Analytical Finance: Volume I

Jan R. M. Röman

Analytical Finance: Volume I

The Mathematics of Equity Derivatives, Markets, Risk and Valuation

palgrave
macmillan

Jan R. M. Röman
Västerås, Sweden

ISBN 978-3-319-34026-5 ISBN 978-3-319-34027-2 (eBook)
DOI 10.1007/978-3-319-34027-2

Library of Congress Control Number: 2016956452

Printed on acid-free paper

This Palgrave Macmillan imprint is published by Springer Nature
The registered company is Springer International Publishing AG
The registered company address is: Gewerbestrasse 11, 6330 Cham, Switzerland

To my soulmate, supporter and love –
Jing Fang

Preface

This book is based upon lecture notes, used and developed for the course *Analytical Finance I* at Mälardalen University in Sweden. The aim is to cover the most essential elements of valuing derivatives on equity markets. This will also include the maths needed to understand the theory behind the pricing of the market instruments, that is, probability theory and stochastic processes. We will include pricing with time-discrete models and models in continuous time.

First, in Chap. 1 and 2 we give a short introduction to trading, risk and arbitrage-free pricing, which is the platform for the rest of the book. Then a number of different binomial models are discussed. Binomial models are important, not only to understand arbitrage and martingales, but also they are widely used to calculate the price and the Greeks for many types of derivative. Binomial models are used in trading software to handle and value several kinds of derivative, especially Bermudan and American type options. We also discuss how to increase accuracy when using binomial models. We continue with an introduction to numerical methods to solve partial differential equations (PDEs) and Monte Carlo simulations.

In Chap. 3, an introduction to probability theory and stochastic integration is given. Thereafter we are ready to study continuous finance and partial differential equations, which is used to model many financial derivatives. We focus on the Black–Scholes equation in particular. In the continuous time model, there are no solutions to American options, since they can be exercised during the entire lifetime of the contracts. Therefore we have no well-defined boundary condition. Since most exchange-traded options with stocks as underlying are of American type, we still need to use descrete models, such as the binomial model.

We will also discuss a number of generalizations relating to Black–Scholes, such as stochastic volatility and time-dependent parameters. We also discuss a number of analytical approximations for American options.

A short introduction to Poisson processes is also given. Then we study diffusion processes in general, martingale representation and the Girsanov theorem. Before finishing off with a general guide to pricing via Black–Scholes we also give an introduction to exotic options such as weather derivatives and volatility models.

As we will see, many kinds of financial instrument can be valued via a discounted expected payoff of a contingent claim in the future. We will denote this expectation $E[X(T)]$ where $X(T)$ is the so-called contingent claim at time T. This future value must then be discounted with a risk-free interest rate, r, to give the present value of the claim. If we use continuous compounding we can write the present value of the contingent claim as

$$X(t) = e^{-r(T-t)}E[X(T)].$$

In the equation above, T is the maturity time and t the present time.

Example: If you buy a call option on an underlying (stock) with maturity T and strike price K, you will have the right, but not the obligation, to buy the stock at time T, to the price K. If $S(t)$ represents the stock price at time t, the contingent claim can be expressed as $X(T) = \max\{S(T) - K, 0\}$. This means that the present value is given by

$$X(t) = e^{-r(T-t)}E[X(T)] = e^{-r(T-t)}E[max\{S(T) - K, 0\}].$$

The max function indicates a price of zero if $K \geq S(T)$. With this condition you can buy the underlying stock at a lower (same) price on the market, so the option is worthless.

By solving this expectation value we will see that this can be given (in continuous time) as the Black–Scholes–Merton formula. But generally we have a solution as

$$X(t) = S(0) \cdot Q_1(S(T) > K) - e^{-r(T-t)}K \cdot Q_2(S(T) > K),$$

where $Q_1(S(T) > K)$ and $Q_2(S(T) > K)$ make up the risk neutral probability for the underlying price to reach the strike price K in different "reference systems". This can be simplified to the Black–Scholes–Merton formula as

$$X(t) = S(0)\cdot N(d_1) - e^{-r(T-t)}K\cdot N(d_2).$$

Here d_1 and d_2 are given (derived) variables. $N(x)$ is the standard normal distribution with mean 0 and variance 1, so $N(d_2)$ represent the probability for the stock to reach the strike price K. The variables d_1 and d_2 will depend on the initial stock price, the strike price, interest rate, maturity time and volatility. The volatility is a measure of how much the stock price may vary in a specific period in time. Normally we use 252 days, since this is an approximation of the number of trading days in a year.

Also remark that by buying a call option (i.e., going long in the option contract), as in the example above, we do not take any risk. The reason is that we cannot lose more money than what we invested. This is because we have the right, but not the obligation, to fulfil the contract. The seller, on the other hand, takes the risk, since he/she has to sell the underlying stock at price K. So if he/she doesn't own the underlying stock he/she might have to buy the stock at a very high price and then sell it at a much lower price, the option strike price K. Therefore, a seller of a call option, who have the obligation to sell the underlying stock to the holder, takes a risky position if the stock price becomes higher than the option strike price.

Acknowledgements

I would like to thank all my students for their comments and questions during my lectures. Special thanks go to Mai Xin who asked me to translate my notes into English. I would also like to thank Professor Dmitrii Silvestrov, who encouraged me to teach Analytical Finance, and Professor Anatoly Malyarenko for his assistance and advice.

Contents

Notations

$B(t)$	The value of the money market account at time t
r	The risk-free interest rate
R	A short notation of $1 + r$
Ω	A sample space
ω_i	Outcome i from a sample space Ω
$S(t)$	Price of a security (financial instrument, equity, stock) at time t
$F(t)$	The forward price of a security (financial instrument, equity, stock) at time t
q	The risk neutral (risk-free) probability of an increase in price
p	The objective (real) probability or the risk-free probability of an decreasing price
Q	The risk neutral probability measure
P	The objective (real) probability measure
$E^Q[\cdot]$	The expectation value with respect to Q
$Var^Q[\cdot]$	The variance with respect to Q
ρ	The risk premium
$X(t)$	A stochastic value/process
I_t	The information set at time t
u	The binomial "up" factor with risk-neutral probability p_u or q
d	The binomial "down" factor with risk-neutral probability p_d or p
Z	A stochastic variable
$V(t)$	A value (process)
μ,α	The drift in a stochastic process
σ	The volatility in a stochastic process
t	Time
T	Time to maturity

K	The option strike price
λ	The market price of (volatility) risk (the sharp ratio)
C	A (call) option value
Δ	The change in the option value w.r.t. the underlying price, S
Γ	The change in the option Δ w.r.t. the underlying price, S
ν	The change in the option value w.r.t. the volatility, σ
Θ	The change in the option value w.r.t. time, t
ρ	The change in the option value w.r.t. the interest rate, r
d_1, d_2	Coefficients (variables) in the Black–Scholes model
VaR	Value-at-Risk
$\mathcal{F}$	A set or subsets to the sample space Ω
μ	A finite measure on a measurable space
$W(t)$	A Wiener process
$N[\mu, \sigma]$	A Normal distribution with mean μ and variance σ
τ	A stopping time (usually for American options)
L_t	A likelihood function of time t

List of Figures

List of Tables

1

Trading Financial Instruments

Financial instruments can be traded on an exchange or over the counter (OTC). Exchange trades securities are standardized instruments. A clearinghouse in connection to a marketplace clears most securities. In such a way the clearinghouse is counterparty to both the seller and the buyer.

1.1 Clearing and Settlement

Clearing is the process of settling a trade including the deposit of any necessary collateral with the clearing organization and exchange of any necessary cash and paperwork. The term clearing usually implies that the clearing organization becomes a party in contracts, rather than merely putting other parties in contact with each other. For example, A wishes to sell to B. In practice, A sells to C, the clearinghouse, and B buys from C.

Settlement is used to refer to the completion of any required payment between two parties to fulfil an obligation. Settlement also refers to the process by which a trade is entered onto the books and records of all the parties to the transaction including brokers or dealers, a clearinghouse, and any other financial institution with a stake in the trade.

How settlement and clearing take place depends on what kinds of instrument are traded and the type of trade process, for example at an exchange or over the counter.

J.R.M. Röman, *Analytical Finance: Volume I*,
DOI 10.1007/978-3-319-34027-2_1

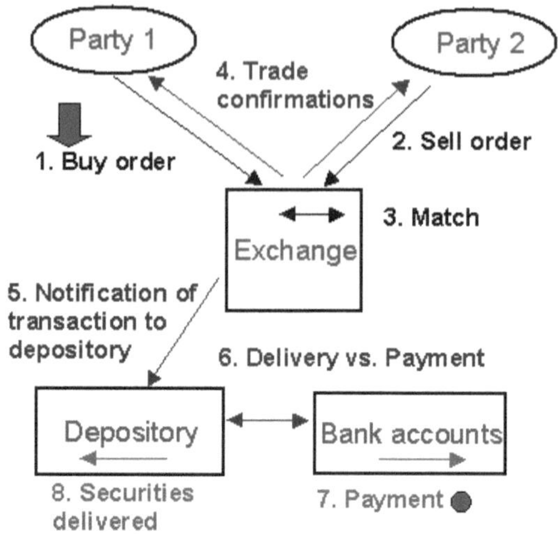

Fig. 1.1 The flows in a typical trade between two parties who place their orders to an exchange

1.1.1 Exchange Traded Securities

In Fig. 1.1 we illustrate a typical trade with exchange-traded instruments.

As seen in Fig. 1.1 the two parties are anonymous to each other. The trade-flow follow includes the following steps:

1. The buyers place their orders in the market.
2. The sellers place their orders in the market. Orders are offers to either buy or sell a particular security at a specified price.
3. Buy orders are matched with suitable sell orders. This may be done electronically or by traders making agreements verbally in a trading pit
4. When a trade has been agreed, confirmations are sent to each party, confirming the details of the trade
5. At the same time as sending confirmation to each party, the exchange notifies the depository of the transaction
6. Delivery vs. payment. The depository sends instructions for money to be transferred from one account to another. This may be in the form of SWIFT transfers between accounts held at banks or the depository may have its own money holding accounts. As this transaction is confirmed, ownership of the securities is transferred.

Table 1.1 Service providers on some exchanges

Exchange services	Depository services	Connected bank accounts
NasdaqOMX (Sweden)	VPC (Värdepappercetralen)	VPC RIX account and other accounts connected to the Central Banks RIX clearing system
London Stock Exchange	CREST	CREST and other accounts
EUREX	EUREX	EUREX

7. Payments are made simultaneously with
8. Delivery of the securities. The credit risk has then been minimized.

In Table 1.1 we show the different service providers at the Sweden Stock and Derivative Exchange, at London Stock Exchange and at EUREX.

A *depository* is an organization that acts as a custodian of securities on behalf of account holders. When Party 1 buys a security from Party 2, instead of physically transferring the securities, the depository simply moves ownership from one account to another. This is similar to the way a bank transfers money from one account to another without physically moving any cash.

1.1.2 Exchange-Traded Derivatives

In Fig. 1.2 we illustrate a typical trade with exchange-traded derivatives.

As seen in Fig. 1.2 the trade-flow follow of exchange-traded derivatives includes the following steps:

1. The buyers place their order in the market.
2. The sellers place their order in the market. Orders are offers to either buy or sell a particular derivative at a specified price.
3. Buy orders are matched with suitable sell orders. This may be done electronically or by traders making agreements verbally in a trading pit.
4. When a trade has been agreed, the exchange will confirm a separate agreement with each party.

With exchange-traded derivatives, credit risks occur for each party; for the buyer or seller of the derivative there is a risk that the exchange could default on its obligations. As the exchange does not take a trading position but merely acts as an intermediary this risk is very small.

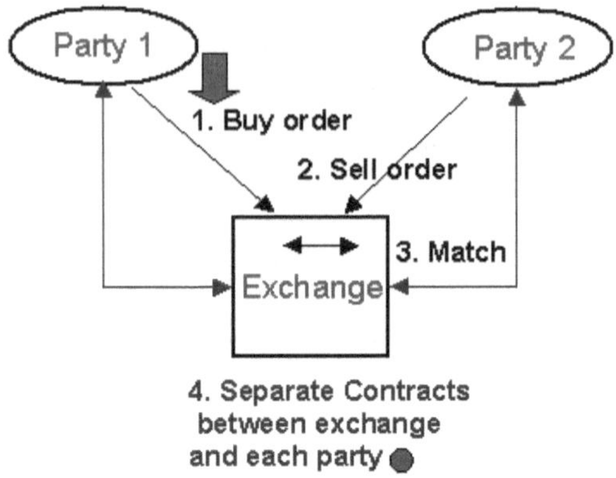

Fig. 1.2 The flows in a typical derivative trade between two parties on an exchange

For the exchange there is a risk that each party to a trade could default on its obligations. To minimize this risk, margining agreements are used. An *initial margin* agreement requires that the counterparty deposit collateral in the form of cash or securities with the exchange (or sometimes a third party). The size of the margin is usually related to the total size of the counterparty's obligations (or potential obligations) to the exchange. A *variation margin* agreement requires cash payments to be made, typically at the end of each day so that outstanding long and short positions are *marked to the market*. This means that, as the market price of a derivative varies, payments are made to reflect that day's gain or loss and prevent any debt or credit building up over time.

The management of margin payments and all other administration is handled by the exchange or a clearinghouse used by the exchange. Other administrative tasks include:

- Exercise/assignment
  When for example an option buyer exercises their option, this action must be assigned to the seller of a matching option. The selection of counterparty is made (at random) by the exchange.
- Expiry
  When derivatives expire, margining agreements and procedures must be terminated.
- New contracts
  The exchange is responsible for defining new contracts.

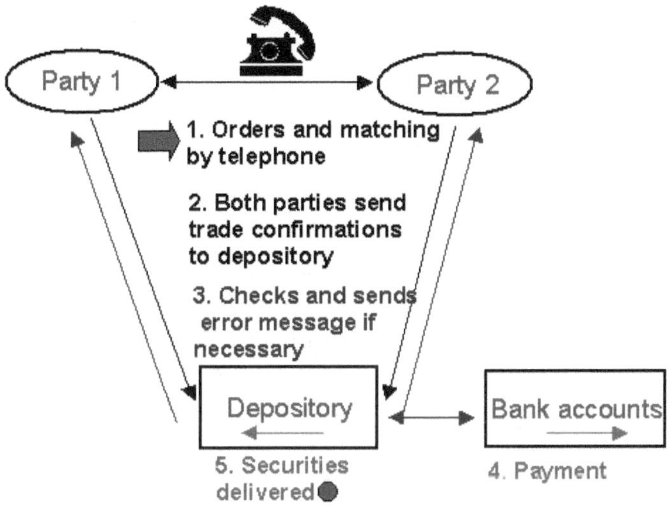

Fig. 1.3 The flows in a typical trade between two parties on an exchange

1.1.3 OTC-Traded Securities

In Fig. 1.3 we illustrate a typical trade on OTC-traded securities.

As seen in Fig. 1.3 the trade-flow follow of OTC-traded securities include the following steps:

1. Buyers and sellers negotiate a trade over the telephone. Conversations are tape recorded to resolve any possible disputes as to what was agreed.
2. When a trade has been agreed, both parties send a confirmation to the depository of
   - The instrument traded, usually defined by a standard code such as ISIN code or VKN number or similar
   - The quantity
   - The agreed price

Dates are usually determined by the choice of instrument according to convention.

3. The depository checks that confirmations from both parties carry the same information and then arrange for delivery versus payment.
4. The payment is made.
5. Ownership of the securities is transferred at the same time as payment is made.

There are several trading codes. The most common is the *International Securities Identification Number* (ISIN) which uniquely identifies a security. Securities for which ISINs are issued include bonds, commercial paper, debt securities, futures, shares, options, warrants and other derivatives. The ISIN code is a 12-character alpha-numerical code that consists of three parts, a two-letter country code, a nine-character alpha-numeric national security identifier, and a single check digit. International securities cleared through Clearstream or Euroclear, which are worldwide, use XS as the country code.

In the United Kingdom and Ireland, SEDOL, which stands for Stock Exchange Daily Official List, are used for clearing purposes. The numbers are assigned by the London Stock Exchange on request by the security issuer. SEDOLs are also part of the security's ISIN. The SEDOL Masterfile (SMF) provides reference data on millions of global multi-asset securities each uniquely identified at the market level.

A CUSIP is a nine-character alphanumeric code that identifies a North American financial security for the purposes of clearing and settlement. The CUSIP system is owned by the American Bankers Association, and is operated by S&P Capital IQ.

The Wertpapierkennnummer (WKN, WPKN, WPK or simply Wert), is a German securities identification code. It comprises six digits or capital letters (excluding I and O) and no check digit. WKNs may become obsolete in the future, since they may be replaced by ISINs.

1.1.4 OTC-Traded Derivatives

In Fig. 1.4 we illustrate a typical trade on OTC-traded derivatives.

As seen in Fig. 1.4 the trade-flow follow of OTC-traded derivatives include the following steps:

1. Buyers and sellers negotiate a trade over the telephone. Conversations are tape recorded to resolve any possible disputes as to what was agreed.
2. When a trade has been agreed, the parties must confirm their agreement in writing. This process will begin typically with a signed contract based on a standard contract, for example one set up by the International Swaps and Derivatives Association (ISDA) (see http://www.isda.org/).
3. All payments and administration (including daily mark to market payments in some cases) must be managed by each party. This may involve a considerable amount of work and may continue for 10–25 years in some cases.

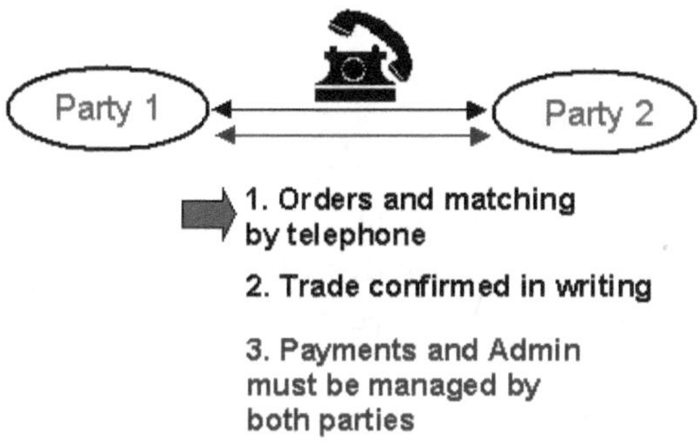

Fig. 1.4 The flows in a typical trade between two parties on OTC derivatives

Payments are usually made in the form of SWIFT transfers. SWIFT stands for the Society for Worldwide Interbank Financial Telecommunications. It is a messaging network that financial institutions use to securely transmit information and instructions through a standardized system of codes.

With OTC derivatives there is a bilateral credit risk. If one party should default, there is little to protect the other party. Various methods exist to reduce the amount of credit exposure, such as netting agreements.

The principle of netting agreements is that when a party fails to honor its obligations due to bankruptcy, then any losses you incur as a result can be offset by any obligations you have toward that party, within the terms of the agreement. This means that two parties can do many trades with each other, but the total credit liability is related to the net position of one party to the other instead of the total credit amount of the defaulting party. Standard agreements to facilitate this are prepared by ISDA, for example.

1.2 About Risk

We will not discuss financial risk in general in this book, but, since we will calculate different risk measures used on the market we will briefly describe the most common risks in the perspective of a bank or another financial institution. Risk can be divided into several main classes:

Market Risk refers to the risk that changes in interest rates, exchange rates and equity prices will lead to a decline in the value of a bank's net assets, including derivatives.

Liquidity Risk refers to the risk that a bank cannot fulfil its payment commitments on any given date without significantly raising the cost. Most institutions face two types of liquidity risk. The first relates to the depth of markets for specific products and the second to funding the financial trading activities. When dealing with OTC market, risks may also rise from the early termination of contracts.

Currency Risk refers to the risk that the value of the assets, liabilities and derivatives may fluctuate due to changes in exchange rates.

Interest Rate Risk refers to the risk that the value of the assets, liabilities and interest-related derivatives may be negatively affected by changes in interest levels.

Equity Price Risk refers to the risk that the value the holdings of equities and equity-related derivatives may be affected negatively as a consequence of changes in prices for equities.

Credit Risk is defined as the risk that the counterparty fails to meet the contractual obligations and the risk that *collateral* will not cover the claim. Credit risk also arises when dealing in financial instruments, but this is often called *counterparty risk*. The risk arises as an effect of the possible failure by the counterparty in a financial transaction to meet its obligations. This risk is often expressed as the current market value of the contract adjusted with an *add-on* for future potential movements in the underlying risk factors. Therefore, counterparty risk usually refers to trading activities. Connected to counterparty risk is also *sovereign risk*, which is the risk that a government action will interfere with repayment of a loan or security. This is measured by the past performance of the nation and present default rate and political, social and economic conditions. Credit risk also includes *concentration risk*, which refers, for example, to large exposures or concentrations in the credit portfolio to certain regions or industries.

Correlation Risk refers to the probability of loss from a disparity between the estimated and actual correlation between assets, currencies, derivatives, instruments or markets.

Model Risk refers to the possibility of loss due to errors in mathematical models, often models of derivatives. Since these models contain parameters such as volatility, we can also speak of *parameter risk*, *volatility risk* and so forth.

Operational Risk refers to the risk of losses resulting from inadequate or failed internal processes or routines, human error, incorrect systems or external events.

Legal or *Compliance risk* refers to the risk of legal consequences, major economic damage or the loss of reputation that a bank could suffer due to failure to comply with laws, regulations or other external policies and instructions. This also includes internal rules such as ethical guidelines that govern how the group conducts its operations.

1.2.1 Risk and Randomness

Before looking at the mathematics of risk we should understand the difference between risk, randomness and uncertainty. When measuring risk we often use probabilistic concepts. But this requires having a distribution for the randomness in investments, a probability density function, for example. With enough data or suitable model we may have a good idea of the distribution of returns. However, without the data, or when embarking into unknown territory, we may be completely in the dark as so the probabilities. This is especially true when looking at scenarios that are incredibly rare or have never even happened before. For example, we may have a good idea of the results of an alien invasion—after all, many scenarios have been explored in the movies—but what is the probability of this happening? When you do not know the probabilities then you have uncertainty.

We have two situations of how to use probabilities:

1. Where the probabilities that specific events will occur in the future are measurable and known—that is, where we have randomness but with known probabilities. This can be further divided:
   i. A priori risk, such as the outcome of the roll of a dice, tossing coins, etc.
   ii. Estimable risk, where the probabilities can be estimated through statistical analysis of the past, for example, the probability of a one-day fall of 10 % or more in a stock index.
2. With uncertainty the probabilities of future events cannot be estimated or calculated.

In finance we tend to concentrate on risk with probabilities that we are able to estimate. We then have all the tools of statistics and probability for quantifying various aspects of that risk. In some financial models we do attempt to address the uncertain, for example the uncertain volatility. Here volatility is uncertain, is allowed to lie within a specified range, but the probability of volatility having any value is not given. Instead of working with probabilities we now work with worst-case scenarios. Uncertainty is therefore more associated with the idea of stress-testing portfolios.

A starting point for a mathematical definition of risk is simply standard deviation. This is essential because of the results of the *central limit theorem*: if you add up a large number of investments what matters as far as the statistical properties of the portfolio are concerned are just the expected return and the standard deviation of individual investments, and the resulting portfolio returns are normally distributed. As the normal distribution is symmetrical about the mean, the potential downside can be measured in terms of the standard deviation.

However, this is only meaningful if the conditions for the central limit theorem are satisfied. For example, if we only have a small number of investments, or if the investments are correlated, or if they don't have finite variance, then standard deviation may not be relevant.

In the following, when we say risk we mean the risk in volatility terms—that is, the change in the underlying stock when we calculate the value of a derivative.

1.3 Credit and Counterparty Risk

Credit risk managers try to estimate the likelihood of default by the borrower or counterparty due to a default, losses in loans, bonds or other obligations that will not be repaid on time or in full. The counterparty can also fail to perform an obligation to the institution trades in OTC derivatives.

The likelihood of this happening is measured through the repayment record/default rate of the borrowing entity, determination of market conditions, default rate, for example.

With loans or bonds, the amount of the total risk is determined by the outstanding balance that the counterparty has yet to repay. However, the credit risk of derivatives is measured as the sum of the current replacement cost of a position plus an estimate of the firm's potential future exposure from the instrument due to market moves and what it may cost to replace the position in the future.

Senior managers must establish how the firm calculates replacement cost. The Basel Committee indicates that it prefers the current mark-to-market price to determine the cost of current replacement. An alternative approach would be to determine the present value of future payments under current market conditions.

The measurement of potential future exposure is more subjective as it is primarily a function of the time remaining to maturity and the expected volatility of the asset underlying the contract. The Basel Committee for Banking Supervision indicates that it prefers multiplying the notional principal of a transaction by an appropriate add-on factor/percentage to determine the potential replacement value of the contract (simply percentages of the notional value of the financial instrument).

Senior management may also determine whether this potential exposure should be measured by using simulation (or other modelling techniques such as Monte Carlo, probability analysis or option valuation models). By modelling the volatility of the underlying stock price it is possible to estimate an expected exposure.

Credit risk limits are part of a well-designed limit system. They should be established for all counterparties with whom an institution conducts business, and no dealings can begin before the counterparty's credit limit is approved. The credit limit for counterparty must be aggregated globally and across all products (i.e. loans, securities, derivatives) so that a firm is aware of its aggregated exposure to each counterparty. Procedures for authorizing credit limit excesses must be established and serious breaches reported to the supervisory board. These limits should be reviewed and revised regularly. Credit officers should also monitor the usage of credit risk by each counterparty against its limits.

Once a counterparty exceeds the credit exposure limits, no additional deals are allowed until the exposure with that counterparty is reduced to an amount within the established limit.

Senior managers should try to reduce counterparty risks by putting in place master netting as well as collateral agreements. Under a master netting agreement, losses associated with one transaction with a counterparty are offset against gains associated with another transaction so that the exposure is limited to the net of all gains and losses related to the transactions covered by the agreement.

The Basle Committee for Banking Supervision estimates that netting reduces current (gross) replacement value on average by 50 % per counterparty. However, board members, senior management and line personnel must be aware that netting agreements are not yet legally enforceable in several European and Asian countries, a factor which they must take into consideration in their daily dealings with counterparties in these countries; not doing so will engender a false sense of security. The forms of collateral generally accepted are cash and government bonds.

Another type of counterparty risk is *pre-settlement risk*. This is the risk that a counterparty will default on a forward or derivative contract prior to settlement. The specific event leading to default can range from disavowal of a transaction, default of a trading counterparty before the credit of a clearinghouse is substituted for the counterparty's credit, or something akin to *Herstatt risk*, where one party settles and the other defaults on settlement.

1.4 Settlement Risk

Settlement risk is related to credit risk and is defined as the risk that an expected settlement payment on an obligation will not be made on time due to bankruptcy, inability or time zone differential. A common example involves bilateral obligations in which one party makes a required settlement payment and the counterparty does not.

Settlement risk provides an important motivation to develop netting arrangements and other safeguards. When related to currency transactions, the term Herstatt risk is sometimes used. This is the risk that one party to a currency swap will default after the other side has met its obligation, usually due to a difference in time zones. The settlement of different currencies in different markets and time zones from the moment the sold currency becomes irrevocable until the purchased currency receipt is confirmed. The two parties are paid separately in local payment systems and may be in different time zones, resulting in a lag time of three days and mounting exposure that may exceed a party's capital. The risk is reduced by improved reconciliation and netting agreements.

The Herstatt risk is named after an incident in Bankhaus Herstatt, a private German bank on June 26, 1974. The bank was then closed by German financial regulators (Bundesaufsichtsamt für das Kreditwesen) who ordered it into liquidation after the close of the interbank payments system in Germany.

Prior to the announcement of Herstatt's closure, several of its counterparties had irrevocably paid approximately $620 million in Deutsche Marks to Herstatt. Upon the termination of Herstatt's at 10.30 a.m. New York time, 3.30 p.m. in Frankfurt, Herstatt's New York correspondent bank suspended outgoing US dollar payments from Herstatt's account.

This action left Herstatt's counterparty banks exposed for the full value of the Deutsche Mark deliveries made. Moreover, banks which had entered into forward trades with Herstatt not yet due for settlement lost money in replacing the contracts in the market, and others had deposits with Herstatt.

1.5 Market Risk

Some of the risks above can be aggregated into a more general risk, the market risk. Market risk deals with all kinds of change in market data that affect prices of assets contained in a portfolio. This includes stock, bond, commodity and other prices. It also includes market data such as interest rates and exchange rates, volatilities and liquidity. Such changes in prices can destroy a financial institution's capital base.

Market risk is different from an asset's mark-to-market calculation, which is the current value of the financial instruments. Market risk represents what we could lose if prices or volatility change in the future. Therefore, we need to measure the market risks in portfolio of financial instruments. For active portfolios we need to calculate their exposure on a daily basis, while those with small portfolios could be analysed less frequently.

The total market risk can be measured as the potential gain or loss in a portfolio that is associated with price movements of given probability over a specified time horizon. This is the *Value-at-Risk* (VaR) approach. VaR can be measured by different models, as we will discuss in Chapter 2. The chosen model is a decision taken by the board of directors on the advice of senior managers and depends on requirements from the supervisory authorities.

Interest rate risk is related to market risk and arises from changes in interest rates. This will result in financial losses related to asset/liability management. It is measured by past and present interest rates and market volatility. It is controlled by hedging the assets and liabilities by swaps, futures and options, and accurately makes changes in possible future scenarios.

Foreign exchange risk is also a part of market risk. This is the risk that changes in the foreign exchange rate will cause assets to fall in value or that foreign exchange denominated liabilities will rise in expense. It is measured by marking-to-market the value of the asset, or increase of the liability. This is done by actual movement of the exchange rate between the currency of the asset/liability and the currency of the booked or pending asset or liability. It is controlled by hedging the assets and liabilities by swaps, futures or options that can changes possible future scenarios.

1.6 Model Risk

Model risk is a topic of great, and growing, interest in the risk management arena. Financial institutions are obviously concerned about the possibility of direct losses arising from mismarked complex instruments. They are becoming even more concerned about the implications that evidence of model risk mismanagement can have on their reputation, and their perceived ability to control their business.

In July 2009, the Basel Committee on Banking Supervision issued a directive requiring that financial institutions have to quantify their model risk. The committee further stated that two types of risk must be taken into account:

> The model risk associated with using a possibly incorrect valuation, and the risk associated with using unobservable calibration parameters.

On the surface, this seems to be a simple adjustment to the market risk framework, adding model risk to other sources of risk that have already been identified within Basel II. In fact, quantifying model risk is much more complex because the source of risk (using an inadequate model) is much harder to characterize.

Financial assets can be divided into two categories. In the first category, we find the assets for which a price can be directly observed in the financial marketplace. These are the liquid assets for which there are either organized markets (e.g. futures exchanges) or a liquid OTC market (e.g. interest rate swaps). For the vast majority of assets, however, price cannot be directly observed, but needs to be inferred from observable prices of related instruments. This is typically the case for financial derivatives whose price is related to various features of the primary assets, depending on a model. This process is known as *marking-to-model*, and involves both a mathematical algorithm and subjective components, thus exposing the process to estimation error.

There are several distinct possible meanings for the expression model risk. The most common one refers to the risk that, after observing a set of prices for the underlying and hedging instruments, different but identically calibrated models might produce different prices for the same exotic product.

Since, presumably, at most one model can be "true", this would expose the trader to the risk of using a mis-specified model. Sidenius (2000) did a research of model risk in the interest-rate area where he found that significantly different prices were obtained for exotic instruments after the underlying bonds and (a subset of) the underlying plain-vanilla options were correctly priced.

These are interesting questions, and they are the most relevant ones from the trader's perspective. Selling optionality too cheaply is likely to cause an irregular but steady bleeding of money out of the book.

The most relevant question is, if the price of a product cannot be frequently and reliably observed in the market, how can we give a price to it between observation times in such a way as to minimize the risk that its book-and-records value might be proven to be wrong?

In pricing models, model risk is defined as:

The risk arising from the use of a model which cannot accurately evaluate market prices.

In risk measurement models, model risk is defined as:

The risk of not accurately estimating the probability of future losses.

Rebonato (2001) uses the following definition:

> Model risk is the risk of occurrence of a significant difference between the mark-to-model value of a complex and/or illiquid instrument, and the price at which the same instrument is revealed to have traded in the market.

If reliable prices for all instruments were observable at all times, model risk in valuation would not exist. On the other hand, if different models are used, the hedging will differ. An example of this is when rates get close to zero or below, the standard Black model for swaptions, caps and floors cannot be used. Then, a model with normal distributed forward rates must be used to allow zero or negative interest rates.

Sources of model risk in pricing models include:

- use of wrong assumptions,
- errors in estimations of parameters,
- errors resulting from discretization, and
- errors in market data.

Sources of model risk in risk measurement models include:

- the difference between assumed and actual distribution[1], and
- errors in the logical framework of the model.

Derman (1996) refers to the following types of model risk:

- inapplicability of modelling,
- incorrect model,
- correct model but incorrect solution,
- correct model but inappropriate use.

---

[1] For instance, the Black–Scholes model assumes that underlying asset prices fluctuate according to a lognormal process, whereas actual market price fluctuations do not necessarily follow this process.

- badly approximated solutions,
- software and hardware bugs,
- unstable data.

Complex financial products require sophisticated financial engineering capabilities for proper risk control, including accurate valuation, hedging, and risk measurement.

Model risk has often been associated with complex derivatives products, but a deeply out-of-the money call and an illiquid corporate bond can both present substantial model risk. What both these instruments have in common is that the value at which they would trade in the market cannot be readily ascertained via screen quotes, intelligence of market transactions or broker quotes.

Model risk arises not because of a discrepancy between the model value and the "true" value of an instrument (whatever that might mean), but because of a discrepancy between the model value and the value that must be recorded for accounting purposes.

Model validation is usually meant to be the review of the assumptions and of the implementation of the model used by the front office for pricing deals, and by finance to mark their value.

The absence of computational mistakes is clearly a requirement for a valid valuation methodology. Rejecting a model because 'it does not allow for stochastic volatility' or because 'it neglects the stochastic nature of discounting' can be totally inappropriate, from a risk perspective. If we require that a product should be marked to market, using a more sophisticated model can be misguided.

From risk perspective the first and foremost task in model risk management is identification of the model ("right" or "wrong" as it may be) currently used by the market in order to arrive at the observed traded prices. In order to carry out this task, it is very important to be able to use reverse-engineering to match observed prices using a variety of models in order to "guess" which model is currently most likely to be used in order to arrive at the observed traded prices. In order to carry out this task we will need a variety of properly calibrated valuation models, and information about as many traded prices as possible.

The next important task is to surmise how today's accepted pricing methodology might change in the future. Notice that the expression 'pricing methodology' makes reference not just to the model, but also to the valuation of the underlying instruments, to its calibration, and possibly, to its numerical implementation. We should not assume that this dynamic process of change should necessarily take place in an evolutionary sense towards better and more realistic models and more liquid and efficient markets. An interesting question could be: "How would the price of a complex instrument change if a particular hedging instrument (say, a very-long-dated FX option) were no longer available tomorrow?"

1.6.1 Some Examples of Model Risk Failure

Index Swaps

Index swaps are swap transactions in which floating interest rates are based on indices other than LIBOR. It is therefore necessary to manage the position and the risks in line with the relevant index. This requires a full understanding of various types of indices, as well as the structure of index swap markets.

A certain financial institution accumulated a substantial position in a special type of index swaps. At the time, the market participants were using several types of models for the valuation of this index swap. One financial institution began trading in this product using what was recognized at the time as the leading mainstream model. As the market for this index swap shrank, some participants left the market. Thereafter, another model, which was being used by some of the remaining participants, became the dominant model in the market.

While maintaining a very large position in this swap index, this financial institution fell behind in research of the most dominant pricing model for this product in the market. Consequently, it failed to recognize that a switch had been made in the dominant model until adjusting its position. As a result, it registered losses amounting to several billions when it finally adopted the new model and made the necessary adjustments in its current price valuations.

Caps

Caps are generally an OTC product with relatively high liquidity. The broker screen displays the implied volatility for each strike price and time period as calculated for cap prices using the Black model. The volatility exhibits a certain skew structures by strike prices and by time periods. To calculate the current price of any given cap, the volatility corresponding to the time period and strike price of the cap is first estimated (interpolated) on the basis of the skew, which is normally observed in the market.

A certain Japanese financial institution was engaged in German cap transactions. At the time, the number of time periods and strike prices for which volatility could be confirmed on the screen was relatively small compared with yen caps. The estimation of volatility was particularly difficult for caps with significant differences between market interest rates and strike interest rates.

The financial institution was using the Black model as its internal pricing model for caps. This institution uses the broker-screen volatility of the closest strike price as the volatility of far-out strikes. Some cap dealers attempted to capitalize on the inevitable difference between market prices and valuation prices by trading aggressively in far-out strikes. This strategy generated internal valuation profits.

The financial institution fell behind in improving its pricing model and failed to minimize the gap between market prices and valuation prices. Continued cap dealer transactions under an unimproved model resulted in the accumulation of substantial internal valuation profits. However, when the internal pricing model was finally revised, the financial institution reported several tens of billion in losses.

LCTM

Long-term capital management (LCTM) was a hedge fund in Greenwich, Connecticut that used absolute-return trading strategies combined with high financial leverage to accumulated a credit spread position, which combined emerging bonds, loans and other instruments. The position was structured to generate profits as spreads narrowed. LTCM suffered huge losses as a result of the sudden increase in spreads following the Russian crisis in 1998.

Various reasons have been given for these huge losses. For instance, LTCM was unable to hedge or cancel its transactions because its liquidity had dried up in the market. On this point, it has been said that LTCM had not taken liquidity into account when building its model. Others have pointed to internal problems in LTCM's risk measurement model. Specifically, problems with wrong assumptions concerning the distribution of underlying asset prices and errors in data used in estimating the distribution of underlying asset prices have been pointed out. Both would lead to fatal errors in risk measurement.

1.6.2 Measurement of Model Risk

If we try to get any kind of measure for model risk to be formulated in a mathematical perspective, we can use the analogy with the VaR method for computing market risk. The calculation of VaR involves two steps:

- The identification of the market risk factors and the estimation of the dynamic of these risk factors (the classical VaR framework assumes a multivariate log-normal distribution for asset prices).
- The definition of a risk measure, for example the 99.5 % confidence interval for a 10-day holding period.

What would be the equivalent when considering model risk? In this case, the risk factors include the risk of model mis-specification (leaving out important sources of risk, mis-specifying the dynamic of the risk factors), and the risk of improper calibration, even though the chosen model may be perfectly calibrated to a set of liquid instruments.

The second step involves defining a reasonable family of models over which the risk should be assessed. The family of models is restricted to the models that can be precisely calibrated to a set of liquid instruments. This constraint alone still defines such a large set of models than further restrictions need to be applied. Intuitively, one needs to define a meaningful notion of "distance" between models, in order to define a normalized measure of model risk.

Let I be a set of liquid instruments, with $H_{i\in I}$ being the corresponding payoffs, and $C_{i\in I}$ the mid-market prices, with $C_i \in \left[C_i^{\text{bid}}, C_i^{\text{ask}}\right]$. Let Ω be a set of models, consistent with the market prices of benchmark instruments

$$Q \in \Omega \Rightarrow E^Q[H_i] \in \left[C_i^{\text{bid}}, C_i^{\text{ask}}\right], \quad \forall i \in I$$

Define next the upper and lower price bounds over the family of models, for a payoff X

$$\overline{\pi}(X) = \sup_{j=1,\dots,n} E^{Q_j}[X], \qquad \underline{\pi}(X) = \inf_{j=1,\dots,n} E^{Q_j}[X]$$

The risk measure is finally the range of values caused by model uncertainty:

$$\mu_\Omega = \overline{\pi}(X) - \underline{\pi}(X)$$

The crucial aspect of this procedure is the definition of a suitable set, Ω. There are many ways of defining it:

- Choose a class of models, and construct a set of models by varying some unobservable parameters, while ensuring that each model calibrates to the set of benchmark instruments.

- Select several types of model (local volatility, stochastic volatility, etc.), and calibrate each model to the same set of benchmark instruments.

It is clear that the variability of models forming the set Ω, needs to be normalized. In the same way as one computes "99.5 % VaR for a 10 day holding period", one needs a normalizing factor to qualify model risk. In other words, one needs to define the aforementioned notion of "distance" between models.

2

Time-Discrete Models

2.1 Pricing via Arbitrage

To study arbitrage-free pricing, we start with a simple financial market containing two instruments, a money-market account instrument (in some literature referred as a bond) B and another security S. The other security can be a stock (equity) or some kind of derivative, such as an option. We want to study a portfolio (B, S) today (at time $t = 0$) and at a future time t. The money-market account has the following simple property

$$B(0) = 1, \;\; B(t) = 1 + r.$$

This means that the value of the money-market account instrument today is 1 (in some currency) and at the future time t, the value is given by $1 + r$, where r is the *risk-free interest rate*. An important property of the money-market account is that the interest rate is the same for borrowing as for lending.

On this market, two events may occur at time t: ω_1 and ω_2. We say that we have a *sample space* Ω with two possible *outcomes* $\Omega = \{\omega_1, \omega_2\}$. On event ω_1 the price of the security S will be $S_1(t)$ and on ω_2 the price of S will become $S_2(t)$. For simplicity, no other outcomes (events) exist.

We then have the following situation in matrix representation[1]

[1] This is only an abstract representation of the situation with two events that might be true in the future.

J.R.M. Röman, *Analytical Finance: Volume I*,
DOI 10.1007/978-3-319-34027-2_2

$$\begin{pmatrix} B(0) \\ S(0) \end{pmatrix} = \begin{pmatrix} B(t) & B(t) \\ S_1(t) & S_2(t) \end{pmatrix} \begin{pmatrix} \omega_1 \\ \omega_2 \end{pmatrix}.$$

As we know, the only outcome for B at time t is $B(t) = 1 + r$, which simplify the first equation

$$\begin{pmatrix} B(0) \\ S(0) \end{pmatrix} = \begin{pmatrix} 1+r & 1+r \\ S_1(t) & S_2(t) \end{pmatrix} \begin{pmatrix} \omega_1 \\ \omega_2 \end{pmatrix}.$$

The first equation can be written as

$$1 = B(0) = (1+r)\omega_1 + (1+r)\omega_2 = q_1 + q_2,$$

where we have defined q_1 and q_2. Since the sum of q_1 and q_2 is equal to 1, we can interpret them as they were probabilities. We do not allow them to be less than zero. The second equation can then be written

$$\begin{aligned} S(0) = S_1(t) \cdot \omega_1 + S_2(t) \cdot \omega_2 &= \frac{1}{1+r} q_1 \cdot S_1(t) + \frac{1}{1+r} q_2 \cdot S_2(t) \\ &= \frac{1}{1+r} [q_1 \cdot S_1(t) + q_2 \cdot S_2(t)]. \end{aligned}$$

We then say that under the *probability measure* $Q = (q_1, q_2)$, the value of S today (at time $t = 0$) is given by the *discounted expected payoff*. We write this as

$$S(0) = \frac{1}{1+r} E^Q[S(t)].$$

Remarks

1. These probabilities have nothing to do with the real probability for the outcome in Ω. Therefore, we call these probabilities, *risk-adjusted probabilities*.
2. If we have other securities, also depending on the same outcome, they should also be given by the same expression. The reason is that the probabilities are given by the risk-free interest rate and the sample space Ω.

If we had used the true (objective) probabilities, P for the outcomes $\{\omega_1, \omega_2\}$, then

$$S(0) < \frac{1}{1+r} E^P[S(t)].$$

The reason is that those probabilities are not risk-free. If we are willing to buy a stock, which is riskier than the money-market account (which pays a risk-free interest rate) we must be compensated for the higher risk. We say that we have a *risk premium* ρ to go into the position of S:

$$S(0) = \frac{1}{1+r+\rho} E^P[S(t)].$$

This is the reason why we buy equities instead of putting the same amount of money into a risk-free money-market account. We take the risk, since we hope we will get a better payoff. The expected payoff increases with the level of risk. Options have better payoff than stocks, since they are more risky.

2.2 Martingales

Expressions such as the expectation value above will be frequently used in this course, especially when dealing with martingales. A martingale with respect to a given probability measure Q, is defined by

$$E^Q[X(t+s)|I_t] = X(t)$$

for all $s > 0$. I_t is the information set that affects the value of the stochastic process X. In other words, this expectation value is saying:

> *Standing at a time t, with a stochastic process X, under a given probability measure Q and a given information set I_t, (with information known up to time t), the calculated expected future value of $X(t+s)$ (where $s > 0$) is equal to $X(t)$.*

This is the same value for X as the value today. A martingale is said to represent a *fair game.*

Example 2.1

If we are tossing a coin, we will get a head or a tail. Suppose we win one cash unit on head, and lose one on tail. This is a fair game since the probability to win money when the number of tosses $\rightarrow \infty$ is zero. We will lose as much as we win.

Example 2.2

We can also construct a martingale measure, that is, a fair game, by making a deal where two parties make an agreement. Say that John said to Lisa that he will give her $100 if it rains tomorrow. He asks Lisa how much she is willing to pay for this agreement (contract). Suppose Lisa is willing to pay $45 and John wants $55. If they finally compromise and agree that Lisa will pay John $50, and then, after making some additional restrictions in the agreement, such as it has to rain in their home town and at least 1 mm,they have an agreement. Now, both John and Lisa feels that they are risk neutral and both believe that it will rain tomorrow with a probability of 50 %. Then, the risk-neutral (martingale) probability to rain next day is 50 %. Remark that this has nothing to do with the real (objective) probability.

The conclusion of this example is that, as soon we know the (possible) price (or prices), the risk-neutral probabilities are known. We can also state that, as soon as we know the risk-neutral probability measure, we also know the possible prices. This will be clear when we study the binomial model below.

When we build binomial models in finance we are creating a situation like tossing a coin, where the stock price goes up if we get a head and down if we get a tail. The only difference is that the probabilities for heads and tails are not the same. Such financial processes are therefore not martingale. But, as we will see, such a process can be transformed into a martingale by changing the probability measure.

If we have

$$E^P[X(t+s)|I_t] \leq X(t),$$

where I_t is the information-set at time t, we say that X is a *super-martingale* and if

$$E^P[X(t+s)|I_t] \geq X(t)$$

X is said to be a *sub-martingale*. If we return to the expression

$$S(0) = \frac{1}{1+r} E^Q[S(t)],$$

the process S is martingale, but since $r > 0$

$$E^P[S(t)]$$

is a sub-martingale because of

$$E^P[S(t)] > \frac{1}{1+r} E^P[S(t)].$$

Here P represents the objective probability measure. This is a fundamental concept in finance. We can transform the process for the stock price to a martingale, just by multiplying with the discounting factor $1/(1+r)$. This means, under the probability measure Q and with the discounting factor $1/(1+r)$ the process of S is martingale. As we will see later, the process above can be transformed to a martingale in two ways; by changing the probability measure or by multiplying with a discount factor.

2.3 The Central Limit Theorem

We will now study the game of tossing a coin and calculate the possible outcomes. Let heads be the outcome u (winning one cash unit) and tails, d (losing one cash unit). We study the total outcome of tossing 1, 2, 4, 8, 16 and 32 rimes.

After tossing the coin twice we have the possible outcomes: $\{uu, ud, du, dd\}$ giving the total profit $\{2, 0, 0, -2\}$. Since we have equal probabilities we can plot the possible payoffs in this game. In Fig. 2.1 we see the outcome of 1 and 2 tosses in Fig. 2.2 we see the outcome of 4 and 8 tosses and in Fig. 2.3 we see the outcome of 16 and 32 tosses.

As we can observe, the coin-tossing game seems to lead in the limit to the normal distribution. If we change the probabilities, we will in the limit reach a

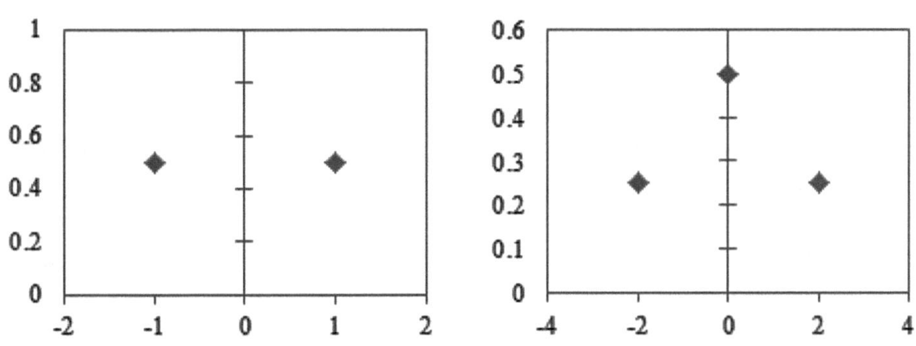

Fig. 2.1 When tossing the coin one time we have two outcomes, −1 or 1, both with probability 1/2. When tossing the coin two times we have three outcomes, −2 with probability 1/4, outcome 0 with probability 1/2

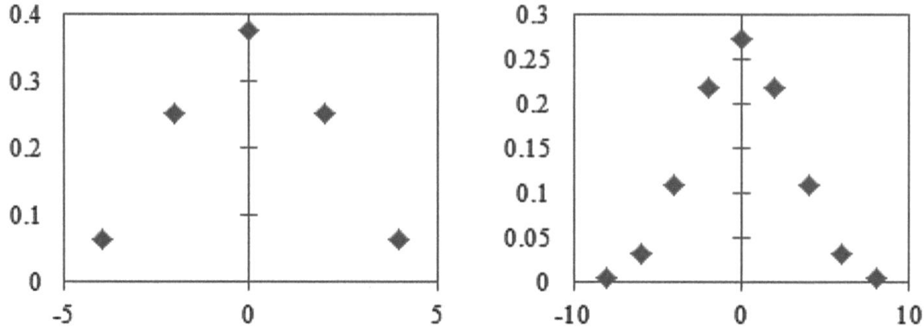

Fig. 2.2 When tossing the coin four and eight times we have five and nine different outcomes with the probability distributions as above

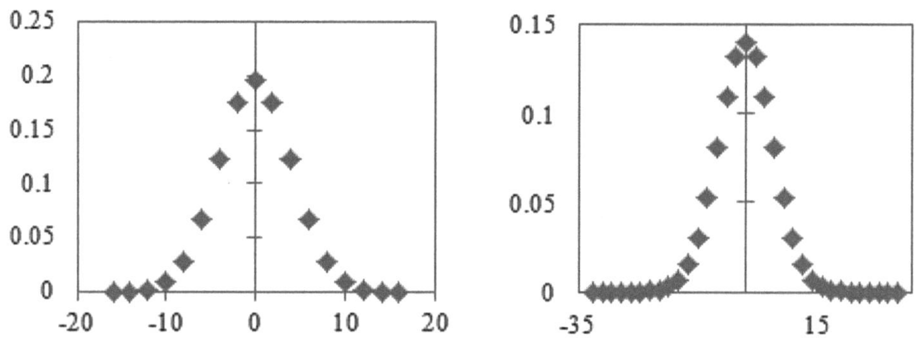

Fig. 2.3 When tossing the coin 16 and 32 times we have 17 and 33 different outcomes with the probability distributions as above

normal distribution with a mean not at zero. If we have three different outcomes with different probabilities, we still reach a normal distribution.

In general, the central limit theorem states that, given certain conditions, the arithmetic mean of a sufficiently large number of iterates of independent random variables, each with a well-defined expected value and well-defined variance, will be approximately normally distributed, regardless of the underlying distribution.

We will see in Sect. 2.5 that when we create a tree model, called the binomial model, the solution when making infinite number of infinitesimal small steps will converge to a normal distributed model in continuous time.

2.4 A Simple Random Walk

Before we start to study stochastic processes we will study a simple random walk. In a random walk we can take a step forwards or a step backwards dependent on some random event, $Z = \{-1, +1\}$. On fix time intervals we can take a step forwards ($Z = +1$) with probability p or a step backwards ($Z = -1$) at probability $q = 1 - p$. During an arbitrary time interval, the given displacement ($E[Z]$) and its variance ($\mathrm{Var}(Z)$) are given by

$$
\begin{aligned}
E[Z] &= (+1)p + (-1)q = p - q \\
\mathrm{Var}(Z) = E\left[Z^2\right] - (E[Z])^2 &= (+1)^2 p + (-1)^2 q - (p-q)^2 = 1 - (p-q)^2 \\
&= (p+q)^2 - (p-q)^2 = 4pq
\end{aligned}
$$

We are now interested in the position, X_n of the process after n such steps. The outcome of this event obeys the *Markov properties*, namely that every event is independent of earlier events, and so we can scale up the displacement by a linear factor

$$
\begin{aligned}
E[X_n] &= \{(+1)p + (-1)q\} = n(p-q) \\
\mathrm{Var}(X_n) &= 4npq,
\end{aligned}
$$

where

$$X_n = \sum_{i=1}^{n} Z_i.$$

With the Markov property, we can put the expectation inside the summation

$$E\left[X_n^m\right] = E\left[\sum_{i=1}^{n} Z_i^m\right] = \sum_{i=1}^{n} E\left[Z_i^m\right].$$

We now ask for the probability distribution $P(X_n)$ to reach the position X_n after n steps. This position can be reached by many different paths, but we have to take f steps forwards and b steps backwards, so we have $X_n = f - b$. The probability to reach this point is given by $p^f q^b$ and to get the probability to get here, we have to multiply with the number of different paths. Then

$$P(X_n) = \binom{f+b}{f} \cdot p^f \cdot q^b = \binom{n}{\frac{n+X_n}{2}} \cdot p^{\frac{n+X_n}{2}} \cdot q^{\frac{n-X_n}{2}},$$

where

$$\binom{n}{m} = \frac{n!}{m!(n-m)!}$$

The *binomial distribution* is given by

$$\begin{aligned} B(n,m) &= \binom{n}{m} p^m \cdot q^{n-m} \\ E[B] &= n \cdot p \\ Var(B) &= n \cdot p \cdot q \end{aligned}$$

so we see that we have a binomial distribution. A *diffusion process* is a *Brownian motion* and behaves like such a random walk with $p = q(= \frac{1}{2})$ in the continuous limit. To see this, let's assume that our walker takes steps of length r between each time interval t. Since $p = q$ his expected position at the next time is his current position (see martingale property) and the variance of his displacement is r^2. To go to the continuous time limit we split the time interval t into n subintervals of length τ and between each subinterval we allow the walker to take steps of $\pm\, r'$ with equal probability. After a time t, the position of the walker is found by summing the n independent identical random variables Z. According to the central limit theorem, as n gets large the probability distribution of the positions will begin to resemble a Gaussian distribution with zero mean and variance nr'^2 (Fig. 2.4).

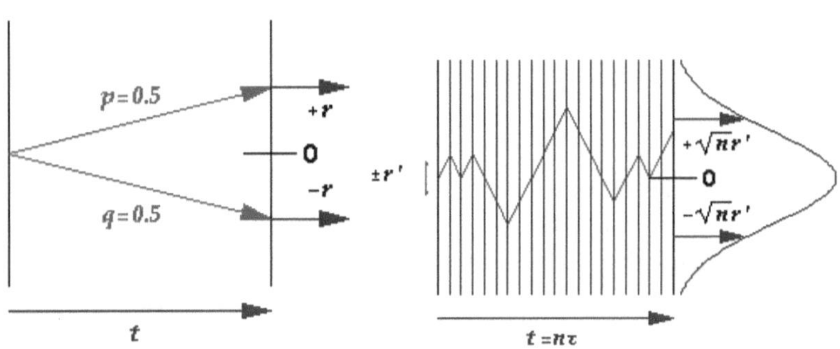

Fig. 2.4 In the continuous limit, a random walk with equal probabilities converges to a Gaussian probability distribution

To prevent the walker from having an infinite or a zero variance as n goes to infinity our only possible choice for r' which also preserves the characteristics of our original random walker is to set

$$r' = \frac{r}{\sqrt{n}}.$$

Hence if time gets rescaled by factor n then the space is rescaled by $\sqrt{n}$ and this preserves the physical properties of walker. The continuous probability distribution $P(x, t)$ of being at position x at time t, given that $P(x_0, 0) = 1$ evolves according to a parabolic partial differential equation called the *Fokker–Planck* equation. However, since $p = q$ there is no drift term and this equation reduces to the Diffusion equation

$$\frac{\partial P(x,t)}{\partial t} = D\frac{\partial^2 P(x,t)}{\partial x^2},$$

where D is called the diffusion constant given by

$$D = \frac{2 \cdot r^2}{t}.$$

2.5 The Binomial Model

We will now discuss the most common model for American options. These options have nothing to do with USA, the country or the continent. American options are options that allows the holder to exercise at any time of the option life-time. European options can only be exercised at maturity. There are also options, especially in the interest rate theory, that can be exercised on specific days. This kind of option is said to be of Bermudan type. We will also study so-called Asian options. These are of European type, but the final value at maturity depends not on the final underlying price, but instead on the average price during a time-period.

2.5.1 Background and Theory

Consider a financial market during one period in time, from $t = 0$ to $t = 1$ with two possible investments (or two different securities), B and S. Here

B represents a deterministic money-market account (or in some literature, a bond) with the price process

$$\begin{cases} B(0) = 1 \\ B(1) = 1 + r \end{cases}$$

where r represent the interest rate. S is considered to be a stock with a stochastic price process given by

$$\begin{cases} S(0) = s \\ S(1) = \begin{cases} u \cdot s & \text{with probability } p_u \\ d \cdot s & \text{with probability } p_d \end{cases} \end{cases}$$

At time $t = 1$ the stock can reach two possible value $u \cdot s$ where $u > 1$ or $d \cdot s$ where $d < 1$. In other words, the stock price can either increase or decrease with probability p_u and p_d respectively. Here $p_u + p_d = 1$.

Furthermore, we suppose that we can buy (*going long* in *S*) or sell (*going short* in *S*) the stock and we can invest (put money, i.e., go long in *B*) or lend (borrow money, i.e., go short in *B*) in the money-market account. The interest rate for saving and lending money from the money-market account is for simplicity the same, *r*.

Now, we write $S(t) = Z \cdot s$ where Z is a *stochastic variable* and consider a *portfolio h* on the (*B*, *S*)-market, as a vector $h = (x, y) \in R^2$ where x is the number of money-market securities and y the number of stocks. x and y may take any number, including negative and fractions where negative values represent short positions. We also suppose that the market is 100 % liquid, that is, we can trade whenever we want.

Definition 2.3 The *value process* of the portfolio h is defined as

$$V(t, h) = x \cdot B(t) + y \cdot S(t); \quad t = 0, 1$$

i.e.

$$\begin{cases} V(0, h) = x + y \cdot s \\ V(1, h) = x \cdot (1 + r) + y \cdot s \cdot Z \end{cases}.$$

Definition 2.4 An *Arbitrage portfolio* of h is defined as

$$\begin{cases} V(0,h) = 0 \\ V(1,h) > 0 \quad \text{with probability } 1 \end{cases}.$$

This means that we can borrow money at time $t = 0$ and buy the stock, or we can sell the stock and put the money in the money-market account. The total value of our portfolio h is then at time $t = 0$ is equal zero. If for sure (with probability 1) our portfolio at time $t = 1$ have a value greater than zero we have made *arbitrage.*

The portfolio in the binomial model above is *free of arbitrage* if and only if $d \leq 1 + r \leq u$. The reason for this is that; If $d \leq u \leq 1 + r$ we can go short in the stock and invest in the risk-free interest rate. If on the other hand, $1 + r \leq d \leq u$ we can go short in the risk-free interest rate and invest in the stock. In both situations we will make arbitrage.

From now on we denote the objective (true or market) probabilities as $P = (p_u, p_d)$ and the risk-free (martingale) probabilities as $Q = (q_u, q_d)$. If the portfolio is risk-free we must have probabilities such as

$$1 + r = u \cdot q_u + d \cdot q_d; \quad q_u + q_d = 1$$

We say that we have a *probability measure* Q defined as

$$Q : \begin{cases} Q(Z = u) = q_u \\ Q(Z = d) = q_d \end{cases}$$

We then have

$$\begin{aligned} \frac{1}{1+r} E^Q[S(1)] &= \frac{1}{1+r}(u \cdot S(0) \cdot q_u + d \cdot S(0) \cdot q_d) \\ &= \frac{1}{1+r} \cdot S(0) \cdot (1+r) = S(0) \end{aligned}$$

i.e.

$$S(0) = \frac{1}{1+r} E^Q[S(1)]$$

This is called the *risk-neutral valuation formula.* Q is called the *risk-neutral probability measure* or the *martingale measure.* If we use continuous compounded interest rate for a security with maturity T, we use the following approximation

$$\frac{1}{(1+r)^{T-t}} \cong e^{-r(T-t)}$$

and get the following general pricing formula for all kinds of securities under the money-market account as a numeraire

$$S(t) = e^{-r(T-t)} \cdot E^Q[S(T)].$$

In later chapters we will return to the meaning if this.

On a multi-period market we have

$$\begin{cases} B_0 = 1 \\ B_{n+1} = (1+r)B_n \end{cases}$$

and

$$\begin{cases} S_0 = s \\ S_{n+1} = Z_n S_n \end{cases}.$$

Definition 2.5 A *portfolio strategy* $h: \{h_t = (x_t, y_t); \quad t = 0, 1, 2..., T\}$ is a stochastic process with a *value process*

$$V_t^h = x_t B_t + y_t S_t.$$

Definition 2.6 A portfolio in discrete time is said to be *self-financing* if

$$x_t B_t + y_t S_t = x_{t-1} B_t + y_{t-1} S_t.$$

Definition 2.7 A portfolio in continuous time is said to be *self-financing* if

$$dV_t = x dB_t + y dS_t.$$

Consider a one-period binomial model (Fig. 2.5).

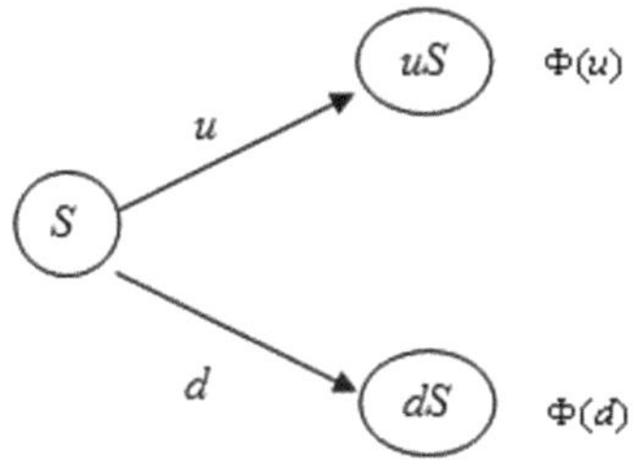

Fig. 2.5 In the on-step binomial model, the stock price may take two different prices, *uS* or *dS*. A derivative on the stock, e.g., a call option can therefore also take two different values, $\Phi(u)$ or $\Phi(d)$

In the Black–Scholes world, the stock prices S follow a stochastic process (the same as for a *Geometrical Brownian Motion*, GBM):

$$dS = \mu S dt + \sigma S dW,$$

where μ and σ are constants, which represents the *drift* and the *volatility* respectively. A binomial model is characterized by the constants u and d, describing how much the price can increase or decrease in each step in time and the probabilities that the price goes up and down. Since volatility measures the changes of the price with respect to time, u and d are functions of the volatility. The simplest model used is the *Cox–Ross–Rubinstein* model. In their model, the factors u and d is given by

$$\begin{aligned} u &= e^{\sigma \cdot \sqrt{dt}} \\ d &= e^{-\sigma \cdot \sqrt{dt}} . \end{aligned}$$

Here dt is the time interval between observations of the prices and σ the volatility of the underlying security. We will understand these formulas after defining the Wiener process dW above.

From Example 2.2 we can conclude that as soon as we know the volatility we also know the possible prices, or as soon the prices are known, we know the volatility. Therefore there exist a one-to-one relationship between the volatility and the prices. We also call the volatility, estimated from known prices, implied volatility.

With continuous compounding of interest rate r we have

$$S_0 = e^{-r \cdot dt}(q_u \cdot u \cdot S_0 + q_d \cdot d \cdot S_0) = e^{-r \cdot dt}(q_u \cdot u \cdot s + q_d \cdot d \cdot s),$$

where the risk-neutral probabilities are given by

$$\begin{cases} q_u + q_d = 1 \\ q_u \cdot u + q_d \cdot d = e^{r \cdot dt} \end{cases}$$

then

$$q_u = \frac{1}{u-d} \cdot \left[e^{r \cdot dt} - d\right] = \frac{a-d}{u-d} \qquad q_d = 1 - q_u.$$

Here we have defined a as e^{rdt}. To have an arbitrage-free market we must have $d < a < u$.

2.5.2 The Risk-Free Probability

One should ask what kind of information is offered from risk-neutral probability and where we can find this measure in the real world.

The first question leads to an equivalent definition of risk-neutral probability. A risk-neutral probability is the probability of a future event or state that both trading parties in the market agree upon.

Let us return to Example 2.2 where John and Lisa made an agreement based upon the likelihood or not of rain tomorrow. Both agreed that the probability of the event that it will rain tomorrow is 50 %, otherwise they wouldn't have reached that agreement and signed the contract. So this price reflects the common beliefs of both parties towards the probability that the event happens. 50 % is the risk-neutral probability of the event that happens. It is not a historical or statistical prediction of any kind. Nor is it a true probability. Simply put, it is just a belief that is shared between the two trading parties in the market.

For the simple example mentioned above, once the price is established, the risk-neutral measure is also determined. Whenever you have a pricing problem in which the event is measurable under this measure, you have to use this measure to avoid arbitrage. If you don't, it's as if you are simply giving out another price for the same event at the same time, which is an obvious arbitrage opportunity.

A more complicated example is the Black–Scholes world, in which we assume the stock follows a Brownian motion. In this setting, the stock price itself is enough to reveal the common belief between the trading parties towards the stock return distribution. The argument is similar to Example 2.2. And as a result, we have the famous Black–Scholes formula for European options. In the real world, the stock dynamics is not a Brownian motion, so the price given by Black–Scholes formula is just a reference price to the risk-free interest rate.

A more accurate information source for risk-neutral probability is the market prices of the stock options. In practice, people use options prices to get the risk-neutral measure and further price more complicated *contingent claims*, such as *exotic options*.

2.5.3 The Replicated Portfolio

Let us use $\Phi(u)$ to denote the value of the option if the stock price increases and $\Phi(d)$ to denote the option value if the stock price decreases. Also let x be the amount of money in our money-market account and y the number of stocks in our portfolio. We then have from the value process two equations with two unknowns, x and y,

$$\begin{cases} (1+r)\cdot x + uS_0 \cdot y = \Phi(u) \\ (1+r)\cdot x + dS_0 \cdot y = \Phi(d) \end{cases}.$$

We can solve this system of equations to find the *replicated* (balanced) *portfolio* in each node in the binomial tree:

$$\begin{cases} x = \dfrac{1}{1+r}\dfrac{u\cdot\Phi(d) - d\cdot\Phi(u)}{u-d} \\ y = \dfrac{1}{S_0}\dfrac{\Phi(u)-\Phi(d)}{u-d} \end{cases}.$$

This proves that the binomial model is *complete* since we can always replicate the option value with the money-market account and the value if the stock (see Definition 2.9 below). The price $\Pi[X, 0]$ of a contingent claim, X, is then, at time $t = 0$ given by

$$\Pi[X, 0] = x + y\cdot S_0 = \frac{1}{1+r}\{q_u\Phi(u) + q_d\Phi(d)\} = \frac{1}{1+r}E^Q[S_1],$$

where the risk-neutral probabilities is given as

$$q_u = \frac{(1+r)-d}{u-d}, \qquad q_d = \frac{u-(1+r)}{u-d}.$$

Here we have used simple compounding of the interest rate. With continuous compounding we write this for a single period of time as

$$\Pi[X, 0] = x + y \cdot S_0 = e^{-r}\{q_u\Phi(u) + q_d\Phi(d)\} = e^{-r}E^Q[S_1],$$

where

$$q_u = \frac{e^r - d}{u - d}, \qquad q_d = \frac{u - e^r}{u - d}.$$

Definition 2.8 A *contingent claim* (a financial derivative) is a stochastic variable $X = F(Z)$, where Z is a stochastic variable that is driving the stock price.

We interpret the contingent claim as a contract that generates X cash units at maturity. In other words, a contingent claim is a security (a financial instrument) whose value is dependent on the outcome of another underlying instrument.

Definition 2.9 A given contingent claim X is said to be *reachable* if there exist a portfolio h so that $V(h, 1) = X$ with probability one. $\left(\exists h \mid V_1^h = X,\right.$ with prob.1$\left.\right)$ Then, h is called a *hedging portfolio* or a *replicating portfolio* that *generates X.*

Definition 2.10 If all contingent claims are reachable, the market is said to be *complete.*

We have seen that the binomial model is complete, since we can replicate all contracts using the money-market account and the stock. The reason for completeness is that we have two securities, which solve the two equations. We can handle many periods in the binomial tree just because we can rebalance the tree with help of *intermediate trading.* For the same reason, a general market is complete if the number of securities is equal to the number of possible outcomes.

This can be stated as the following theorem.

Theorem 2.11 *The Meta Theorem. If we let N be the number of underlying securities on the market (excluding the risk-free) and K the number of random sources. Then*

(1) *The market is free of arbitrage if* $N \leq K$.
(2) *The market is complete of* $N \geq K$.
(3) *The market is complete and free of arbitrage if* $N = K$.

Theorem 2.12 *If the market is free of arbitrage there exists one (or many) martingale probability measure(s).*

Theorem 2.13 *If a martingale measure exists, the market is free of arbitrage.*

Remember, a martingale is a fair game.

Example 2.14

Consider a European call option with the strike price K, $(dS < K < uS)$. The contingent claim X is then, at maturity given by

$$X = \begin{cases} uS - K & \text{if } Z = u \\ 0 & \text{if } Z = d \end{cases}.$$

Expressed in option prices this is $\Phi(u) = uS - K$ and $\Phi(d) = 0$.

Example 2.15

Consider an American call option with the following data.

Current stock price	$S_0 = 100$
Volatility	$\sigma = 20$ %
Risk-free interest rate	$r = 5$ %
Strike	$K = 110$
Time to maturity $s = 20$ %	$T = 180$ days

Using a single iteration $\Delta t = 180/365 \cong 0.5$ year we get $(a = e^{r \cdot \Delta t})$:
$u = 1.152$
$d = 0.868$
$a = 1.025$
$q = 0.553$ (Fig. 2.6)

(continued)

Example 2.15 (continued)

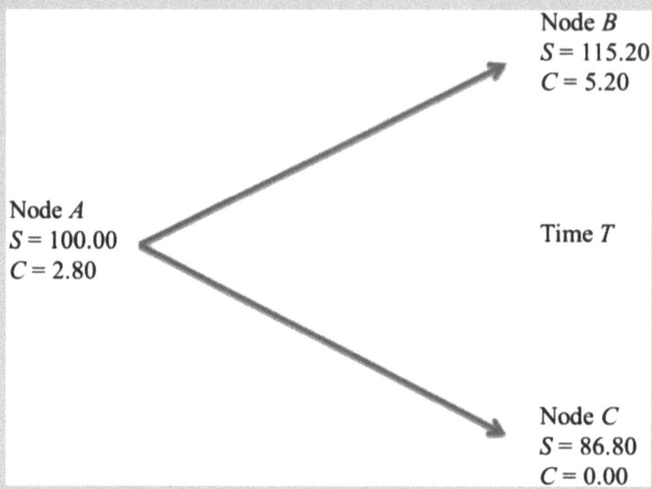

Fig. 2.6 In a one-step binomial model for an American call option, the stock price may take two different prices

The option value (fair value) in node A is then given by:

$$\begin{aligned} C &= \max(S_A - K, e^{-r\Delta t}(qC_B + (1-q)C_C)) \\ &= \max\left(100 - 110, e^{-0.05 \cdot 0.5}(0.553 \cdot 5.20 + (1 - 0.553) \cdot 0)\right) \\ &= 2.80 \end{aligned}$$

2.6 Modern Pricing Theory Based on Risk-Neutral Valuation

There are different kinds of investor; those who like to trade in a risk-neutral world and those who like to trade in a *risk-averse* world. As we have seen, in a risk-neutral world, prices are based on expected values of future payoffs. In a risk-averse world, investors choose the security with less risk if they have the same expected return. This leads to a *risk-return trade-off*.

We can illustrate this on a market where the interest rate is 10 % as (Fig. 2.7):

We use the risk-neutral valuation to value contingent claims. Contingent claims are securities which prices depends on the outcome from other sources. Bonds depend on interest rates and equity options depend on the outcome of the underlying equity. The pricing tool is always arbitrage conditions.

Fig. 2.7 A one-step binomial model in a risk-neutral and a risk averse world. The value 23.64 is calculated as $(30 \times 0.8 + 10 \times 0.20)/1.10 = 23.64$. A higher risk aversion leads to a lower price

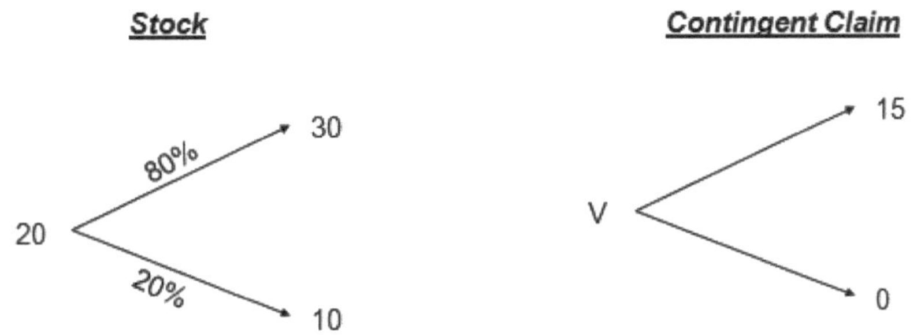

Fig. 2.8 A one-step binomial model for an underlying stock and an option

In arbitrage theory, equivalent securities (or portfolio of securities) should sell at equivalent prices. If not, arbitrage possibilities can be made from misaligned market prices. The no arbitrage requirement, lead to the law of one price. Therefore, we use the concept of arbitrage for pricing contingent claims.

We illustrate in Fig. 2.8 a simple arbitrage strategy for a contingent claim of a stock with an interest rate of 10 %.

Consider the portfolio by buying 0.75 shares the stock and sell one option. On an up movement, we get for the total portfolio, a value $30 \times 0.75 - 15 = 7.50$ and on a down movement $10 \times 0.75 - 0 = 7.50$. Therefore the strategy has a risk-less payoff of 7.50 in one period. But, since the risk-free security returns 10 %, arbitrage theory forces the return to be the same. Therefore the following must hold

$$(20 \times 0.75 - V) \times 1.10 = 7.50.$$

This gives an option price V, equal to 8.18, see Fig. 2.9. We then observe that the probability of the move in stock price where not used in the valuation of the option. But we can calculate the implied probabilities from the option

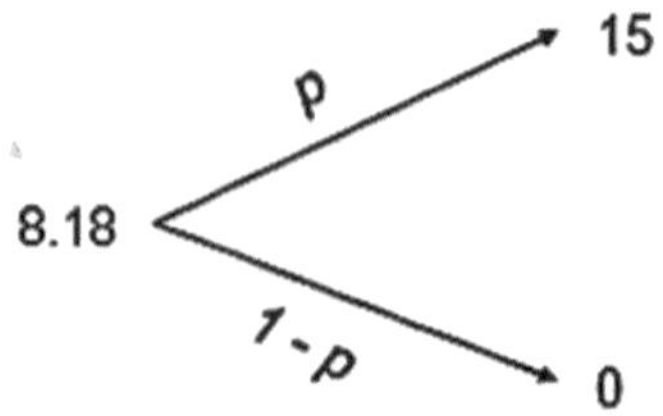

Fig. 2.9 The arbitrage-free price of the option let us calculate the risk-neutral probabilities

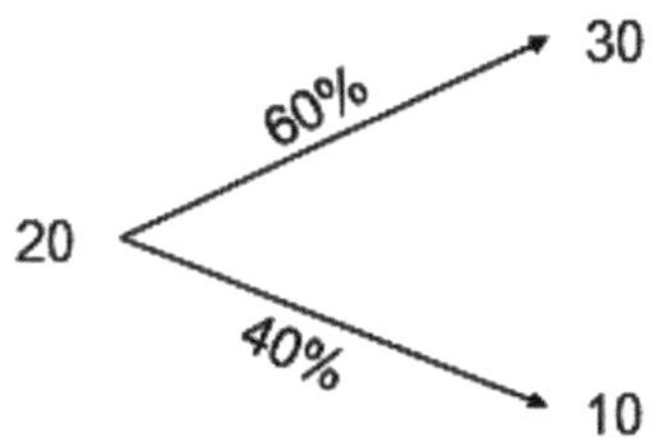

Fig. 2.10 The arbitrage-free price of the option gives the risk-neutral probabilities where $p = 0.6$. As we see, we have a relationship between the prices and probabilities

price. We therefore pretend that there are only risk-neutral probabilities. Then we have:

The risk-neutral probabilities can now be used to value *any* contingent claim of this stock. The risk-neutral probabilities are given by (Fig. 2.10)

$$\frac{15 - 8.18}{8.18}p + \left(\frac{0 - 8.18}{8.18}\right)(1 - p) = 10\ \%.$$

From $30p + 10(1 - p) = 30 \times 0.60 + 10 \times 0.40 = 22$ we also observe that in the risk-neutral world, also the return on the stock is 10 %.

We now introduce the *market price of* (volatility) *risk*. The market price of risk is defined as the extra compensation (per risk units) needed to take the higher risk. The stock and the option have the same source of risk, but the risk exposure is higher for the contingent claim. If we require, as in Fig. 2.9 that return of the stock, we get $30 \times 0.8 + 10 \times 0.2 = 26$. That is an expected return of 30 %. The risk (volatility) is given by $u = \frac{30}{20} = e^{\sqrt{t} \cdot \sigma} = e^{\sigma}$ i.e., $\sigma = \ln(1.5) = 40.5\%$.

Similarly, the option payoff we get is $15 \times 0.8 + 0 \times 0.2 = 12$. That is an expected return of $12/8.18 = 46.67\%$. Since the sharp ratio, market price of risk per volatility unit must be the same we must have

$$\lambda = \frac{E[r] - r_f}{\sigma} = \frac{30\ \% - 10\ \%}{40\ \%} = \frac{46.67\ \% - 10\ \%}{73.33\ \%} = 0.5,$$

where we have calculated the volatility for the option. We summarize this in Table 2.1.

Table 2.1 The stock and the option have difference return and risk

Security	Exp. Return	Risk (σ)
Stock	30%	40%
Contingent Claim	46.67%	73.33%

The conclusions of this are:

- Risk-neutral valuation is useful for contingent pricing
- For the real-world returns (what we observe) we have to include the market price of risk
- Shifting to the risk-neutral world will eliminate the extra return for accepting risk. This is usually a lower return
- All securities' returns are identical in the risk-neutral world and equal to the risk-free interest rate
- All securities that depend on the same underlying return earn the same risk premium, per unit of risk

2.6.1 An Example of Arbitrage

If the conclusions above do not hold, we can have a free lunch by making arbitrage. We always want to buy at a low and sell at a high price. Suppose we have a stock at 100 CU (cash units) and we want to buy an option with strike 110. We suppose that $u = 1.2$ and , $q_u = q_d = 0.5$ and $r = 0$. This gives us $S_0 = 100$, $uS_0 = 120$, $dS_0 = 80$ and

$$\begin{aligned} \Phi(u) &= \max(uS_0 - X, 0) = 10 \\ \Phi(d) &= \max(dS_0 - X, 0) = 0. \end{aligned}$$

The option value at $t = 0$ is then g $d = 0.8$ iven by $(0.5 * 10 + 0.5 * 0) = 5$ since $r = 0$.

Now, suppose someone on the market is trading the option for 8 CU (with the same price for bid and ask). We then take a short position in the option, invest 5 in shares and borrow 20 at the risk-free interest rate. We can then put 3 CU in our pocket to use for a free lunch.

$$\text{At time} = 0: \quad \begin{array}{r l} 8 & \text{sell the option} \\ 20 & \text{borrow from the bank} \\ -25 & \text{invest in a } 1/4 \text{ of a stock} \end{array}$$

This gives us 3 CU in our pocket.

If the stock price increases to 120, we can sell the shares to the price of $1/4 \times 120 = 30$, pay back the loan, 20 and pay the buyer 10 for the option

$$\text{At time} = 1: \quad \begin{array}{r l} -10 & \text{pay the buyer of the option} \\ -20 & \text{pay back to the bank} \\ 30 & \text{sell the } 1/4 \text{ of a stock} \end{array}$$

If the stock price decreases to 80, we can sell the shares to the price of $1/4 \times 80 = 20$ and pay back the loan, 20 to the bank. The option is worth nothing.

$$\text{At time} = 1: \quad \begin{array}{r l} 20 & \text{pay back to the bank} \\ 20 & \text{sell the } 1/4 \text{ of a stock} \end{array}$$

We still have our free lunch.

Suppose, on the other hand, that someone is trading the option at 3. Then we take a long position in the option, go short in the share, receiving $1/4 \times 100 = 25$ and put 20 at the bank. We then have 2 CU for the free lunch.

$$\text{At time} = 0: \quad \begin{array}{r l} -3 & \text{buy in the option} \\ -20 & \text{put money into the bank} \\ 25 & \text{sell a } 1/4 \text{ of a stock} \end{array}$$

This gives us 2 cash units in our pocket.

If the stock price increases, we will get 10 for the option, take the money from the bank and buy back the shares at 30.

$$\text{At time} = 1: \quad \begin{array}{r l} 10 & \text{payoff from the option} \\ 20 & \text{take back the money from the bank} \\ -30 & \text{buy back the } 1/4 \text{ of a stock} \end{array}$$

If the shares decrease, the option is worth nothing. We then take the money from the bank and buy the shares.

$$\text{At time} = 1: \quad \begin{array}{r l} 20 & \text{take back the money from the bank} \\ -20 & \text{buy back the } 1/4 \text{ of a stock} \end{array}$$

Remark We buy one option and want to hedge the change in the option value by Δ number of stocks where Δ is calculated as

$$\Delta = (C(u) - C(d))/(S(u) - S(d)) = (10 - 0)/(120 - 80) = 10/40 = \tfrac{1}{4}.$$

2.7 More on Binomial Models

Before we describe some other binomial models, we will discuss some general principles on building such models. First we define a *growth factor g*. This factor is the risk-free interest rate if we use the underlying instrument to value options and zero if we use the underlying forward/future in the valuation. The reason is the relation between the forward price F and the stock price S: $F = S \cdot e^{rT}$.

Furthermore, we can use the stochastic process for the stock price or its natural logarithm. For this reason we introduce the following variables

$$X = \frac{S_{i+1}}{S_i} \quad \text{and} \quad Y = \ln\left(\frac{S_{i+1}}{S_i}\right),$$

where $S_i = S(t)$ and $S_{i+1} = S(t + Dt)$. If we study a change in the stock price with the stochastic variable Y, we have the first order momentum in the normal distribution and if we use X we have a lognormal momentum. In the Black–Scholes world we have a price processes where:

$$S(t) = S_0 \cdot e^{\left(r - \frac{1}{2}\sigma^2\right) \cdot t + \sigma\sqrt{t} \cdot z(t)} \quad \text{where} \quad z(t) \sim N(0, 1).$$

This will be further explained in a later section. In the Black–Scholes world X follow a Brownian motion with the following expectation values and variances

$$E[X] = e^{g \cdot \Delta t}$$
$$Var(X) = E^2[X] \cdot \left(e^{\sigma^2 \cdot \Delta t} - 1\right) = e^{2 \cdot g \cdot \Delta t} \cdot \left(e^{\sigma^2 \cdot \Delta t} - 1\right)$$

and

$$E[Y] = \left(g - \frac{1}{2}\sigma^2\right) \cdot \Delta t.$$
$$Var(Y) = \sigma^2 \cdot \Delta t$$

When we approximate the Brownian motion with a binomial process we get

$$E[X] = q \cdot u + (1-q) \cdot d$$
$$E[Y] = q \cdot \ln u + (1-q) \cdot \ln d$$

and

$$Var(X) = q \cdot u^2 + (1-q) \cdot d^2 - E^2[X]$$
$$Var(Y) = q \cdot (\ln u)^2 + (1-q) \cdot (\ln d)^2 - E^2[Y].$$

For different choices of q, u and d we get different binomial models.

2.7.1 Normal Distribution with $q = ½$

With $Y = \ln(S_{i+1}/S_i)$ and $q = ½$ we get

$$E[Y] = q \cdot \ln u + (1-q) \cdot \ln d = \left(g - \frac{1}{2}\sigma^2\right) \cdot \Delta t$$
$$Var(Y) = E[Y^2] - E^2[Y] = q \cdot (\ln u)^2 + (1-q) \cdot (\ln d)^2 - E^2[Y] = \sigma^2 \Delta t$$

$$u = e^{\left(g - \frac{1}{2}\sigma^2\right) \cdot \Delta t + \sigma\sqrt{\Delta t}}$$
$$d = e^{\left(g - \frac{1}{2}\sigma^2\right) \cdot \Delta t - \sigma\sqrt{\Delta t}}.$$

2.7.2 Normal Distribution with $u = 1/d$

With $Y = \ln(S_{i+1}/S_i)$ and $u = 1/d$ we get

$$E[Y] = \left(g - \frac{1}{2}\sigma^2\right) \cdot \Delta t$$
$$Var(Y) = \sigma^2 \Delta t$$
$$q = \frac{1}{2} + \frac{g - \frac{1}{2}\sigma^2}{2 \cdot \sqrt{\sigma^2 + \left(g - \frac{1}{2}\sigma^2\right)^2}}$$
$$u = e^{\sqrt{\sigma^2 \cdot \Delta t + \left(g - \frac{1}{2}\sigma^2\right)^2 \cdot \Delta t^2}}.$$

2.7.3 Log-Normal Distribution with $q = ½$

With $X = S_{i+1}/S_i$ and $q = ½$ we get

$$\begin{aligned}
&E[X] = q \cdot u + (1-p) \cdot d = e^{g \cdot \Delta t}\\
&Var(Y) = E\left[Y^2\right] - E^2[Y] = q \cdot u^2 + (1-q) \cdot d^2 - E^2[Y] = e^{2 \cdot g \cdot \Delta t}\left(e^{\sigma^2 \Delta t} - 1\right)\\
&u = e^{g \cdot \Delta t}\left\{1 + \sqrt{\left(e^{\sigma^2 \cdot \Delta t} - 1\right)}\right\}\\
&d = e^{g \cdot \Delta t}\left\{1 - \sqrt{\left(e^{\sigma^2 \cdot \Delta t} - 1\right)}\right\}.
\end{aligned}$$

2.7.4 Log-Normal Distribution with $u = 1/d$

With $X = S_{i+1}/S_i$ and $u = 1/d$ we get

$$\begin{aligned}
&E[X] = e^{g \cdot \Delta t}\\
&Var(Y) = e^{2 \cdot g \cdot \Delta t}\left(e^{\sigma^2 \Delta t} - 1\right)\\
&q = \frac{e^{g \cdot \Delta t} - d}{u - d}\\
&u = \frac{1}{2} e^{-g \cdot \Delta t}\left(e^{\left(2 \cdot g + \sigma^2\right) \cdot \Delta t} + 1\right) + \sqrt{\frac{1}{4} e^{-2 \cdot g \cdot \Delta t}\left(e^{\left(2 \cdot g + \sigma^2\right) \cdot \Delta t} + 1\right)^2 - 1}.
\end{aligned}$$

2.7.5 Mixed Normal/Log-Normal Distribution

With $X = S_{i+1}/S_i$ and $Y = \ln\left(S_{i+1}/S_i\right)$ we get

$$\begin{aligned}
&E[X] = e^{g \cdot \Delta t}\\
&E[Y] = \left(g - \frac{1}{2}\sigma^2\right) \cdot \Delta t\\
&Var(Y) = \sigma^2 \cdot \Delta t\\
&u = e^{\left(g - \frac{1}{2}\sigma^2\right) \cdot \Delta t} \cdot e^{\sigma \cdot \sqrt{\left(\frac{1}{q} - 1\right)\Delta t}}\\
&d = e^{\left(g - \frac{1}{2}\sigma^2\right) \cdot \Delta t} \cdot e^{-\sigma \cdot \sqrt{\left(1 - \frac{1}{q}\right)\Delta t}},
\end{aligned}$$

where q is solved numerically by the equation

$$q \cdot e^{\sigma^2 \sqrt{\frac{1-q}{q}}} + (1-q) \cdot e^{\sigma^2 \sqrt{\frac{q}{1-q}}} = e^{\frac{\sigma^2}{2}\Delta t}.$$

2.7.6 The Cox–Ross–Rubinstein Model

This is perhaps the most common model

$$\begin{cases} u = e^{\sigma \cdot \sqrt{\Delta t}} \\ d = 1/u = e^{-\sigma \cdot \sqrt{\Delta t}} \end{cases}$$

$$q = \frac{e^{r \cdot \Delta t} - d}{u - d}.$$

2.7.7 The Second Order Cox–Ross–Rubinstein

This model is a variant of the model above and gives almost the same result.

$$\begin{cases} u = \dfrac{a^2 + b^2 + 1 + \sqrt{(a^2 + b^2 + 1)^2 - 4a^2}}{2a} \\ d = 1/u \end{cases}$$

$$\begin{cases} a = e^{r \cdot \Delta t} \\ b^2 = a^2 \cdot (e^{\sigma^2 \cdot \Delta t} - 1) \end{cases}$$

where

$$q = \frac{e^{r \cdot \Delta t} - d}{u - d}.$$

2.7.8 The Jarrow–Rudd Model

Also this model is a minor modification to the CCR with almost the same behavior.

$$\begin{cases} u = e^{\left(r - \frac{1}{2}\sigma^2\right) \cdot \Delta t + \sigma \cdot \sqrt{\Delta t}} \\ d = e^{\left(r - \frac{1}{2}\sigma^2\right) \cdot \Delta t - \sigma \cdot \sqrt{\Delta t}} \end{cases}$$

where

$$q = \frac{e^{r \cdot \Delta t} - d}{u - d}$$

For this model we have the following expectation and variance

$$E\left[\ln\left(\frac{S_{i+1}}{S_i}\right)\right] = q \cdot \ln u + (1-q) \cdot \ln d = \left(r - \frac{1}{2}\sigma^2\right)\Delta T$$

$$E\left[\left\{\ln\left(\frac{S_{i+1}}{S_i}\right)\right\}^2\right] = q \cdot (\ln u)^2 + (1-q) \cdot (\ln d)^2 = \sigma^2 \Delta T.$$

2.7.9 The Tian Model

If we also use the second order moments for the normal distribution we get this model with somewhat better accuracy:

$$\begin{cases} u = \frac{M \cdot V}{2}\left[V + 1 + \sqrt{V^2 + 2V - 3}\right] \\ d = \frac{M \cdot V}{2}\left[V + 1 - \sqrt{V^2 + 2V - 3}\right] \end{cases}$$

where

$$\begin{cases} M = e^{r \cdot \Delta t} \\ V = e^{\sigma^2 \cdot \Delta t} \end{cases}$$

$$q = \frac{e^{r \cdot \Delta t} - d}{u - d}.$$

2.7.10 The Tigori Model

In this model we model the logarithm of the stock price and define u as dx and d as $-dx$. Instead of multiplying with u and d we add dx and $-dx$

$$\begin{cases} dx = \sqrt{\sigma^2 \Delta t + \left(r - \frac{1}{2}\sigma^2\right)^2 \cdot (\Delta t)^2} \\ p = \frac{1}{2} + \frac{1}{2}\left(r - \frac{1}{2}\sigma^2\right)\frac{dx}{\Delta t} \end{cases}$$

2.7.11 The Leisen–Reimer Model

One of the latest binomial models is the Leisen–Reimer model. This model has an advantage against the other models. The model has quadratic convergency in the number of time steps, at least for European options and American call options, while the other models have a linear convergence. Therefore the accuracy is much better. Furthermore, since there are no (or small) oscillations in this model, we can use Richardson extrapolation to increase the accuracy even more. The Richardson extrapolation is, however, not always recommended for American put options (depending on the strike), since the early exercise will modify the tree in such a way that the extrapolation doesn't give any extra accuracy. First we define

$$a = e^{r \cdot \Delta t}$$

$$d_1 = \frac{\ln\left(\frac{S}{K}\right) + \left(r + \frac{1}{2}\sigma^2\right) \cdot (T - t)}{\sigma\sqrt{T - t}}$$

$$d_2 = d_1 - \sigma\sqrt{T - t},$$

where we recognize d_1 and d_2 from Black–Scholes equations. We then introduce:

$$p = B(d_2, N)$$
$$\overline{p} = B\left(d_2 + \sigma \cdot \sqrt{T - t}, N\right),$$

where B is the inverse of the binomial distribution and N the number of time steps. We use the Peizer–Pratt method to invert the binomial distribution $[j + ½ = n - (j + ½), n = 2j + 1]$

$$p = B(z, n) = \frac{1}{2} \mp \left[\frac{1}{4} - \frac{1}{4} \cdot \exp\left\{-\left(\frac{z}{n + ⅓}\right)^2 \cdot \left(n + \frac{1}{6}\right)\right\}\right]^{\frac{1}{2}},$$

where the sign is the sign of z. We get

$$\begin{cases} u = a \cdot \dfrac{\overline{p}}{p} \\ d = a \cdot \dfrac{1 - \overline{p}}{1 - p} \end{cases}$$

Remark With the inverse as above, we must have an odd number of time steps.

2.7.12 Black–Scholes Smoothing

There exists a method to get less oscillation in many models. Without oscillations we can use Richardson extrapolation Sect. 2.7.18 to get a more accurate result in price. The convergence is still linear before the extrapolation. The method is called Black–Scholes smoothing. We use the Black–Scholes formula discussed in Sect. 4.3 to calculate the values in three of the nodes, closest to the strike price (marked with an arrow), at the last time step, as in the Fig. 2.11.

The reason that the Black Scholes smoothing (also called mollification for dealing with ill-posed problems) minimizes the oscillations is that we get a much smoother distribution one step from maturity.

At maturity, the option value converges as to a "hockey stick"—that is, we can approximate the call option with a function:

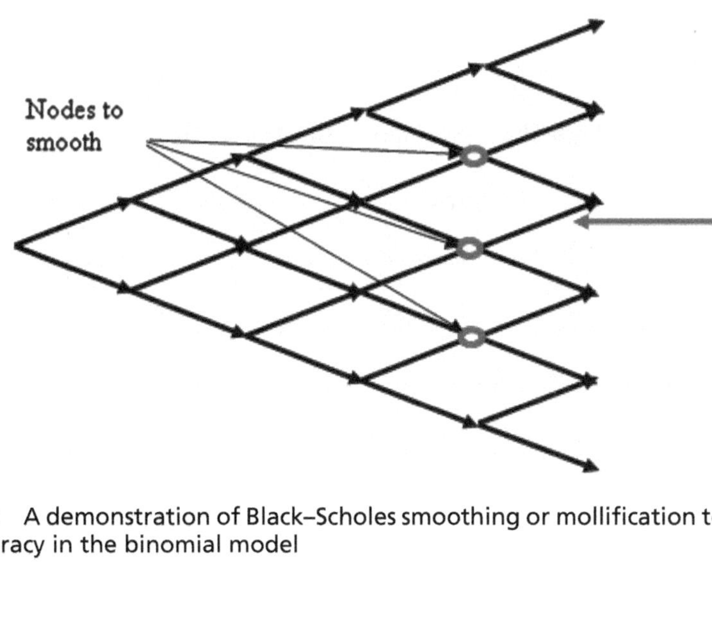

Fig. 2.11 A demonstration of Black–Scholes smoothing or mollification to increase the accuracy in the binomial model

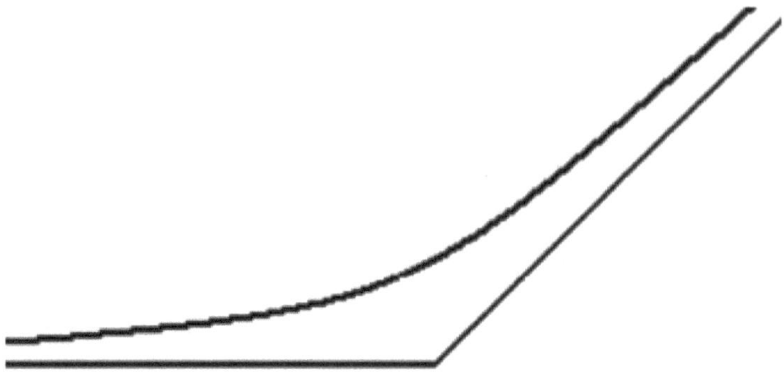

Fig. 2.12 This illustrates how the price of a call option as function of the underlying price behaves before maturity, where the price converges to the shape of a hockey stick

$$y = \begin{cases} 0 & \text{if } x < 0 \\ x & \text{if } x \geq 0 \end{cases}.$$

This function has a "knee" in $x = 0$. But, using the Black–Scholes formula at the last nodes, we add the time value and get a nice smooth curve (Fig 2.12).

When we build trees of different sizes we have no singularity in the curve, as we have if we use the two lines.

2.7.13 Pegging the Strike

Another method to get rid of the oscillations in the solutions in the binomial model is to "peg" the strike. If we start with the Cox–Ross–Rubinstein model

$$\begin{cases} u = e^{\sigma \cdot \sqrt{\Delta t}} \\ d = 1/u = e^{-\sigma \cdot \sqrt{\Delta t}} \end{cases}$$

$$q = \frac{e^{r \cdot \Delta t} - d}{u - d}$$

and replace the factors, u and d to

$$\begin{cases} u = e^{\sigma \cdot \sqrt{\Delta t} + \Delta t \cdot \ln(K/S)} \\ d = 1/u = e^{-\sigma \cdot \sqrt{\Delta t} + \Delta t \cdot \ln(K/S)} \end{cases}$$

we will dramatically reduce the oscillations since we always hit the strike in the tree. This gives a "sloped" tree and a result on which we can apply Richardson extrapolation to increase the accuracy even more.

To compare the models above, we will study the following American call option.

• Underlying price:	100
• Strike price:	110
• Time to maturity	183 days
• Risk-free interest rate	2 %
• Volatility:	40 %
• Number of time step	[25, 250]

The Black–Scholes value is 7.836944. CCR gives the result as in Fig. 2.13.

If we apply Black–Scholes smoothing we get the result as in Fig. 2.14.

If we also use Richardson extrapolation we finally get the result as in Fig. 2.15.

In Figs. 2.16 and 2.17 we see the Leisen–Reimer model without and with Richardson extrapolation respectively.

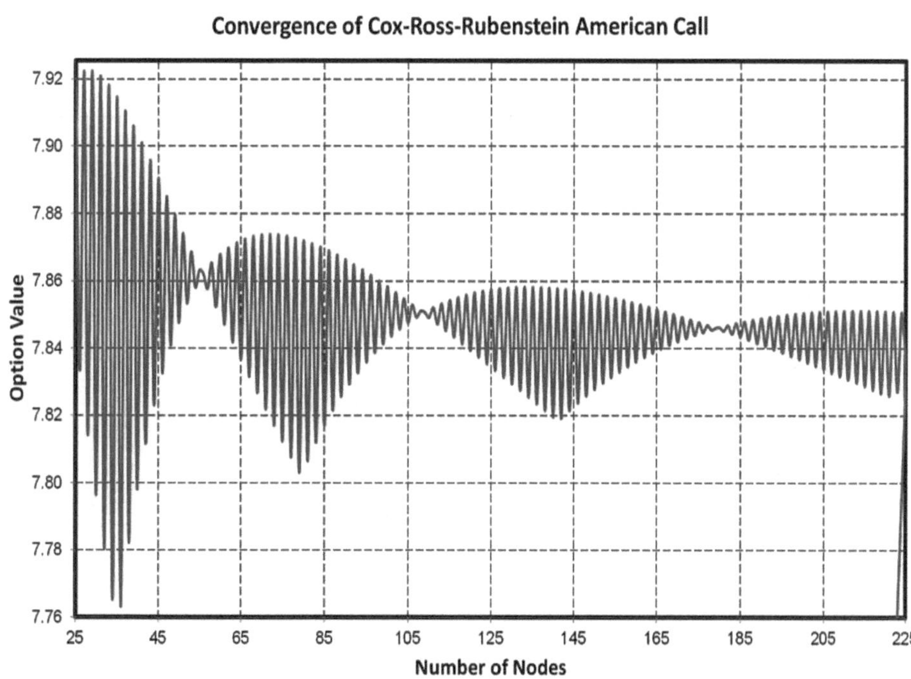

Fig. 2.13 The CCR convergence with oscillations

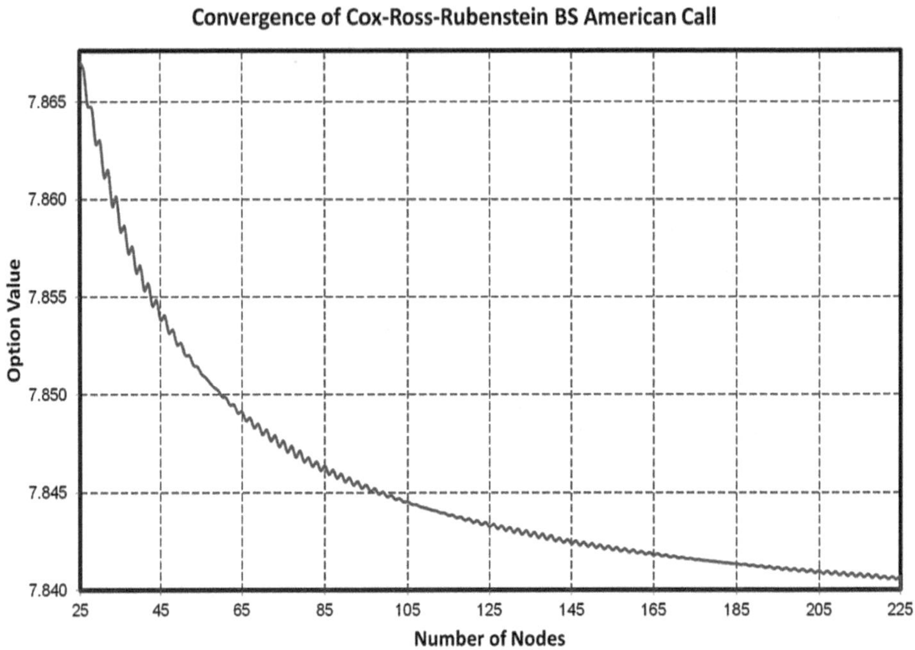

Fig. 2.14 The CCR convergence with Black–Scholes smoothing

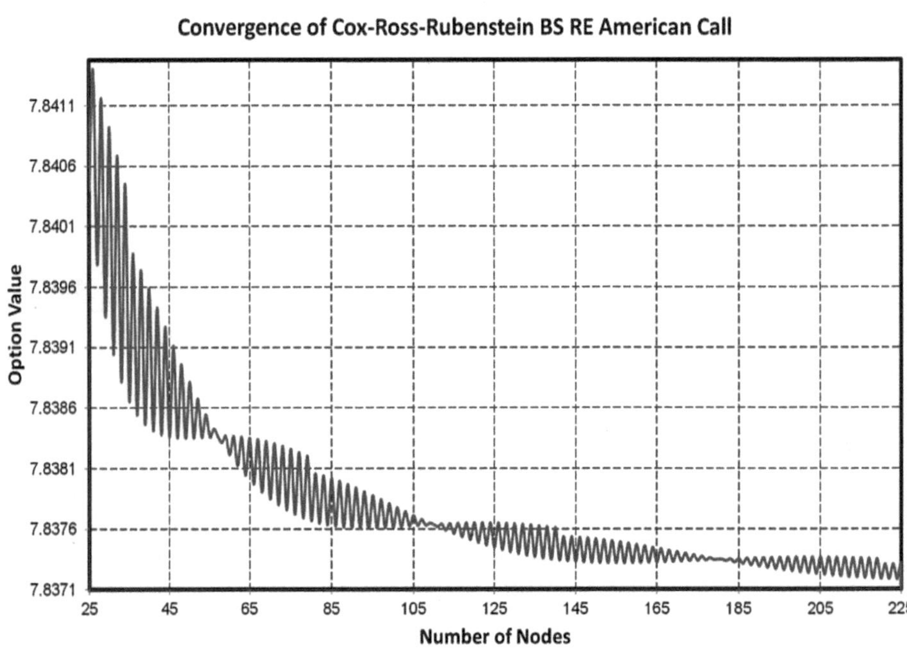

Fig. 2.15 The CCR convergence with Black–Scholes smoothing with Richardson extrapolation. Note the increasing accuracy in the option price

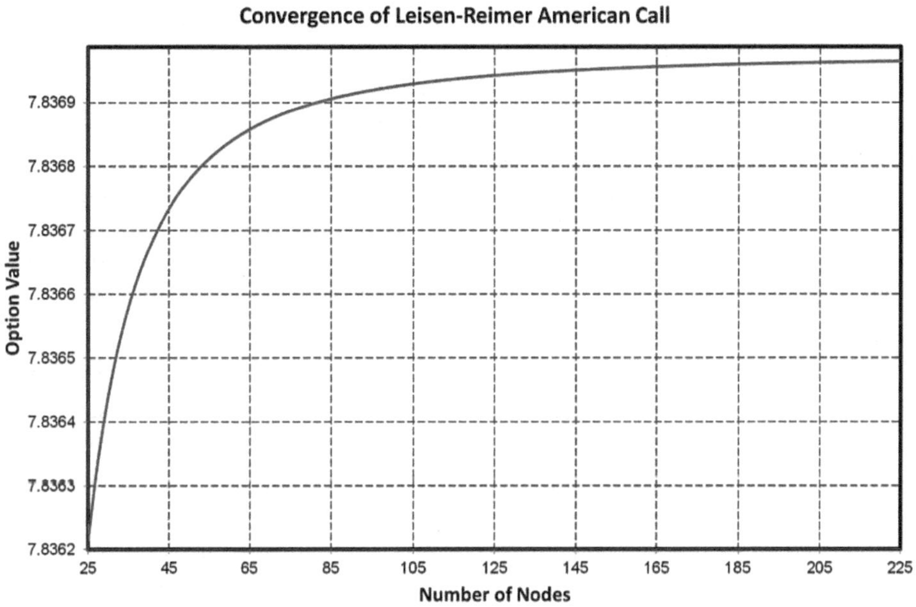

Fig. 2.16 The convergence using the Leisen–Reimer model

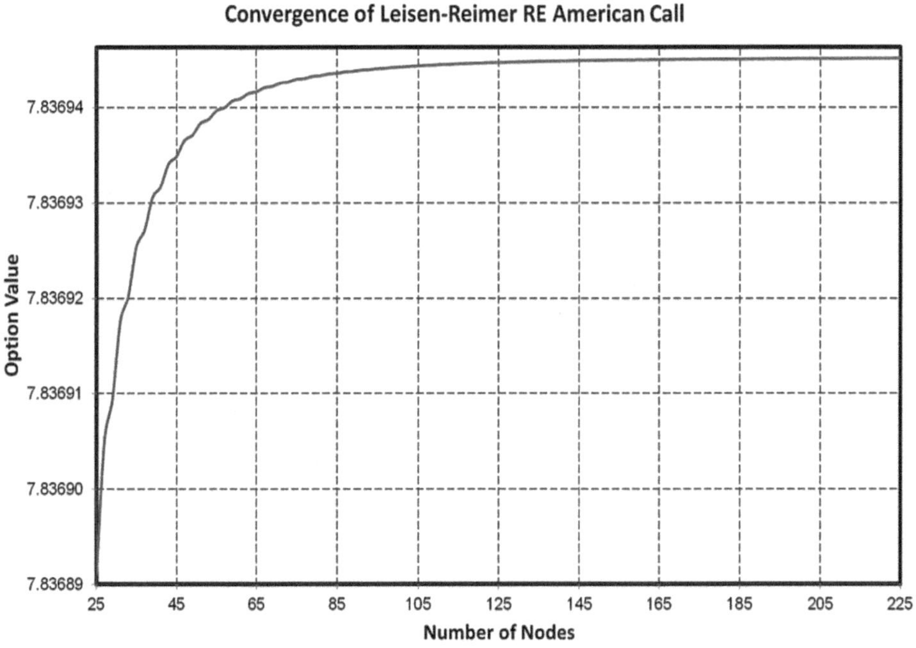

Fig. 2.17 The convergence using the Leisen–Reimer model with Richardson extrapolation. As we see, we need to use five decimal places on the y-axis

Note that how the number of decimals increases on the option value axis when we succeed to increase the accuracy of the option price.

Let's study an American put option to see how good the different binomial models will converge. We also use only odd nodes to minimize the oscillations due to jumps between odd and even numbers of nodes. In Fig. 2.18 we use the Cox–Ross–Rubinstein model (CCR), CCR with Black–Scholes smoothing (CRR+BS), CCR with Richardson extrapolation (CRR+BS+RE). We also using the method of pegging the strike price (PEG, PEG+BS and PEG+BS +RE). Finally, we use the Leisen–Reimer model, with and without Richardson extrapolation (LR and LR+RE).

In Fig. 2.19 we zoom in to see how the best binomial models behave. As we can see, the Cox–Ross–Rubinstein and the model with pegging the strike are very accurate when we combine Black–Scholes smoothing and Richardson extrapolation. The Leisen–Reimer model behaves very well both with and without Richardson extrapolation.

When we use the Richardson extrapolation we assume an error of second order. So we use two calculations for each value. This means that when we use

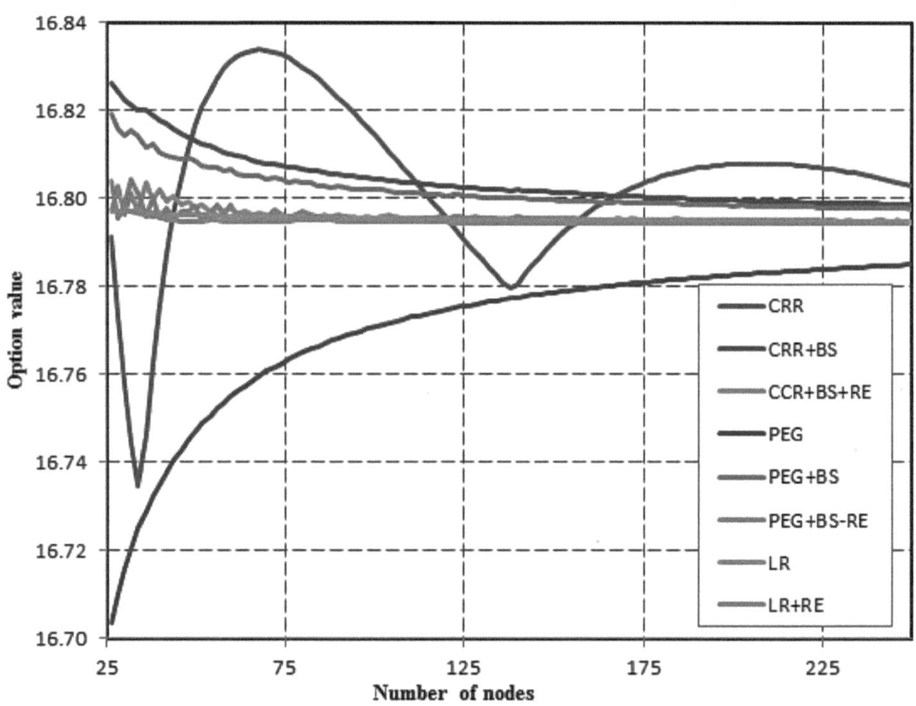

Fig. 2.18 Convergences in the different binomial models for a European call option

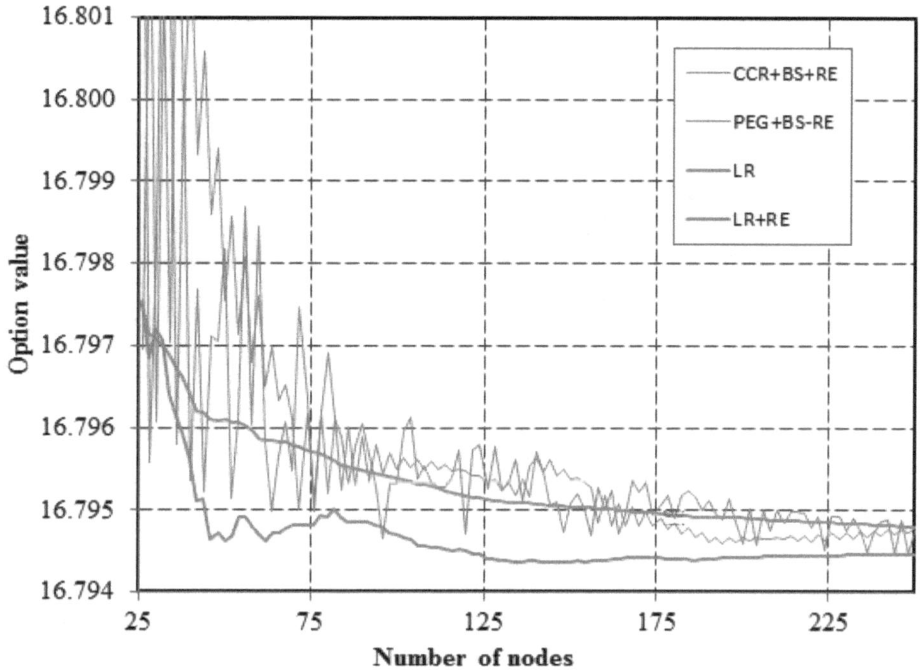

Fig. 2.19 A closer look at convergences in the different binomial models for a European call option

a calculation with N nodes, we also use the value by using $N/2$ nodes to eliminate the error to second order. Richardson extrapolation is explained in Sect. 2.7.17.

2.7.14 Binomial Model: The Numerical Algorithm

We will now briefly describe the algorithm for the binomial model. In the tree below, we study a bought American put option with strike K and a current stock price S_0.

The calculations of the option price, C_0 can be made as follows:

1. Start at the end of the tree (at time T). The lowest node has the value: $S_0 \cdot d^N$ where N is the number of time-steps. Set the boundary condition in these nodes with respect to the option type (see Sect. 2.7.15).
2. For the remaining nodes at the same time, move upwards and calculate the price by multiplying with u/d and use the same boundary condition.

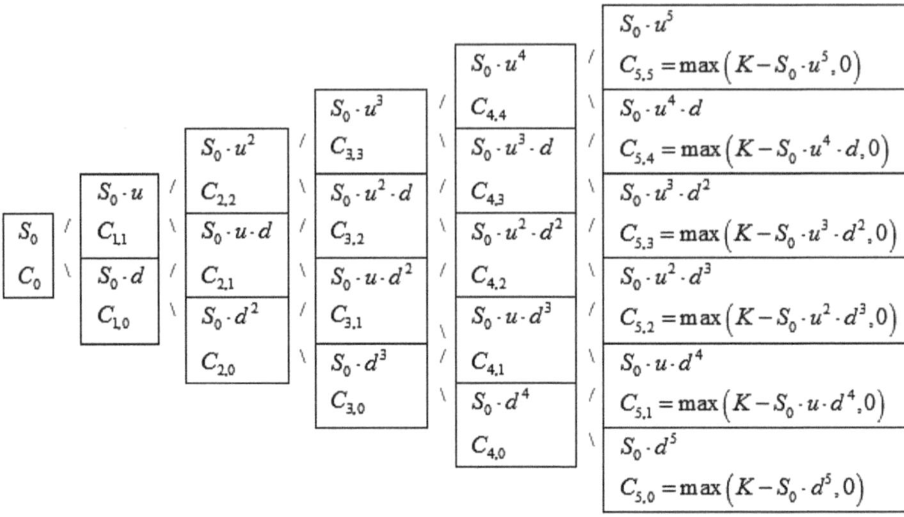

Fig. 2.20 How to implement a binomial model for an American put option

3. Go backwards in the tree and calculate all possible stock prices as in Fig. 2.20. Then calculate the option values $C_{i,j}$. For an American option this is done as

$$\begin{aligned} C_{4,4} &= \max\left\{K - S_0 u^4, e^{-r\cdot\Delta t}(q_u \cdot C_{5,5} + q_d \cdot C_{5,4})\right\} \\ C_{4,3} &= \max\left\{K - S_0 u^3 d, e^{-r\cdot\Delta t}(q_u \cdot C_{5,4} + q_d \cdot C_{5,3})\right\} \\ C_{4,2} &= \max\left\{K - S_0 \cdot u^2 \cdot d^2, e^{-r\cdot\Delta t}(q_u \cdot C_{5,3} + q_d \cdot C_{5,2})\right\} . \\ &\cdots\cdots\cdots \\ C_{4,0} &= \max\left\{K - S_0 \cdot d^4, e^{-r\cdot\Delta t}(q_u \cdot C_{5,1} + q_d \cdot C_{5,0})\right\} \end{aligned}$$

Since the option is of American type, and therefore can be exercised at any time, we need to calculate both the intrinsic and the discounted values. The intrinsic value is given by the strike price minus the stock value. If we instead have a European option, we do not need to calculate the intrinsic value, since we do not have the right to exercise. Therefore, we just have to calculate the discounted values

$$\begin{aligned} C_{4,4} &= e^{-r\cdot\Delta t} \cdot (q_u \cdot C_{5,5} + q_d \cdot C_{5,4}) \\ C_{4,3} &= e^{-r\cdot\Delta t} \cdot (q_u \cdot C_{5,4} + q_d \cdot C_{5,3}) . \\ &\cdots\cdots\cdots \\ C_{4,0} &= e^{-r\cdot\Delta t} \cdot (q_u \cdot C_{5,1} + q_d \cdot C_{5,0}) \end{aligned}$$

The American option will always have a value greater or equal the corresponding European option. The reason is that the American option is more flexible since it can be exercised at any time during the lifetime. When we are finished, the price of the option is given by C_0. The Greeks—that is, the hedge parameters—can be calculated using the values in the binomial trees:

$$\Delta = \frac{C_{1,1} - C_{1,0}}{S_0 \cdot u - S_0 \cdot d} \qquad \left(= \frac{\partial C}{\partial S}\right)$$

$$\Gamma = \frac{\dfrac{C_{2,2} - C_{2,1}}{S_0 \cdot u^2 - S_0 \cdot u \cdot d} - \dfrac{C_{2,1} - C_{2,0}}{S_0 \cdot u \cdot d - S_0 \cdot d^2}}{\dfrac{1}{2} \cdot \left(S_0 \cdot u^2 - S_0 \cdot d^2\right)} \qquad \left(= \frac{\partial^2 C}{\partial S^2}\right)$$

$$\Theta = \frac{C_{2,1} - C_0}{2 \cdot \Delta t} \qquad \left(= \frac{\partial C}{\partial t}\right).$$

To calculate Vega and Rho, we have to build two new trees where we use another volatility and risk-free interest rate, respectively. We can then use

$$\upsilon = \frac{C_0(\sigma) - C_0(\sigma + \Delta\sigma)}{\Delta\sigma} \qquad \left(= \frac{\partial C}{\partial \sigma}\right)$$

$$\rho = \frac{C_0(r) - C_0(r + \Delta r)}{\Delta r} \qquad \left(= \frac{\partial C}{\partial r}\right).$$

The hedge parameters in continuous time are defined by the partial derivatives

$$\Delta = \frac{\partial P}{\partial S}, \quad \Gamma = \frac{\partial^2 P}{\partial S^2}, \quad \Theta = \frac{\partial P}{\partial T}, \quad \upsilon = \frac{\partial P}{\partial \sigma} \quad \text{and} \quad \rho = \frac{\partial P}{\partial r}.$$

We use the hedge parameters to calculate the sensitivities in the option price with respect to the underlying price, the time to maturity, the volatility and the risk-free interest rate. With good accuracy it is also possible to build trees with different initial stock prices.

2.7.15 Boundary Conditions

At maturity we use the following conditions, depending on the option type:

$$\begin{array}{ll}
X(T) = \max(S(T) - K, 0) & \text{Bought call option.} \\
X(T) = -\max(S(T) - K, 0) = \min(K - S(T), 0) & \text{Sold call option.} \\
X(T) = \max(K - S(T), 0) & \text{Bought put option.} \\
X(T) = -\max(K - S(T), 0) = \min(S(T) - K, 0) & \text{Sold put option.}
\end{array}$$

2.7.16 More on Probabilities in the Binomial Model

When we build the Cox–Ross–Rubinstein tree, we use

$$u = e^{\sigma \cdot \sqrt{dt}}$$
$$d = e^{-\sigma \cdot \sqrt{dt}}.$$

The maximum stock price in the tree at maturity therefore becomes

$$S_{\max} = S_0 \cdot u^n = S_0 \cdot e^{n \cdot \sigma \cdot \sqrt{dt}}.$$

Similarly, the lowest stock price is given by

$$S_{\min} = S_0 \cdot d^n = S_0 \cdot e^{-n \cdot \sigma \cdot \sqrt{dt}}.$$

We also have the maximum and minimum probabilities at time to maturity

$$q(S_{\max}) = q_u^n \quad resp. \quad q(S_{\min}) = q_d^n.$$

The number of paths reaching the nodes at maturity is shown in the Fig. 2.21. This can be used to calculate the probability to reach a certain stock price.

If we let N represent the time-node and n the nodes for the level of the price then we can denote nodes in the tree by (N, n). We can now calculate the number of paths reaching a specific node by

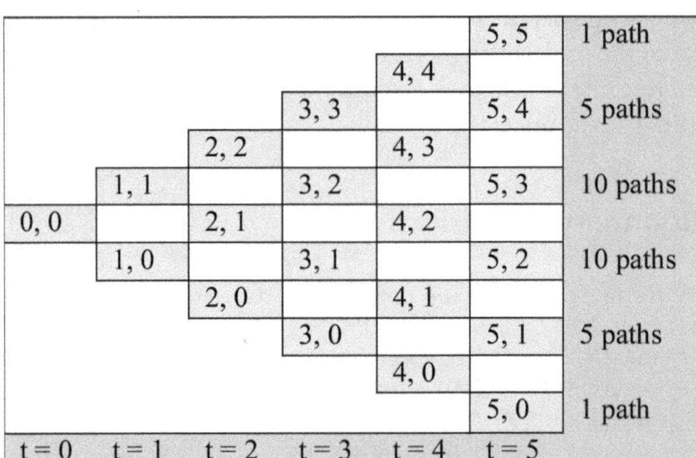

Fig. 2.21 The number of paths reaching the nodes at maturity in a binomial tree

$$\binom{N}{n} = \frac{N!}{n!(N-n)!}$$

The numbers of paths which reach node (5, 3) are given by

$$\binom{5}{3} = \frac{5!}{3!2!} = \frac{1\cdot 2\cdot 3\cdot 4\cdot 5}{(1\cdot 2\cdot 3)\cdot(1\cdot 2)} = \frac{4\cdot 5}{2} = 10.$$

The 10 paths to reach node (5, 3) can be expressed using up (*u*) and down (*d*) as {*uuudd, uudud, uuddu, uduud, ududu, udduu, duuud, duudu, duduu and dduuu*}. The probability of reaching the node is:

$$P = \binom{5}{3} q_u^3 \cdot q_d^2 = \left\{\text{if } q_u = q_d = \frac{1}{2}\right\} = 2^{-5}\cdot\binom{5}{3} = \frac{10}{32}.$$

We can use this to calculate the probability to get a profit (reaching the strike price) from our option, just by adding the probability for the nodes, which have a positive value (is in-the-money)

$$P(n) = \sum_{i=n}^{N} \binom{N}{i} q_u^i \cdot q_d^{N-i}.$$

In the limit when the number of time-steps goes to infinity, the probabilities will, according to the Central Limit Theorem, converge to a normal distribution.

2.7.17 Cox–Ross–Rubinstein Formula

From the above discussion we see that, for a European call option, we can write the price, with continuous interest rate as:

$$C(q,N,S,K) = e^{-N\cdot r\cdot\Delta t}\sum_{i=0}^{N}\binom{N}{i} q^i(1-q)^{N-i}\max\left(u^i d^{N-i} S - K, 0\right),$$

where

$$q = \frac{e^{N\cdot r\cdot\Delta t} - d}{u-d}.$$

We see that we don't need to sum from $i = 0$, since many of the low values will be zero due to $\max(u^i d^{N-i}S - K, 0)$. We can find the low index i_0 from

$$\max\left(u^i d^{N-i}S - K, 0\right) = \begin{cases} u^i d^{N-i}S - K & \text{if} \quad i \geq i_0 \\ 0 & \text{else} \end{cases}.$$

We can find i_0 by taking the logarithm and using $u = 1/d = e^{\sigma\sqrt{\Delta t}}$

$$i_0 = \frac{\ln(K/S) + \sigma\sqrt{\Delta T}}{2\sigma\sqrt{\Delta T}}.$$

We then get

$$C = C(q, N, S, K) = e^{-N \cdot r \cdot \Delta t} \sum_{i=i_0}^{N} \binom{N}{i} q^i (1-q)^{N-i} \left(u^i d^{N-i} S - K\right).$$

So

$$\begin{aligned} C &= S \cdot \sum_{i=i_0}^{N} \binom{N}{i} (q \cdot u \cdot e^{-r \cdot \Delta t})^i \left((1-q) \cdot d \cdot e^{-r \cdot \Delta t}\right)^{N-i} - e^{-N \cdot r \cdot \Delta t} K \cdot \sum_{i=i_0}^{N} \binom{N}{i} q^i (1-q)^{N-i} \\ &= S \cdot \sum_{i=i_0}^{N} b(i, N, q \cdot u \cdot e^{-r \cdot \Delta t}) - e^{-N \cdot r \cdot \Delta t} K \cdot \sum_{i=i_0}^{N} b(i, N, q) \\ &= S \cdot \Phi(i_0, N, q \cdot u \cdot e^{-r \cdot \Delta t}) - e^{-r \cdot T} K \cdot \Phi(i_0, N, q) \end{aligned}$$

where $\Phi(i, n, q)$ is the binomial probability distribution. In the above calculations we have used that

$$q \cdot u \cdot e^{-r \cdot \Delta t} + (1-q) \cdot d \cdot e^{-r \cdot \Delta t} = 1$$

To find the formula for a put option, we can use the put-call parity (see Sect. 4.5).

$$\begin{aligned} P &= e^{-r \cdot T} K \cdot \left(1 - \Phi(i_0, N, q)\right) - S \cdot \left(1 - \Phi(i_0, N, q \cdot u \cdot e^{-r \cdot \Delta t})\right) \\ &= e^{-r \cdot T} K \cdot \Phi(N - i_0 + 1, N, 1 - q) - S \cdot \Phi(N - i_0 + 1, N, 1 - q \cdot u \cdot e^{-r \cdot \Delta t}) \end{aligned}$$

The formulas for call and put options above can be compared by the famous Black–Scholes–Merton formula in continuous time. We will discuss the continuous time and Black–Scholes in Chap. 4.

Example 2.16

Compute the price of an American put option with strike price $K = 100$ and exercise time $T = 2$ years, using a binomial tree with two trading dates $t_1 = 0$ and $t_2 = 1$ (your portfolio at time $t_3 = 2$ is the same as your portfolio at time $t_2 = 1$) and parameters $s_0 = 100$, $u = 1.4$, $d = 0.8$, $r = 10\%$, and $p = 0.75$.

Solution

First of all, we have to calculate the risk neutral probabilities. With a simple discounting the probability for an increasing price is given by

$$q = q_u = \frac{1 + r - d}{u - d} = 0.5 \quad \text{and} \quad q_d = 1 - q_u = 0.5\,.$$

Using them we obtain the binomial tree as in Fig. 2.22, where the value of the stock is written in the nodes and the value of the option in the adjacent boxes. The value 20 adjacent to the node with stock price 80 is obtained as max{*exercise value, discounted binomial value*}

$$C' = \max\left\{100 - 80, \frac{1}{1 + 0.10} \cdot \left(\frac{1}{2} \cdot 0 + \frac{1}{2} \cdot 36\right)\right\} = \max\{20, \quad 16.36\} = 20.$$

Thus, an early exercise of the option is optimal in this node. The total price of the option is then given by

$$C = \max\left\{100 - 100, \frac{1}{1 + 0.10} \cdot \left(\frac{1}{2} \cdot 0 + \frac{1}{2} \cdot 20\right)\right\} = \max\{0, \quad 9.09\} = 9.09.$$

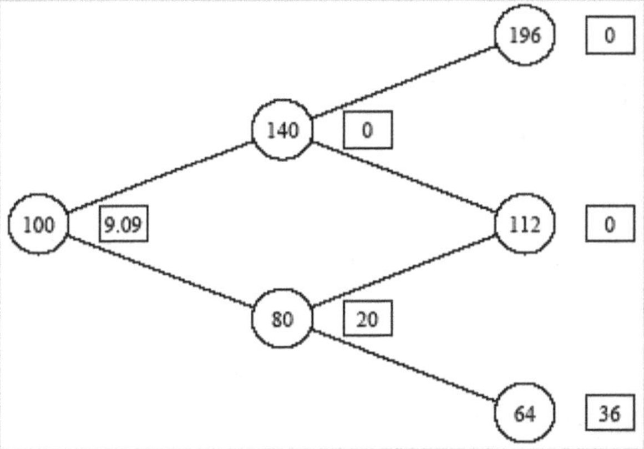

Fig. 2.22 The binomial tree, given the parameters $s_0 = 100$, $u = 1.4$, $d = 0.8$ and $r = 10\%$

(continued)

Remark 2.17

We never use the objective probability, $p = 0.75$. In a risk-neutral world we use the risk-neutral martingale probabilities.

Remark 2.18

As soon as we know the volatility of a specific model, the possible prices are known. In the binomial model, u and d are given by the volatility. In this example u and d were explicitly given. As soon we know the possible prices, we also know the probability distribution, here given by q_u and q_d. This means that there is a one-to-one relationship between the volatilities and the prices. We will discuss this in more detail later.

The price of the American option is thus 9.09. The corresponding price of a European put with the same parameters as above, is given by

$$C^E = \frac{1}{1+0.10} \cdot \left(\frac{1}{2} \cdot 0 + \frac{1}{2} \cdot 16.36 \right) = 7.44.$$

This price is lower than the American option since we cannot make an early exercise.

Example 2.19

Compute the price of an European binary asset-or-nothing call option with strike price $K = 120$ and exercise time $T = 2$ years, using a binomial tree with two trading dates $t_1 = 0$ and $t_2 = 1$ (your portfolio at time $t_3 = 2$ is the same as your portfolio at time $t_2 = 1$) with parameters $s_0 = 80$,

$u = 1.5$, $d = 0.5$, $r = 0\%$, and $q = 0.5$.

Solution

An asset-or-nothing call means that, if we reach the strike, we will get the asset—that is, no payments are made for the underlying asset as it is for a plain vanilla call option where we have the right to buy the underlying asset at the strike price. We use the tree in Fig. 2.23.

We get the price of the option as 45.

We can use the values in the tree to calculate the replicating portfolio. At $t = 0$ the following must hold:

$$\begin{cases} x + y \cdot 120 = 90 \\ x + y \cdot 40 = 0 \end{cases}.$$

This means that the value process for the replicating portfolio, consisting of the money-market account (B) and the underlying asset must be equal to the option value. In other words, regardless if the stock price increase or decrease, the value of the portfolio should equal the value of the option. This yields that $x = -45$ and $y = 9/8$. We can also use the formula we derived in Sect. 2.5.3:

(continued)

Remark 2.17 (continued)

$$\begin{cases} x = \dfrac{1}{1+r}\dfrac{u\cdot\Phi(d)-d\cdot\Phi(u)}{u-d} = \dfrac{1}{1}\dfrac{1.5\cdot 0-0.5\cdot 90}{1.5-0.5} = -45 \\ y = \dfrac{1}{S_0}\dfrac{\Phi(u)-\Phi(d)}{u-d} = \dfrac{1}{80}\dfrac{90-0}{1.5-0.5} = \dfrac{9}{8} \end{cases}.$$

The same calculations can be made to find the replicated portfolio in all the nodes, e.g., where $S=120$

$$\begin{cases} x = \dfrac{1}{1}\dfrac{1.5\cdot 0-0.5\cdot 180}{1.5-0.5} = -90 \\ y = \dfrac{1}{120}\dfrac{180-0}{1.5-0.5} = \dfrac{3}{2} \end{cases}.$$

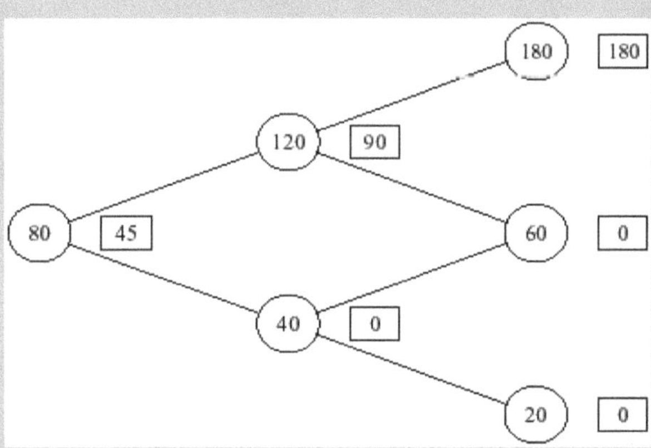

Fig. 2.23 The binomial tree for an asset-or-nothing call, given the parameters $s_0=80$, $u=1.5$, $d=0.5$, $r=0\%$, and $q=0.5$

2.7.18 Richardson Extrapolation

For those who have not been studying Richardson extrapolation in numerical analysis, we will here give a short introduction. Suppose we have a numerical method with a known error of order p (i.e., *error* ~ h^p where h is small value representing the accuracy in some measurement, e.g. in time, Δt):

$$F = F(h) + O(h^p) = F(h) + c\cdot h^p + O\left(h^{p+1}\right).$$

If we study two such values of h, (h_1, h_2) giving

$$\begin{cases} F = F(h_1) + c \cdot h_1^p + O\left(h_1^{p+1}\right) \\ F = F(h_2) + c \cdot h_2^p + O\left(h_2^{p+1}\right) \end{cases}$$

it is possible to eliminate the constant c by multiplying the first equation with h_2^p and the second wit h_1^p and then subtract them

$$\left(h_2^p - h_1^p\right)F = h_2^p F(h_1) - h_1^p F(h_2) + O\left(h^{2p+1}\right).$$

We then have

$$F = \frac{h_2^p F(h_1) - h_1^p F(h_2)}{h_2^p - h_1^p} + O\left(h^{p+1}\right).$$

In that way, we have increased the accuracy from order $O(h^p)$ to $O(h^{p+1})$. Typically, we have $h_1 = h$ and $h_2 = h/2$:

$$F_R = \frac{(h/2)^p F(h) - h^p F(h/2)}{(h/2)^p - h^p} = \frac{2^p F(h/2) - F(h)}{2^p - 1}.$$

Example 2.20

Suppose we want to find an approximation of the derivative of the function $f(x) = e^{-x} \sin(x)$ at the point $x = 1.0$ by using a centred divided difference formula and Richardson extrapolation. If we use $h = 0.5$ and $h = 0.25$, we get

$$\frac{e^{-(1,0+0.5)} \cdot \sin(1.0 + 0.5) - e^{-(1.0-0.5)} \cdot \sin(1.0 - 0.5)}{2.0 \cdot 0.5} = -0.068215072$$

and

$$\frac{e^{-(1+0.25)} \cdot \sin(1 + 0.25) - e^{-(1-0.25)} \cdot \sin(1 - 0.25)}{2 \cdot 0.25} = -0.100189411$$

Neither of these approximations is near the correct answer, −0.11079376, however, using one step of Richardson extrapolation we get

$$\frac{4 \cdot (-0.100189411) - (-0.068215072)}{3} = -0.110847524.$$

This value is much closer to the correct value.

2.8 Finite Difference Methods

We will now discuss how to find a numerical solution to partial differential equations (PDEs). In particular, we will consider parabolic boundary value problems of the Black–Scholes type

$$-\frac{\partial C}{\partial t}=\frac{1}{2}S^2\sigma^2\frac{\partial^2 C}{\partial S^2}+(r-\delta)S\frac{\partial C}{\partial S}-rC.$$

This equation can be solved by numerical methods, and we will here discuss some of the most common techniques. Anyone who knows how to numerically approximate derivatives and has some experience in, for example, Microsoft Excel can easily solve the partial differential equation above. If we let $x=\ln(S)$ we can rewrite the PDE above by the use of the chain rule:

$$\begin{aligned}
\frac{\partial C}{\partial S}&=\frac{\partial C}{\partial x}\frac{\partial x}{\partial S}=\frac{1}{S}\frac{\partial C}{\partial x}\\
\frac{\partial^2 C}{\partial S^2}&=\frac{\partial}{\partial S}\left(\frac{1}{S}\frac{\partial C}{\partial x}\right)=-\frac{1}{S^2}\frac{\partial C}{\partial x}+\frac{1}{S}\frac{\partial}{\partial x}\frac{\partial C}{\partial S}=-\frac{1}{S^2}\frac{\partial C}{\partial x}+\frac{1}{S}\frac{\partial}{\partial x}\left(\frac{1}{S}\frac{\partial C}{\partial x}\right)\\
&=-\frac{1}{S^2}\frac{\partial C}{\partial x}+\frac{1}{S^2}\frac{\partial^2 C}{\partial x^2}
\end{aligned}$$

We then get

$$-\frac{\partial C}{\partial t}=\frac{1}{2}\sigma^2\frac{\partial^2 C}{\partial x^2}-\frac{1}{2}\sigma^2\frac{\partial C}{\partial x}+(r-\delta)\frac{\partial C}{\partial x}-rC$$

so

$$-\frac{\partial C}{\partial t}=\frac{1}{2}\sigma^2\frac{\partial^2 C}{\partial x^2}+\upsilon\frac{\partial C}{\partial x}-rC,$$

where $\nu=r-\delta-½\sigma^2$. By doing this we have removed the explicit dependencies of S and thereby get the coefficients independent of the stock price (see the coefficients p_u, p_m and p_d below).

2.8.1 Derivative Approximations

Mathematically, the partial derivative of a function $f(x, y)$ with respect to y is defined by

$$\frac{\partial f(x,y)}{\partial y} = \lim_{\Delta y \to o} \frac{f(x,y+\Delta y) - f(x,y)}{\Delta y}.$$

This can be is approximated with

$$\frac{\partial f(x,y)}{\partial y} \cong \frac{f(x,y+\Delta y) - f(x,y)}{\Delta y},$$

where Δy represent a small change in the variable y. The above approximation is called a forward difference since the difference is in the forward direction. Similarly, the backward difference is defined by

$$\frac{\partial f(x,y)}{\partial y} \cong \frac{f(x,y) - f(x,y-\Delta y)}{\Delta y}.$$

A central difference is therefore given by

$$\frac{\partial f(x,y)}{\partial y} \cong \frac{f(x,y+\Delta y) - f(x,y-\Delta y)}{2\Delta y}.$$

A more stable central difference scheme is

$$\frac{\partial f(x,y)}{\partial y} \cong \frac{f(x,y+\Delta y/2) - f(x,y-\Delta y/2)}{\Delta y}.$$

To find a difference scheme for the second order derivative we use:

$$\begin{aligned}\frac{\partial^2 f(x,y)}{\partial y^2} &= \frac{\partial}{\partial y}\left[\frac{f(x,y+\Delta y/2)}{\partial y} - \frac{f(x,y-\Delta y/2)}{\partial y}\right] \\ &\cong \frac{f(x,y+\Delta y) - 2\cdot f(x,y) + f(x,y-\Delta y)}{(\Delta y)^2}\end{aligned}$$

By substituting these into a partial differential equation we get a scheme to solve it. We will now study three different schemas that are widely used in practice.

The Explicit Finite Difference Method

If we use the following approximations of the derivatives,

$$\frac{\partial C}{\partial x} = \frac{C_{i+1,j+1} - C_{i+1,j-1}}{2 \cdot \Delta x}$$
$$\frac{\partial^2 C}{\partial x^2} = \frac{C_{i+1,j+1} - 2 \cdot C_{i+1,j} + C_{i+1,j-1}}{\Delta x^2},$$

which are called backward differences, we can then write the Black–Scholes PDE as

$$-\frac{C_{i+1,j} - C_{i,j}}{\Delta t} = \frac{1}{2}\sigma^2 \frac{C_{i+1,j+1} - 2 \cdot C_{i+1,j} + C_{i+1,j-1}}{\Delta x^2} + \nu \cdot \frac{C_{i+1,j+1} - C_{i+1,j-1}}{2 \cdot \Delta x} - r \cdot C_{i+1,j}$$

Here i is the time index and j the price index. With some rearrangement we have

$$C_{i,j} = \frac{1}{1 + r \cdot \Delta t}\left(p_u \cdot C_{i+1,j+1} + p_m \cdot C_{i+1,j} + p_d \cdot C_{i+1,j-1}\right),$$

where

$$p_u = \frac{1}{2} \cdot \Delta t \cdot \left(\frac{\sigma^2}{\Delta x^2} + \frac{\nu}{\Delta x}\right)$$
$$p_m = 1 - \Delta t \cdot \frac{\sigma^2}{\Delta x^2}$$
$$p_d = \frac{1}{2} \cdot \Delta t \cdot \left(\frac{\sigma^2}{\Delta x^2} - \frac{\nu}{\Delta x}\right),$$

and where $1/(1 + rDt)$ is the discount factor or an approximation of $e^{-r\Delta t}$. One can show that this method is equivalent to using a trinomial tree (see Sect. 4.12.5). For stability and convergence reasons it has been shown that we should use Δx and Δt such as

$$\Delta x \geq \sigma\sqrt{3 \cdot \Delta t}.$$

As we can see, the result from this method is explicitly given because we know the value (the claim) at the boundary where the option expires. Then, we perform the calculation backwards in time until the valuation date. Since the time dependence (i) only depends on future dates ($i+1$) we can explicitly calculate the change, node by node backward in time.

The Implicit Finite Difference Method

If we instead use the following approximations of the derivatives

$$\frac{\partial C}{\partial x} = \frac{C_{i,j+1} - C_{i,j-1}}{2 \cdot \Delta x}$$
$$\frac{\partial^2 C}{\partial x^2} = \frac{C_{i,j+1} - 2 \cdot C_{i,j} + C_{i,j-1}}{\Delta x^2},$$

which are called forward differences, the stability and convergence will increase considerably. But, for each time-step we now have to solve a system of equations. The Black–Scholes PDE with forward differences is given by

$$-\frac{C_{i+1,j} - C_{i,j}}{\Delta t} = \frac{1}{2} \cdot \sigma^2 \cdot \frac{C_{i,j+1} - 2 \cdot C_{i,j} + C_{i,j-1}}{\Delta x^2} + \nu \cdot \frac{C_{i,j+1} - C_{i,j-1}}{2 \cdot \Delta x} - r \cdot C_{i+1,j}$$

With some rearrangement, we have:

$$p_u \cdot C_{i,j+1} + p_m \cdot C_{i,j} + p_d \cdot C_{i,j-1} = C_{i+1,j},$$

where

$$p_u = \frac{1}{2} \cdot \Delta t \cdot \left(\frac{\sigma^2}{\Delta x^2} + \frac{\nu}{\Delta x} \right)$$
$$p_m = 1 + \Delta t \cdot \frac{\sigma^2}{\Delta x^2} + r \cdot \Delta t$$
$$p_d = \frac{1}{2} \cdot \Delta t \cdot \left(\frac{\sigma^2}{\Delta x^2} - \frac{\nu}{\Delta x} \right).$$

Using the boundary conditions

$$C_{i,N_j} - C_{i,N_{j-1}} = \lambda_U$$
$$C_{i,-N_{j+1}} - C_{i,-N_j} = \lambda_L,$$

we have a system with $2N_j + 1$ equations. The boundary conditions depend on the type of option. For a call option we have

$$\lambda_U = S_{i,N_j} - S_{i,N_{j-1}}$$
$$\lambda_L = 0$$

and for a put option

$$\lambda_U = 0$$
$$\lambda_L = S_{i,-N_j} - S_{i,-N_{j+1}}.$$

The corresponding system of equations can be expressed as

$$\begin{bmatrix} 1 & -1 & 0 & \dots & \dots & \dots & 0 \\ p_u & p_m & p_d & 0 & \dots & \dots & 0 \\ 0 & p_u & p_m & p_d & 0 & \dots & 0 \\ \dots & \dots & \dots & \dots & \dots & \dots & \dots \\ 0 & \dots & 0 & p_u & p_m & p_d & 0 \\ 0 & \dots & \dots & 0 & p_u & p_m & p_d \\ 0 & \dots & \dots & \dots & 0 & 1 & -1 \end{bmatrix} \begin{bmatrix} C_{i,N_j} \\ C_{i,N_{j-1}} \\ C_{i,N_{j-2}} \\ \dots \\ C_{i,-N_{j+2}} \\ C_{i,-N_{j+1}} \\ C_{i,-N_j} \end{bmatrix} = \begin{bmatrix} \lambda_U \\ C_{i+1,N_{j-1}} \\ C_{i+1,N_{j-2}} \\ \dots \\ C_{i+1,N_{j+2}} \\ C_{i+1,N_{j+1}} \\ \lambda_L \end{bmatrix}.$$

As we can see above, in this model the future depends on the past. We don't know the history but we know the value of the option (claim) in the future. This is given by the boundary condition. Therefore, we have to solve this system of equation for all time-steps.

The Crank–Nicholson Method

If we combine the forward and backward differences we can get an even better method, the famous Crank–Nicholson method,

$$-\frac{C_{i+1,j}-C_{i,j}}{\Delta t}=\frac{1}{2}\cdot\sigma^2\cdot\left(\frac{(C_{i+1,j+1}-2\cdot C_{i+1,j}+C_{i+1,j-1})+(C_{i,j+1}-2\cdot C_{i,j}+C_{i,j-1})}{2\Delta x^2}\right)$$
$$+\nu\cdot\left(\frac{(C_{i+1,j+1}-C_{i+1,j-1})+(C_{i,j+1}-C_{i,j-1})}{4\cdot\Delta x}\right)-r\cdot\left(\frac{C_{i+1,j}+C_{i,j}}{2}\right)$$

With some rearrangement we have

$$p_u \cdot C_{i,j+1} + p_m \cdot C_{i,j} + p_d \cdot C_{i,j-1} = -p_u \cdot C_{i+1,j+1} - (p_m - 2) \cdot C_{i+1,j} + p_d \cdot C_{i+1,j-1},$$

where

$$p_u = -\frac{1}{4} \cdot \Delta t \cdot \left(\frac{\sigma^2}{\Delta x^2} + \frac{\nu}{\Delta x} \right)$$
$$p_m = 1 + \Delta t \cdot \frac{\sigma^2}{2 \cdot \Delta x^2} + \frac{r \cdot \Delta t}{2}$$
$$p_d = -\frac{1}{4} \cdot \Delta t \cdot \left(\frac{\sigma^2}{\Delta x^2} - \frac{\nu}{\Delta x} \right).$$

Finally, we can calculate the Greeks as

$$\Delta = \frac{\partial C}{\partial S} \approx \frac{C_{0,j+1} - C_{0,j-1}}{S_{0,j+1} - S_{0,j-1}}$$

$$\Gamma = \frac{\partial^2 C}{\partial S^2} \approx \frac{\left(\frac{C_{0,j+1} - C_{0,j}}{S_{0,j+1} - S_{0,j}} \right) - \left(\frac{C_{0,j} - C_{0,j-1}}{S_{0,j} - S_{0,j-1}} \right)}{\frac{1}{2}\left(S_{0,j+1} - S_{0,j-1}\right)}$$

$$\Theta = \frac{\partial C}{\partial t} \approx \frac{C_{1,j} - C_{0,j}}{\Delta t}$$

$$\upsilon = \frac{\partial C}{\partial \sigma} \approx \frac{C(\sigma) - C(\sigma + \Delta\sigma)}{\Delta \sigma}$$
$$\rho = \frac{\partial C}{\partial r} \approx \frac{C(r) - C(r + \Delta r)}{\Delta r}.$$

The accuracy in the method above is $O(\Delta x + \Delta t)$, $O(\Delta x^2 + \Delta t)$ and $O\Delta x^2 + (\Delta t/2)^2$ respectively. The integration schemas can be illustrated as in the Fig. 2.24.

As we can see in Fig. 2.24, in the explicit method we use the information at time t to calculate the value at $t - \Delta t$. In the implicit method the information passes in the opposite direction so we need to solve a system of equations to find the values in all nodes at time t. If we use Crank–Nicholson we combine the implicit and the explicit method.

The Hopscotch Method

When we solve a partial differential equation, we always create some kind of grid. In the grid shown in Fig. 2.25. we illustrate how we represent the stock price as function of time. At maturity we have the boundary condition representing the contingent claim. The other two, parallel to the time axis represent the minimum and maximum stock price.

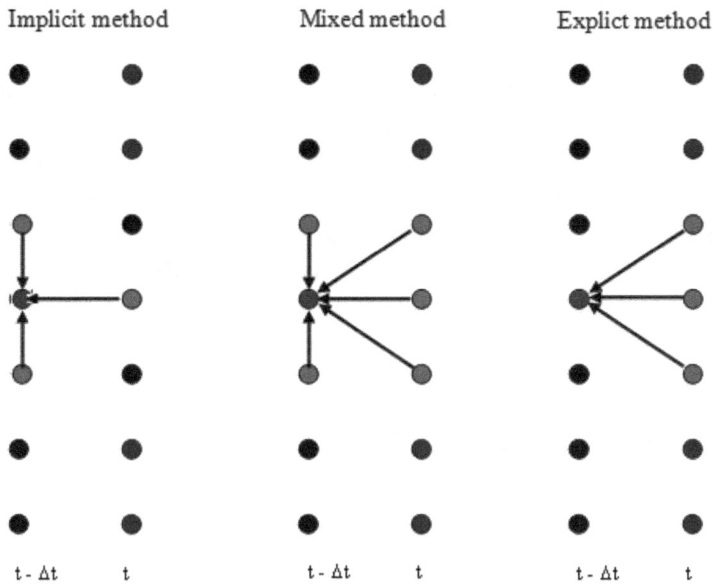

Fig. 2.24 The integration schema can be illustrated like this

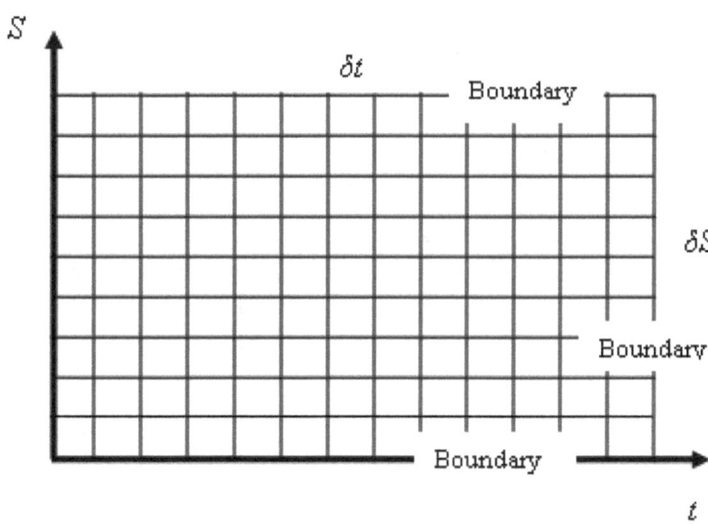

Fig. 2.25 The integration schema can be illustrated like this

If we combine the forward- and backward differences and place the nodes as in Fig. 2.26, we don't have to solve the equations simultaneously as we have to do in the implicit and Crank–Nicholson method. In the Hopscotch model, we start by calculating the explicit nodes, denoted by E for the time before maturity. This is every second node. We then continue with the implicit nodes, denoted by I, who can be calculated by the known explicit nodes and the nodes in the next time step. We continue like this backwards in time and by shifting the explicit and implicit nodes as in Fig. 2.26. By mixing the nodes in this way, we can get almost the same accuracy as the Crank–Nicholson method without having to solve a complete system of equation.

2.8.2 Some Words About Monte Carlo Simulations

In many situations, Monte Carlo simulations can be very useful to price financial instruments. This is especially useful for complex derivatives when no closed form solutions exist. Monte Carlo simulations can also be used when there are many random factors, such as stochastic volatility, stochastic interest rate and more realistic price processes with jumps or for complex boundaries. The disadvantage is the need of extensive and time-consuming calculations, which needs a lot of computer power.

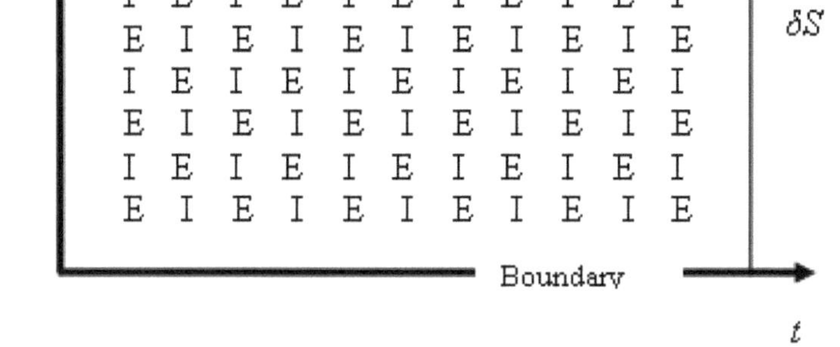

Fig. 2.26 The Hopscotch schema can be illustrated like this. Here, for each time, we always start with the explicit nodes. Thereafter it is possible to calculate the values in the implicit nodes. We continue backwards until the valuation time today

We will first introduce Monte Carlo simulations and then show how we can increase the accuracy by *control variates* and by *quasi-random numbers*. Consider a plain vanilla European call option in the Black–Scholes world with continuous compounding, with a constant risk-free interest rate r. The stock price is following a stochastic process given by

$$dS_t = rS_t dt + \sigma S_t dz_t.$$

For simplicity, we will study the natural logarithm of the stock price, $x_t = \ln(S_t)$ which gives the following dynamics

$$\begin{aligned} dx_t &= \nu dt + \sigma dz_t \\ \nu &= r - \frac{1}{2}\sigma^2. \end{aligned}$$

This process can be simulated as

$$x_{t+\Delta t} = x_t + \nu\Delta t + \sigma(z_{t+\Delta t} - z_t),$$

where the random increment in z is normally distributed with mean zero at variance Δt. Then we can simulate the random process given by $\sqrt{\Delta t} \cdot \varepsilon$ where ε is normally distributed random numbers. We then have

$$\begin{aligned} S_{t_i} &= \exp(x_{t_i}) \\ x_{t_i} &= x_{t_{i-1}} + \nu\Delta t + \sigma\sqrt{\Delta t} \cdot \varepsilon. \end{aligned}$$

In the Fig. 2.27 we show 100 simulations of the stock price during a half of a year divided into 126 trading days. (We suppose there are 252 trading days per year.) At the starting time, the stock price is 100, the volatility 40 % and the risk-free interest rate 2 %. We use Monte Carlo simulation to calculate the price of a European call option with strike price $K = 110$.

A histogram of the stock price at maturity is shown in Fig. 2.28. We observe a typical log-normal distribution. From this histogram we can also calculate the probability that the stock price will be above the strike at maturity. This is done by counting the number of paths ending above K and divide with the total number of simulations.

For each scenario, we calculate the profit of the call options as $\max(S_T - K, 0)$. To find the theoretical option value we then calculate the mean value of the discounted payoff

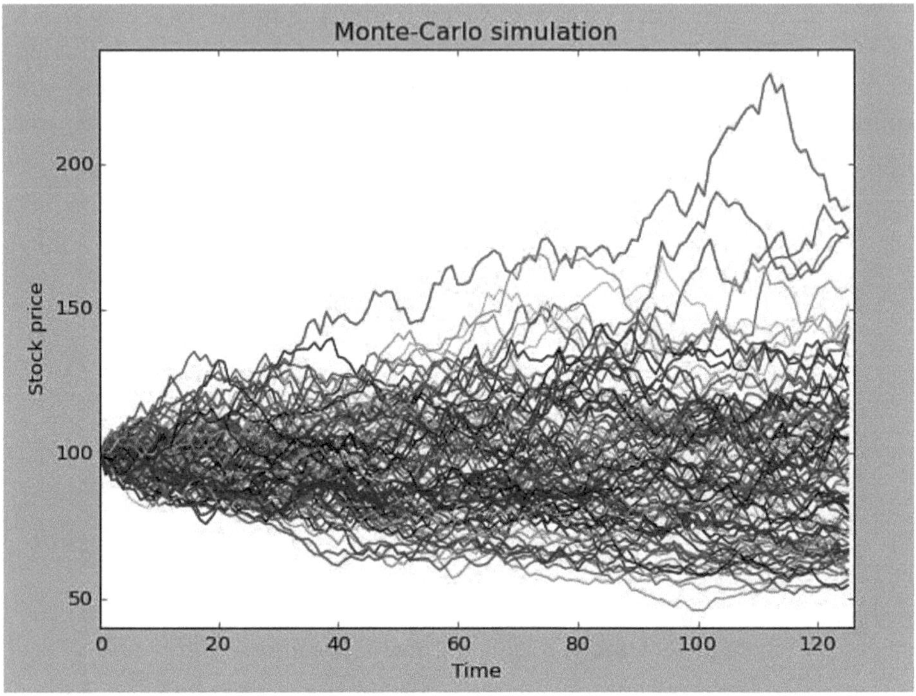

Fig. 2.27 100 Monte Carlo simulations of the stock price starting at 100

$$C_0 = \exp(-rT)\frac{1}{N}\sum_{i=1}^{N}\max(S_{T,i} - K, 0)$$

where K is the strike price of the option. If we make 10 simulations with 10,000 simulations each we get: [7.944, 7.705, 7.373, 7.896, 7.535, 7.781, 7.871, 8.232, 7.991 and 7.953]. As we can see, the simulated values vary very much, also with as many as 10,000 simulations. The average value is 7.828 which can be compared with the Black–Scholes value, 7.836944. As we see with 100,000 simulations, we still have an error of 0.009 or 0.11 %. With one million simulations we get 7.78786, an error of 0.6 %. This was even higher than the previous 100,000 simulations. So the error is very random itself.

The standard deviation (*SD*) of the simulations is given by

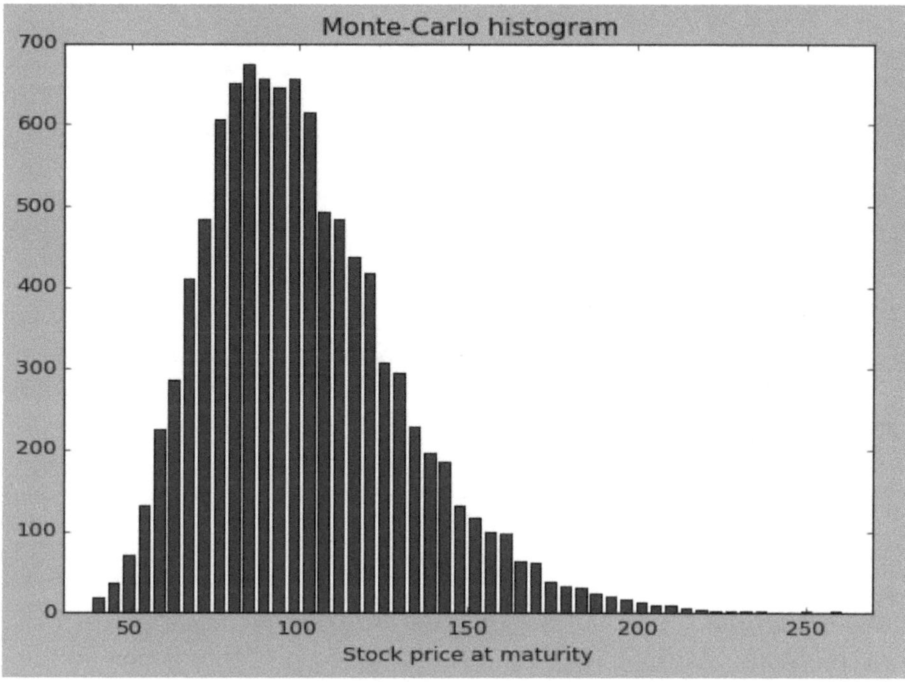

Fig. 2.28 A histogram of 10,000 Monte Carlo simulations

$$SD = \frac{1}{N-1}\sqrt{\sum_{i=1}^{N}(C_{T,i})^2 - \frac{1}{N}\left(\sum_{i=1}^{N}C_{T,i}\right)^2} \cdot \exp(-2rT)$$

and the standard error (SE) is then calculated as

$$SE = \frac{SD}{\sqrt{N}}.$$

Unfortunately, as we have seen, one has to make many simulations to get reasonable accuracy on the option price, usually millions of simulations. But with a different technique, we can increase the accuracy.

In general, when we simulate a portfolio of many instruments, we sum all the expected cash flows, discounted to a present value using the appropriate interest rates. Let s_i denote the discounted cash flow for the i:th path. We have that

$$s_i = \sum_{k=1}^{n} \exp(-r_k t_k) CF_k.$$

If we do n such Monte Carlo simulation and average the results we have

$$\widehat{S} = \frac{1}{n} \sum_{k=1}^{n} s_k$$

The central limit theorem states that $\hat{S}$ will converge to the true expected value $E(s)$ as $n \to \infty$. It is important to realize that $\hat{S}$ only is an approximation of $E(s)$ for any finite n. The central limit theorem states that the averaged mean $\hat{S}$ exhibits a standard error of size

$$\frac{\sigma}{\sqrt{n}}.$$

The standard error is a measure of the insecurity in the estimate of the instruments value. From the size of it we can draw two conclusions. First, we can improve the accuracy of our simulation by performing more simulations. Second, since the error decreases as $O(1/\sqrt{n})$, it is possible that many simulations are needed to provide high accuracy.

Variance Reduction: Control Variates

To increase the accuracy we can study a hedged portfolio of both the stock and the option. This will, in general, give us a much better accuracy of the option price. By creating a hypothetical stock with a perfect negative correlation to the first stock, we have

$$\begin{aligned} dS_{t,1} &= rS_{t,1}dt + \sigma S_{t,1}dz_t \\ dS_{t,2} &= rS_{t,2}dt - \sigma S_{t,2}dz_t \end{aligned}$$

with option prices

$$C_{T,j} = \max\left(0, S \cdot \exp\left(\nu T - \sigma\sqrt{T}\varepsilon_i\right) - X\right)$$

and

$$C^{-}_{T,j} = \max\left(0, S \cdot \exp\left(\nu T + \sigma\sqrt{T}\varepsilon_i\right) - X\right).$$

Both of these should of course have exactly the same value, so we can use the mean value of both. This technique is called variance reduction with opposite variation. To increase the variance even more we can study a delta-hedged position.

The variance without control variance is about 15–25. This is increased, first to 10–15 and with delta hedging to 3–4. If we combine the techniques we can go below 3 and if we also use gamma-hedge we can increase the accuracy even more.

Also the use of random number is of great importance. The best result is given by using quasi-random numbers. They give a better coverage than real random numbers. Therefore they give a better result. (For more details of generating random numbers, see Clewlow and Strickland [2000].)

The use of Monte Carlo methods does not easily handle the pricing of American options due to their early exercise characteristic. Simulation of option prices tends to employ a backwards induction technique, which will tend to overestimate the price of an option. Various algorithms have been put forward to price American options using backwards induction, but many algorithms are computationally intensive adn do not converge readily. A number of authors, including Broadie and Glasserman (1997) and Fu et al. (2000), have suggested that the most flexible and easily implemented procedure is the simulated tree algorithm, but it too has drawbacks, with the primary one being exponential growth in computational with the number of exercise opportunities.

Variance reductions can sometimes also be used to get a better result in the binomial model. Then the more accurate result for an American put option is given by

$$C_{am} = C^{bin}_{am} - C^{bin}_{eur} + C^{bs}_{eur}.$$

The idea is that, for a European option, Black–Scholes gives the exact result and that the difference between the binomial approximation for the European and the exact value is the same for the American option:

$$C^{bin}_{am} - C_{am} = C^{bin}_{eur} - C^{bs}_{eur}.$$

2.9 Value-at-Risk (VaR)

One of the most popular and traditional measure of risk is volatility. The main problem with volatility, however, is that it does not make allowance for the direction of an investment's movement. A stock can be volatile because it suddenly jumps higher. For investors, risk is about the odds of losing money, and VaR is based on that common-sense fact. By assuming investors care about the odds of a really big loss, VaR answers the questions "What is my worst-case scenario?" or "How much could I lose in a really bad month?"

A VaR statistic has three components:

1. The time horizon (period) to be analysed. This may be related to the time period over which a financial institution is committed to holding its portfolio, or to the time required to liquidate the assets. Typical periods using VaR are 1 day, 10 days or 1 year. A 10-day period is used to compute capital requirements under the European Capital Adequacy Directive (CAD) and the Basel II Accords for market risk, whereas a 1-year period is used for credit risk. A problem by using a long time horizon is that the portfolio is not the same in the beginning as at the end of the period.
2. The confidence level in which the VaR would not be expected to exceed the maximum loss. Commonly used confidence levels are 99 % and 95 %. Confidence levels are not indications of probabilities.
3. The loss amount or loss in percentage.

Keep these three parts in mind as we give some examples of variations of the questions that VAR answers:

- What is the most I can—with a 95 % or 99 % level of confidence—expect to lose in dollars over the next month?
- What is the maximum percentage I can—with 95 % or 99 % confidence—expect to lose over the next year?

Institutional investors use VaR to evaluate portfolio risk, but for illustration we will use it to evaluate the risk of a single index.

Three methods are used to calculate VaR: the historical method, the variance-covariance method and using Monte Carlo simulation.

2.9.1 Historical VaR Method

The historical method is the simplest and most transparent method of calculation. This involves running the current portfolio across a set of historical price changes to yield a distribution of changes in portfolio value, and computing a percentile (the VaR).

The benefits of this method are its simplicity to implement, and the fact that it does not assume a normal distribution of asset returns. Drawbacks are the requirement for a large market database and the computationally intensive calculation. By using historical data, we can evaluate VaR as

$$VaR = MV \cdot \sigma_h \cdot \sqrt{d} \cdot 2.3263,$$

where MV is the market value of the portfolio, σ_h the historical volatility of the portfolio and d number of days. The value 2.3263 is a given value used to calculate the level of certainty of 99 %. It can be calculated solving

$$\frac{1}{\sqrt{2\pi}} \int_{-x}^{\infty} e^{-z^2/2} dz = 0.99.$$

By using Excel you can find the value of x as "=NORMSINV(99 %)". In a calculation with a certainty of 95 % we solve

$$\frac{1}{\sqrt{2\pi}} \int_{-x}^{\infty} e^{-z^2/2} dz = 0.95.$$

The value of x will then be =1.6449 (=NORMSINV(95 %) by using Excel).

In the Fig. 2.29 we calculate the daily return of almost 1,400 points and put them in a histogram that compares the frequency of return "buckets". For example, at the highest point of the histogram (the highest bar), there were more than 250 days when the daily return was between 0 % and 1 %. At the far right, you can barely see a tiny bar at 13 %; this represents the one single day within a period of several years when the daily return was 12.4 %.

Notice the red bars (the leftmost 6 bars between −10 % and −5 %) that compose the "left tail" of the histogram. These are the lowest 5 % of daily returns. The worst are always the "left tail". The red bars run from daily losses of 5–10 %. Because these are the worst 5 % of all daily returns, we can say with 95 % confidence that the worst daily loss will not exceed 4 %. Put in another

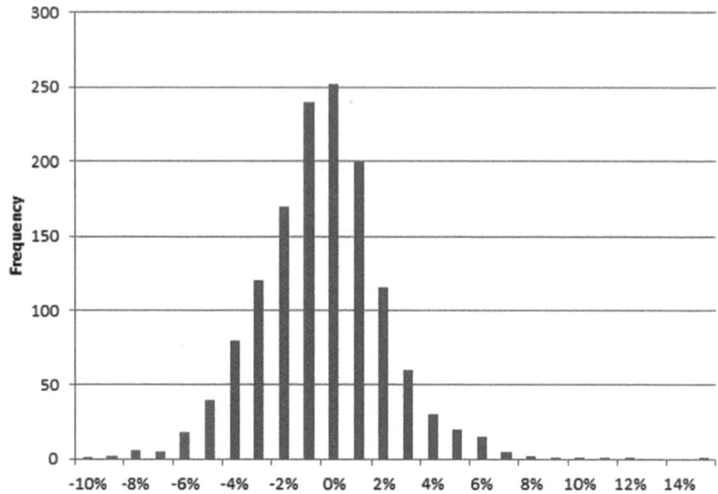

Fig. 2.29 A histogram of 1386 Monte Carlo simulations

way, we expect with 95 % confidence that our gain will exceed −4 %. That is VaR in a nutshell. Let's rephrase the statistic into both percentage and cash terms:

- With 95 % confidence, we expect that our worst daily loss will not exceed 4 %.
- If we invest 100, we are 95 % confident that our worst daily loss will not exceed 4 (100 × −4 %).

You can see that VaR makes a probabilistic estimate. If we want to increase our confidence, we need only to "move to the left" on the same histogram, to where the first two red bars, at −8 % and −7 % represent the worst 1 % of daily returns:

- With 99 % confidence, we expect that the worst daily loss will not exceed 7 %.
- Or, if we invest 100, we are 99 % confident that our worst daily loss will not exceed 7.

2.9.2 The Variance–Covariance Method

This method assumes that stock returns are normally distributed. In other words, it requires that we estimate only two factors - an expected (or average) return and a standard deviation - which allow us to plot a normal distribution curve. Here we plot the normal curve against the same actual return data:

The idea behind the variance–covariance method is similar to the ideas behind the historical method, except that we use the familiar curve instead of actual data. The advantage of the normal curve is that we automatically know where the worst 5 % and 1 % lie on the curve. They are a function of our desired confidence and the standard deviation.

The curve in Fig. 2.30 is based on the actual daily standard deviation of the index, which is 2.6263 %. The average daily return happened to be fairly close to zero, so we will assume an average return of zero for illustrative purposes. Here are the results of plugging the actual standard deviation into the formulas above.

Confidence	# of σ	Calculation	Equals
95 % (high)	1.65 × σ	1.65 × 2.64 %	4.36 %
99 % (very high)	2.33 × σ	2.33 × 2.64 %	6.16 %

We calculate the 99 % VaR as

$$\begin{aligned} \mu_p &= \sum_{i=1}^{N} \omega_i \mu_i \\ \sigma_p &= \sqrt{\boldsymbol{\omega}^{\mathrm{T}} \boldsymbol{\Sigma} \boldsymbol{\omega}} \\ VaR_{99\%} &= -MV\left(\mu_p - 2.3263 \cdot \sigma_p\right), \end{aligned}$$

where

$\omega_i = V_i / V_p$	is the return on asset i in the portfolio.
Σ	the covariance matrix of the N assets.
μ_i	the expected return of asset i, i.e., the mean return.
μ_p	the expected return of the portfolio i.e., the mean.
MV	the market value of the portfolio today.

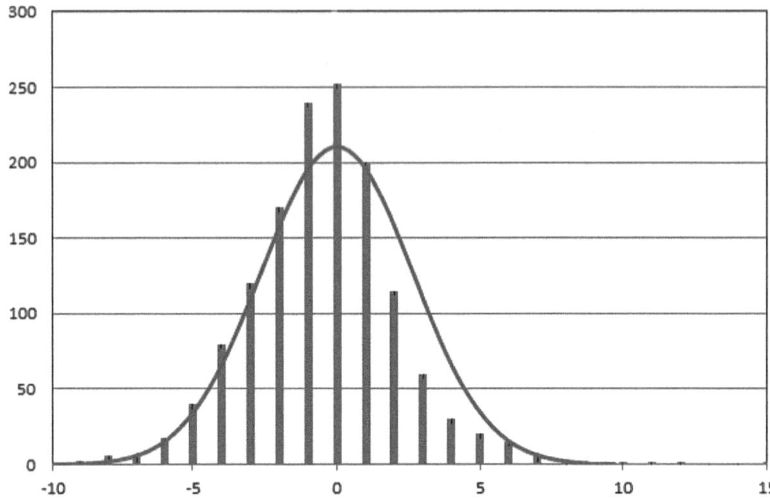

Fig. 2.30 The histogram in Fig. 2.30 fitted to a normal distribution. The mean is 0.0181 % and the standard deviation 2.6263 %

The benefits of the variance–covariance model are the use of a more compact and maintainable data set, which can often be bought from third parties, and the speed of calculation using optimized linear algebra libraries. Drawbacks include the assumption that the portfolio is composed of assets whose delta is linear and the assumption of a normal distribution of asset returns (i.e., market price returns).

2.9.3 Monte Carlo Simulation

The third method involves developing models for future price returns of all financial instruments in the portfolio and running multiple hypothetical scenarios through the models. This is done via Monte Carlo simulations that randomly simulate scenarios generated from historical time series. The result can be arranged into a histogram with monthly returns.

To summarize, we ran 100 hypothetical scenarios of monthly returns. Among them, two outcomes were between −15 % and −20 %; and three were between −20 % and 25 %. That means the worst five outcomes were less than −15 %. The Monte Carlo simulation therefore leads to the following VaR conclusion, with 95 % confidence we do not expect to lose more than 15 % during any given month.

2.9.4 Exponential Weighted VaR

The yearly volatility is calculated as the square root of 252 (the approximate number of trading days on a year) times the standard deviation

$$\sigma = \sqrt{\frac{1}{n-1} \cdot \sum_{i=1}^{n} (x_i - \overline{x})^2}.$$

Here x_i is the logarithmic return on day i and $\overline{x}$ the average return during the time period. In this formula, all the returns are equally weighted. If we use an exponential weight, we can rescale the volatility in such a way that the return of the most nearby return in the history becomes more important than those in the far past. We then use the formula

$$\sigma = \sqrt{(1-\lambda) \cdot \sum_{i=1}^{n} \lambda^{n-i} \cdot (x_i - \overline{x})^2},$$

where λ is a *decay factor*. The value of λ is usually between 0.94 and 0.99. Recall the formulas for mean, variance, co-variance, skewness and kurtosis:

$$\begin{aligned}
\mu &= E[x_i] \\
\sigma &= E\left[(x_i - \mu)^2\right] \\
\sigma_{ij} &= E\left[(x_i - \mu_i)(x_j - \mu_j)\right] \\
s &= E\left[(x_i - \mu)^3\right] / \sigma^3 \\
k &= E\left[(x_i - \mu)^4\right] / \sigma^4.
\end{aligned}$$

Skewness measures the asymmetry of the distribution and is zero for the normal distribution, but non-zero for the lognormal distribution. The kurtosis measures the flatness of the distribution, $k = 3$ for the normal distribution.

2.9.5 Value-at-Risk for Bonds

The value of a bond can be expressed (quoted) in the interest rate yield-to-maturity, ($P = P(y)$). With a simple Taylor expansion we may write

$$P_t(y) \approx P_0 + \frac{dP_t}{dy} \cdot \Delta y + \frac{1}{2} \frac{d^2P_t}{dy^2} \cdot \Delta y^2,$$

giving

$$\Delta P_t \approx -P_0 \cdot D_{\text{mod}} \cdot \Delta y + \frac{1}{2} P_0 \cdot Cnvx \cdot \Delta y^2,$$

where we have defined the *modified duration* and the *convexity* of the bond. If we now define the change in yield as $Y = \Delta y / y_0$ we have

$$\frac{\Delta P_t}{P_0} \approx - \cdot D_{\text{mod}} \cdot Y \cdot y_0 + \frac{1}{2} P_0 \cdot Cnvx \cdot (Y \cdot y_0)^2 = -\delta \cdot y_0 + \frac{1}{2} \gamma \cdot y_0^2,$$

where we have defined δ and γ. This model is called the delta-gamma model. If we put gamma (γ) to zero, we call that model, the delta model. If the changes in the interest rate are normal distributed $Y \sim N(0, \sigma^2)$, then $\Delta P_t / P_0 \sim N(0, \delta^2 \sigma^2)$. If MV is the market value P_0 at $t = 0$, then

$$VaR = MV \cdot \delta \cdot \sigma_h \cdot \sqrt{d} \cdot 2.3263 = D_{\text{mod}} \cdot y_0 \cdot \sigma_h \cdot \sqrt{d} \cdot 2.3263.$$

2.9.6 Portfolio VaR

By using the co-variance between two instruments 1 and 2

$$\sigma_{1,2} = \frac{1}{T-1} \sum_{t=1}^{T} (\Delta P_{1,t} - \mu_1)(\Delta P_{2,t} - \mu_2)$$

and the correlation

$$\rho_{1,2} = \frac{\sigma_{1,2}}{\sigma_1 \sigma_2}$$

we can calculate the VaR of a portfolio of the two assets as

$$VaR_p = \sqrt{(VaR_1)^2 + (VaR_2)^2 + 2 \cdot \rho_{1,2} \cdot VaR_1 \cdot VaR_2}.$$

We see that lower correlation gives better diversification. This formula can be generalized for any number of assets.

Since VaR is not normally linear, we cannot use the superposition principle for VaR, especially when using derivatives. Therefore a bank starts by calculating the VaR at the lowest lever on each trading desks. Then the different trading desks are aggregated into groups, trading on similar markets or instruments. This VaR again aggregates step by step up to the top level in the bank. For each level, the VaR is needed to be simulated as a new part or portfolio.

2.9.7 Conditional Value-at-Risk: Expected Shortfall

Expected shortfall (ES) is an alternative to Value-at-Risk, which is often criticized as not presenting a full picture of the risks a company faces. The "expected shortfall at *q*% level" is the expected return on the portfolio in the worst *q*% of the cases. Expected shortfall is also known as *conditional Value-at-Risk* (CVaR) or *expected tail loss* (ETL).

As we have seen, VaR is defined as the loss level that will not be exceeded with a certain confidence level during a certain period of time. For example, if a bank's 10-day 99 % VaR is \$3 million, there is considered to be only a 1 % chance that losses will exceed \$3 million in 10 days. One problem with VaR is that, when used in an attempt to limit the risks taken by a trader, it can lead to undesirable results.

The VaR can be expressed as

$$VaR_{\alpha}(X) = -\inf\{x : P(X \leq x) \geq \alpha\},$$

where $0 < a < 1$ is the quantile of the distribution of the random variable X. The ES can be expressed as

$$ES_{\alpha}(X) = -E[X|X \leq -VaR_{\alpha}(X)].$$

The meanings of these risk measures are obvious: VaR is a threshold which is fallen short of in $\alpha \cdot 100\%$ of all cases; ES is the expectation (i.e. the mean) of the losses under the condition that this threshold has already been fallen short of. The change of the sign is a matter of interpretation to neutralize losses. Risk capital has to be positive.

Suppose a bank tells a trader that the 1-day 99 % VaR of the trader's portfolio must be kept at less than \$10 million. There is a danger that the trader will construct a portfolio where there is a 99 % chance that the daily loss is less than \$10 million and a 1 % chance that it is \$50 million. The trader is satisfying the risk limits imposed by the bank, but is clearly taking an unacceptable risk. Most traders would, of course, not behave in this way—but some might.

CVaR is a measure that produces better incentives for traders than VaR itself. Where VaR asks the question "how bad can things get?", CVaR asks "if things do get bad, what is our expected loss?"

CVaR, like VaR, is a function of two parameters; N, the time horizon in days and q%, the confidence level. It is the expected loss during an N-day period, conditional that the loss is greater than the qth percentile of the loss distribution. For example, with $q = 99$ and $N = 10$, the expected shortfall is the average amount that is lost over a 10-day period, assuming that the loss is greater than the 99th percentile of the loss distribution. We illustrate the expected shortfall in Fig. 2.31. The expected shortfall is the expectation i.e. the mean of the losses under the condition that we make a loss.

2.9.8 Properties of the Risk Measures

A risk measure that is used for specifying capital requirements can be thought of as the amount of cash (or capital) that must be added to a position to make

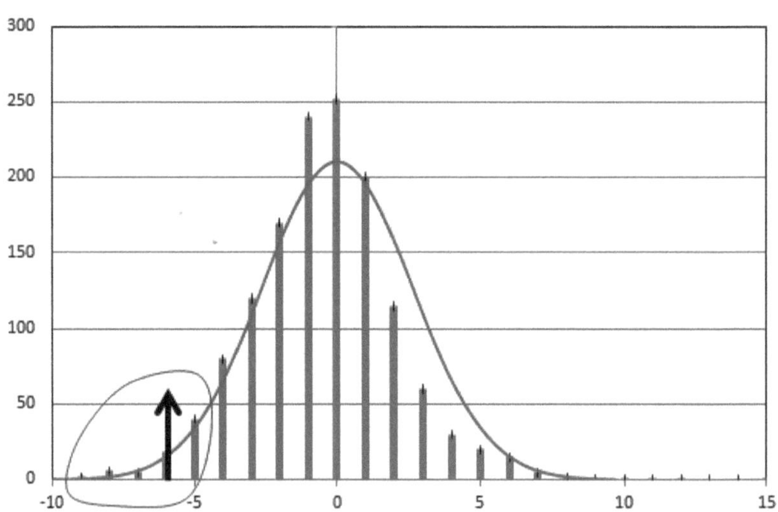

Fig. 2.31 The histogram from Fig. 2.30 illustrating the expected shortfall and the 5 % worst case outcomes

its risk acceptable to regulators. Artzner et al. (1999) have proposed a number of properties that such a risk measure should have. These are

1. *Monotonicity*
   If a portfolio has lower returns than another portfolio for every state of the world, its risk measure should be greater.
2. *Translation invariance*
   If we add an amount of cash C to a portfolio, its risk measure should go down by C.
3. *Homogeneity*
   Changing the size of a portfolio by a factor λ while keeping the relative amounts of different items in the portfolio the same should result in the risk measure being multiplied by λ.
4. *Sub-additivity*
   The risk measure for two portfolios after they have been merged should be no greater than the sum of their risk measures before they were merged.

The first three conditions are straightforward given that the risk measure is the amount of cash needed to be added to the portfolio to make its risk acceptable. The fourth condition states that diversification helps to reduce the risks. When two risks are aggregated, the total of the risk measures corresponding to the risks should either decrease or stay the same.

VaR satisfies the first three conditions, but it does not always satisfy the fourth, as will now be illustrated.

Consider two $10 million one-year loans, each of which has a 1.25 % chance of defaulting. If a default occurs on one of the loans, the recovery of the loan principal is uncertain, with all recoveries between 0 % and 100 % being equally likely. If the loan does not default, a profit of $200,000 is made. To simplify matters, we suppose that if one loan defaults it is certain that the other loan will not default[2]. For a single loan, the one-year 99 % VaR is $2 million. This is because there is a 1.25 % chance of a loss occurring and conditional on a loss, there is an 80 % chance that the loss is greater than $2 million. The unconditional probability that the loss is greater than $2 million is 80 % of 1.25 %, or 1 %.

Consider next the portfolio of two loans. Each loan defaults 1.25 % of the time and they never default together. There is therefore a 2.5 % probability

[2] This is to simplify the calculations. If the loans default independently of each other so that two defaults can occur, the numbers are very slightly different, but the VaR of the portfolio is still greater than the sum of the VaRs of the individual loans.

that a default will occur. The VaR in this case turns out to be \$5.8 million. This is because there is a 2.5 % chance of one of the loans defaulting and conditional on this event, there is a 40 % chance that the loss on the loan that defaults is greater than \$6 million. The unconditional probability that the loss on the defaulting loan is greater than \$6 million is therefore 40 % of 2.5 %, or 1 %. A profit of \$200,000 is made on the other loan, showing that the VAR is \$5.8 million.

The total VaR of the loans considered separately is \$2 million + \$2 million = \$4 million. The total VaR after they have been combined in the portfolio is \$1.8 million greater, at \$5.8 million. This is in spite of the fact that there are very attractive diversification benefits from combining the loans in a single portfolio.

2.9.9 Coherent Risk Measures

Risk measures satisfying all four of the conditions in Sect. 2.9.8 are referred to as coherent. The example illustrates that VaR is not always coherent. It does not satisfy the sub-additivity condition. This is not just a theoretical issue. Risk managers sometimes find that, when they have a portfolio in multiple currencies, the total VaR goes up rather than down as expected.

In contrast, it can be shown that the CVaR[3] (expected shortfall) measure is coherent. Consider again the earlier example. The VaR for a single loan is \$2 million. The expected shortfall from a single loan when the time horizon is one year and the confidence level is 99 % is, therefore, the expected loss on the loan, conditional on a loss greater than \$2 million. Given that losses are uniformly distributed between zero and \$10 million, this is halfway between \$2 million and \$10 million, or \$6 million.

The VaR for a portfolio consisting of the two loans was calculated as \$5.8 million. The expected shortfall from the portfolio is, therefore, the expected loss on the portfolio, conditional on the loss being greater than \$5.8 million. When a loan defaults, the other (by assumption) does not and outcomes are uniformly distributed between a gain of \$200,000 and a loss of \$9.8 million. The expected loss, given that we are in the part of the distribution between \$5.8 million and \$9.8 million, is \$7.8 million. This is therefore the expected shortfall on the portfolio. Because \$6 million + \$6 million > \$7.8 million, the expected shortfall does satisfy the sub-additivity condition for the example.

---

[3] Sometimes CVaR is the acronym for Credit VaR.

A risk measure can be characterized by the weights it assigns to quantiles of the loss distribution. VaR gives a 100 % weighting to the qth quantile and zero to other quantiles. Expected shortfall gives equal weight to all quantiles greater than the qth quantile and zero weight to all quantiles below the qth quantile. We can define what is known as a *spectral risk measure* by making other assumptions about the weights assigned to quantiles. A general result is that a spectral risk measure is coherent (that is, it satisfies the sub-additivity condition) if the weight assigned to the pth quantile of the loss distribution is a non-decreasing function of p. Expected shortfall satisfies this condition. VaR, however, does not because the weights assigned to quantiles greater than q are less than the weight assigned to the qth quantile.

2.9.10 Regulations

Regulators make extensive use of VaR and its importance as a risk measure is therefore unlikely to diminish. However, expected shortfall has a number of advantages over VaR. This has led many financial institutions to use it as a risk measure internally.

3

Introduction to Probability Theory

3.1 Introduction

The study of financial engineering, risk modelling and the valuation of financial instruments requires knowledge in basic statistics and probability theory. We will here provide a short introduction to the basic concepts. Since the main focus of this book is not on statistics and probability theory, we give only the theorems and definitions needed for further reading in financial engineering.

3.2 A Binomial Model

So far we have been studying the binomial model where the price can grow with a factor u from one time to another, or decrease with a factor d during the same time. A stochastic variable, such as tossing a coin, decides whether u or d should be used with some probabilities. We will study such a tree with the following properties:

$u=2 \Rightarrow d=1/u=0.5$, $S_0=4$ and $q_u = q_d = 1/2$ where $S_2(uu)=u^2S_0$, $S_2(ud)=udS_0$ etc. (Fig. 3.1).

If we are tossing a coin once, twice and three times, we get a *sample space* Ω given by

J.R.M. Röman, *Analytical Finance: Volume I*,
DOI 10.1007/978-3-319-34027-2_3

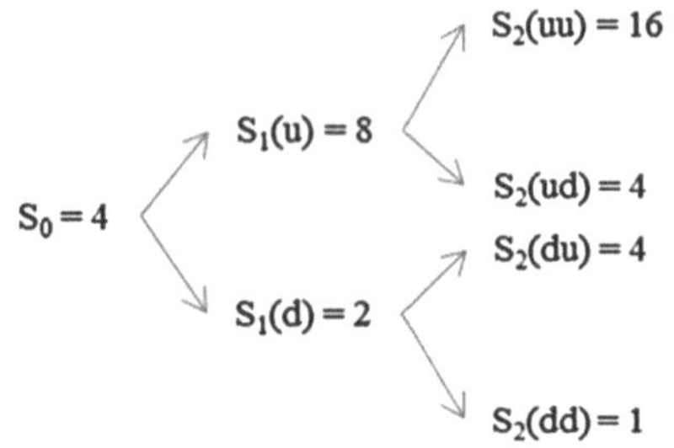

Fig. 3.1 A binomial tree with parameters $u = 2$, $d = 1/u = 0.5$, $S_0 = 4$ and $q_u = q_d = 1/2$

$$\begin{aligned}
\Omega_1 &= \{u, d\} = \{\omega_1\}, \\
\Omega_2 &= \{uu, ud, du, dd\} = \{\omega_2\}, \\
\Omega_3 &= \{uuu, uud, udu, duu, udd, dud, ddu, ddd\} = \{\omega_3\}.
\end{aligned}$$

Here, Ω_i represent the sample space after i tossings. We also introduce the interest rate r, such as for one period in time 1 CU (cash unit) will grow to $(1 + r)$ 1 CU = 1 R CU. Using the no-arbitrage condition we know that the factor R must be in the interval $d \leq R \leq u$ because if $R > u$ nobody would be interested in buying the stock and if $R < d$ then $r < 0$, which is unrealistic.

We say that the model above is free of arbitrage if and only if $d \leq R \leq u$.

Example 3.1

Let's study a European call option with strike K at $t = 1$. On maturity, the value is given by

$$V_1(\omega) = (S_1(\omega) - K)^+ \equiv \max(S_1(\omega) - K, 0).$$

We are now looking for the arbitrage-free price. The two possible outcomes, with u and d are given by

$$V_1(\omega) = \begin{cases} (uS_0 - K)^+ & \text{if} \quad \omega_1 = u \\ (dS_0 - K)^+ & \text{if} \quad \omega_1 = d. \end{cases}$$

To hedge a short position of the option we have to buy Δ_0 stocks. This means that at time $t = 0$ we have sell the option, giving us V_0 cash units. But we also buy Δ_0 stocks at the price of S_0. We then have $(V_0 - \Delta_0 S_0)$ cash units to put in our money-market account. If the sign is negative, that means we have to borrow this amount at a rate of r. The value process gives us two possible values on maturity

$$\begin{aligned}
V_1(u) &= \Delta_0 \cdot S_1(u) + R \cdot (V_0 - \Delta_0 \cdot S_0) \\
V_1(d) &= \Delta_0 \cdot S_1(d) + R \cdot (V_0 - \Delta_0 \cdot S_0).
\end{aligned}$$

We can therefore solve Δ_0 to get

(continued)

Example 3.1 (continued)

$$\Delta_0 = \frac{V_1(u) - V_1(d)}{S_1(u) - S_1(d)} \quad \rightarrow \quad \frac{\partial V}{\partial S}.$$

By inserting Δ_0 into the equation above, we find the price of the option at $t = 0$

$$V_0 = \frac{1}{R}\left\{\frac{R-d}{u-d}V_1(u) - \frac{R-u}{u-d}V_1(d)\right\} = \frac{1}{R}\{q_u \cdot V_1(u) + q_d \cdot V_1(d)\} = \frac{1}{R}E^Q[V_1]$$

Here we have also defined q_u and q_d as the *risk-neutral probabilities* as

$$q_u = \frac{R-d}{u-d} \text{ and } q_u = -\frac{R-u}{u-d}.$$

We also let the expression

$$\Pi[X] = \frac{1}{R}E^Q[X]$$

represent the arbitrage free price on the option on the contingent claim X with respect to the risk-neutral probability measure Q, the martingale measure. Similarly, we get

$$V_1(u) = \frac{1}{R}\{q_u \cdot V_2(uu) + q_d \cdot V_2(ud)\}; \qquad \Delta_1(u) = \frac{V_2(uu) - V_2(ud)}{S_2(uu) - S_2(ud)}$$
$$V_1(d) = \frac{1}{R}\{q_u \cdot V_2(du) + q_d \cdot V_2(dd)\}; \qquad \Delta_1(d) = \frac{V_2(du) - V_2(dd)}{S_2(du) - S_2(dd)}.$$

3.3 Finite Probability Spaces

Let $\mathcal{F}$ be the set of all subsets to the sample space Ω (where $\varnothing$, {*ddd*}, {*uuu, uud, udu, ddd*}, Ω are examples of some) where $\varnothing$ is the empty set. We define a *probability measure P* by a function mapping $\mathcal{F}$ into the interval [0, 1] with $P(\Omega) = 1$ where

$$P\left(\bigcup_{k=1}^{\infty} A_k\right) = \sum_{k=1}^{\infty} P(A_k).$$

Here $A_1, A_2, \ldots$ is a sequence of disjoint sets in $\mathcal{F}$. A probability measures has the following interpretation: let A be a subset of $\mathcal{F}$ and imagine that Ω is the set of all possible outcomes of some random experiment. Then there is a certain probability between 0 and 1 that when the experiment is performed, the outcome will lie in the set A. We think of $P(A)$ as this probability. From now we will use $P_u = 1/3$ and $P_d = 2/3$.

Example 3.2

The probability to get ahead (i.e., a *u*) in the first toss when we are tossing the coin three times, is given by

$$P\{uuu, uud, udu, udd\} = \left(\frac{1}{3}\right)^3 + \frac{2}{3}\left(\frac{1}{3}\right)^2 + \frac{2}{3}\left(\frac{1}{3}\right)^2 + \frac{1}{3}\left(\frac{2}{3}\right)^2 = \frac{1}{3}.$$

Definition 3.3 A σ-*algebra* is a collection $\mathcal{F}$ of subsets in Ω with the following properties

$$\begin{cases} \varnothing \in \mathcal{F} \\ A \in \mathcal{F} \Rightarrow A^C \in \mathcal{F} \\ A_1,\ A_2\ldots \text{ is a sequence of subspaces to } \mathcal{F} \Rightarrow \ \bigcup_k A_k \in \ \mathcal{F} \end{cases}$$

It is essential to understand that, in probabilistic terms, the σ-algebra can be interpreted as *containing all relevant information* about a random variable.

Example 3.4

Some important σ-algebras to Ω above is

$\mathcal{F}_0 = \{\varnothing, \Omega\}$
$\mathcal{F}_1 = \{\varnothing, \Omega, \{uuu, uud, udu, udd\}, \{duu, dud, ddu, ddd\}\}$
$\mathcal{F}_2 = \{\varnothing, \Omega, \{uuu, uud\}, \{udu, udd\}, \{duu, dud\}, \{ddu, ddd\}\}$ and all unions of these}
$\mathcal{F}_3 = \mathcal{F} =$ the set of all subsets of Ω.

We say that $\mathcal{F}_3$ is *finer* than $\mathcal{F}_2$, which is finer than $\mathcal{F}_1$.

If we introduce the terms $A_u = \{uuu, uud, udu, udd\} = \{u^{**}\}$, $A_d = \{d^{**}\}$, $A_{uu} = \{uu^*\}$ etc., we can write

$$\mathcal{F}_1 = \{\varnothing, \Omega, A_u, A_d\}$$
$$\mathcal{F}_2 = \{\varnothing, \Omega, A_u, A_d, A_{uu}, A_{ud}, A_{du}, A_{dd}, A_{uu} U A_{du}, A_{uu} U A_{dd}, A_{ud} U A_{du},$$
$$A_{ud} U A_{dd}, A_{uu}{}^c, A_{ud}{}^c, A_{du}{}^c, A_{dd}{}^c\}$$

We can illustrate $\mathcal{F}_2$ as

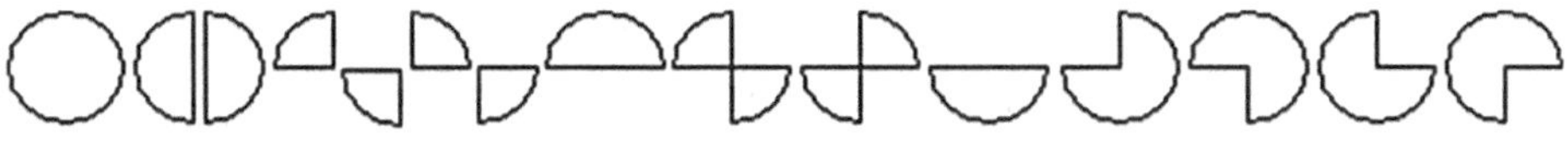

where the circle represent the full set Ω.

Definition 3.5 A pair $(X, \mathcal{F})$ where X is a set and $\mathcal{F}$ a σ-algebra on X is called a *measurable space*. The sub-spaces that exist in $\mathcal{F}$ are called $\mathcal{F}$-*measurable* sets.

In particular, if a random variable Y is a function of X, $(Y = \Phi(X))$, then Y is said to be $\mathcal{F}^X$-measurable.

Definition 3.6 A *finite measure* μ on a measurable space is a function such as

$$\mu(A) \geq 0,$$

$$\mu(\varnothing) = 0,$$

$$\text{If } A_k \in \mathcal{F}\, \forall\, k = 1, 2, \ldots \text{ and } A_i \cap A_j = \varnothing \text{ for } i \neq j, \text{then}$$

$$\mu\left(\bigcup_{k=1}^{\infty} A_k\right) = \sum_{k=1}^{\infty} \mu(A_k)$$

Definition 3.7 A *filtration* $\mathcal{F}_\infty = \underline{\mathcal{F}} = \{\mathcal{F}_t;\, t \geq 0\}$ is a sequence of σ-algebras $\mathcal{F}_0, \mathcal{F}_1, \ldots, \mathcal{F}_n$ such that $\mathcal{F}_t$ contains all sets in $\mathcal{F}_{t-1}$:

$$\begin{cases} \mathcal{F}_t \subseteq \mathcal{F} & \forall\ t \geq 0 \\ s \leq t \quad \Rightarrow & \mathcal{F}_s \in \mathcal{F}_t \end{cases}.$$

We say that the σ-algebra is *generated by* $\mathcal{F}_t$. A finite probability space $(\Omega, \mathcal{F}_t, P)$ with the filtration of σ-algebras is sometimes called σ-*fields*. We also have

$$\mathcal{F}_0 = \{\Omega, \varnothing\} \subseteq \mathcal{F}_1 \subseteq \mathcal{F}_2 \ldots \subseteq \mathcal{F}_T.$$

We say that each σ-algebra $\mathcal{F}_i$ when $0 \leq i \leq T$, are generated by *partitions*. Since $\mathcal{F}_{i-1} \subseteq \mathcal{F}_i$ it follows that the partition generating $\mathcal{F}_i$ is finer than that which generates $\mathcal{F}_{i-1}$. The elements of a partition are sometimes called *cells* or *atoms* since they make up larger objects in the σ-algebras that they generate, just as atoms creates a molecule. Since the partition that creates $\mathcal{F}_i$ is finer than that those who create $\mathcal{F}_{i-1}$ it is clear that the set of partition at time $i-1$ make a *split* to construct the set of the partition at time i. The way in which sets in a partition split to form sets of a new partition turns out to be quite important, so we define the *splitting index* $S(E)$ of a cell E in a partition to be the number of cells that it splits into the new partition.

Definition 3.8 X is $\underline{\mathcal{F}}$*-adapted* if X_t is $\mathcal{F}_t$-measurable for all $t \geq 0$.

Definition 3.9 A function $f: X \rightarrow \boldsymbol{R}$ is said to be $\mathcal{F}$ *–measurable*, if for each interval I the set $f^{-1}(I)$ is $\mathcal{F}$–measurable, i.e.

$$\{x \in X | f(x) \in I\} \in \mathcal{F}.$$

Definition 3.10 A *stochastic variable* X is a mapping of Ω on R such as

$$X : \Omega \rightarrow R \text{ so that } X \, is \, \mathcal{F}\text{-measurable.}$$

Example 3.11

Consider again, the binomial tree above. A *mirror* under S_2 on [4, 27] is given by

$$\{\omega \in \Omega | S_2(\omega) \in [4, 27]\} = \{\omega \in \Omega | 4 \leq S_2(\omega) \leq 27\} = A_{dd}^c.$$

This is all nodes except $S_2(dd) = 1$. The complete list of subset on Ω with mirrors of sets in R is Ø, Ω, A_{uu}, $A_{ud} U A_{du}$, A_{dd} + all unions of these. They form a σ*-algebra generated by* S_2: $\sigma(S_2)$.

The symbol $\mathcal{F}_t^X$ represent the information generated by X on the time interval $[0, t]$, that is, the changes of X on this interval. We base this on the observation of the trajectory $\{X(s) : 0 \leq s \leq t\}$.

If a specific event A occur in this interval, then $A \in \mathcal{F}_t^X$. If the value of a stochastic variable Z can be determined by observation of the trajectory of X, then we write $Z \in \mathcal{F}_t^X$. Furthermore, if a stochastic process $Y \in \mathcal{F}_t^X$ we say that Y is *adapted to the filtration* $\{\mathcal{F}_t^X\}_{t \geq 0}$. Remark, X is always adapted to the *natural filtration* $\underline{\mathcal{F}}^X = \{\mathcal{F}_t^X; t \geq 0\}$

Definition 3.12 A *stochastic process* can be considered as a discrete set of time-indexed random variables $\{X_n\}_{n=1}^{\infty}$ or, as in continuous time, a continuous set $\{X_t\}_{t>0}$. In many situations we consider such a process, containing a *drift* μ and *diffusion* σ

$$X(t + \Delta t) - X(t) = \mu[t, X(t)]\Delta t + \sigma[t, X(t)]Z(t).$$

Sometimes this is interpreted as a random process (a random walk) upon a deterministic drift. In the continuous limit the random process becomes a Wiener process.

Definition 3.13 A stochastic process $\{W(t); t \geq 0\}$ is called a *Wiener process* if

1. $W(0) = 0$
2. $(W(u) - W(t))$ and $(W(s) - W(r))$ is independent (we say that W has independent increments) $r \leq s \leq t \leq u$.
3. $W(t) - W(s)$ is normal distributed $N\left[0, \sqrt{t-s}\right] \ \forall\, s < t$.
4. $W(t)$ has continuous trajectories.

A normal distributed process $N[\mu, \sigma]$ has the mean value given as μ and the variance σ. A very important property of a Wiener process (also called a Brownian motion) is $(dW)^2 = dt$.

From a random process X_n we can construct a continuous process by linear interpolation between the distinct points

$$Y(t) = X_i + (t - ndt)(X_{i+1} - X_i),$$

where $idt \leq t \leq (i+1)\, dt$. This process has the following properties:

1. $Y(t)$ is said to have the *Markov property* if given $Y(t)$ and $s > t\, Y(s)$ is independent of $Y(u)$ for all $u < t$.
2. $E[Y(T)] = 0$
3. $E[Y(T)^2] = T$

If we define a Wiener process as $X(t) = \lim_{dt \to 0} Y(t)$ with $X(0) = 0$ it follows from the central limit theorem that the probability distribution of the increments $X(t + a) - X(t)$ are normal distributed around 0 with the variance a. Furthermore, the Markov property gives

$$\int_0^T dX(t)^2 = \left\langle \lim_{dt \to 0} \left\{ \sum_{i=1}^{n} (X_i - X_{i-1})^2 \right\} \right\rangle = \left\langle \lim_{dt \to 0} (X_n)^2 \right\rangle = T = \int_0^T dt$$

or

$$dX(t)^2 = dt.$$

Here the brackets $\langle x \rangle$ mean the mean value of x. We also have that

$$E_t[X(t + a) - X(t)] = 0,$$

where E_t is the expected value at time t, which gives the martingale property

$$E_t[X(t+a)] = E_t[X(t)]$$

such as

$$E_s[X(t)] = E_s[E_t[X(t)]],$$
$$E_s[X(t)X(t)] = \min(t,s) = s. \quad \text{for } s \leq \text{t}.$$

In the risk-neutral valuation we use in continuous time a risk-free money-market account and where the stock is following the process:

$$dS(t) = \alpha(t)S(t)dt + \sigma(t)S(t)dW(t)$$

Here $W(t)$ is a Wiener process on $(\Omega, \mathcal{F}, P)$ with filtration $\mathcal{F}(t)$ and where $\alpha(t)$, $\sigma(t)$ and the interest rate $r(t)$ are adapted to $\mathcal{F}(t)$.

Definition 3.14 A σ-*algebra generated by* X is the complete list of all sets

$$\{\omega \in \Omega | X(\omega) \in A\},$$

where $A \subseteq R$.

Let $\mathcal{G}$ be a sub σ-algebra of $\mathcal{F}$. We then say that X is $\mathcal{G}$-*measurable* if all sets in $\sigma(X) \in \mathcal{G}$.

Definition 3.15 Given $(\Omega, \mathcal{F}, P, X)$. If $A \subseteq R$ we define the *distribution measure* as

$$\mu_X(A) = P(X \in A).$$

Example 3.16

From the binomial tree; $\mu_{S_2}(\varnothing) = P(\varnothing) = 0$, $\mu_{S_2}(R) = P(\Omega) = 1$ $\mu_{S_2}[0,3] = P(S_2 = 1) = P(A_{dd}) = (2/3)^2$. The distribution measure of S_2 place the mass $(1/3)^2 = 1/9$ on $S_2 = 16$, the mass $2*(1/3)*(2/3) = 4/9$ on $S_2 = 4$ and $(2/3)^2 = 4/9$ on $S_2 = 1$. Hereby we have the following *distribution function* on S_2

$$F_{S_2}(x) = P(S_2 \leq x) = \begin{cases} 0 & x < 1 \\ 4/9 & 1 \leq x < 4 \\ 4/9 & 4 \leq x < 16 \\ 1 & x \leq 16. \end{cases}$$

Remark 3.17 A stochastic variable can have many distributions since they depend on the choice of probabilities. For the same reason, different random variables can have the same distribution function.

Definition 3.18 2^X is defined as the set of all subsets of X in Ω, i.e. as

$$2^X = \{A | A \subseteq X\}.$$

Remark 3.19 2^X is a set which elements are subsets of X. We illustrate this with the following example:

$$X = \{1, 2, 3\} \Rightarrow 2^X = \{\emptyset, X, \{1\}, \{2\}, \{3\}, \{1, 2\}, \{1, 3\}, \{2, 3\}\}.$$

A *partition* $\mathcal{P}$ on Ω, can be written as $\mathcal{P} = \{A_i, i = 1, 2, \ldots, k\}$.

Example 3.20

We can create partitions of $\Omega = [0, 1]$, $\mathcal{P}_1 = \{A_1, A_2, A_3, A_4\}$, $\mathcal{P}_2 = \{B_1, B_2, B_3\}$ as $A_1 = [0, 1/3)$, $A_2 = [1/3, 1/2)$, $A_3 = [1/2, 3/4)$, $A_4 = [3/4, 1]$ and $B_1 = [0, 1/3)$, $B_2 = [1/3, 3/4)$, $B_3 = [3/4, 1]$.

Definition 3.21 A partition S says to be *finer* than another partition $\mathcal{P}$ if all components in $\mathcal{P}$ are a union of components in S.

Definition 3.22 If $\mathcal{P}$ is a partition of Ω and f: $\Omega \to R$ is a given map. Then, we say that the function f is $\mathcal{P}$ *–measurable* if $\mathcal{P}$ is finer than $\sigma(f)$.

Theorem 3.23 *If f is $\mathcal{P}$ –measurable, then f is $\sigma\{\mathcal{P}\}$-measurable.*

This is obvious since $\mathcal{P} \subseteq \sigma\{\mathcal{P}\}$. We interpret this as the function f is constant on each of the components of $\mathcal{P}$.

We now have:

1. $\mathcal{P}$ is generating a natural σ-algebra $\sigma\{\mathcal{P}\}$.
2. Given $\sigma\{\mathcal{P}\}$ we can recreate $\mathcal{P}$ via $A \in \mathcal{P}$ if and only if $A \neq \emptyset$, and $A \in \sigma\{\mathcal{P}\}$ and no subset of A belong to $\sigma\{\mathcal{P}\}$.
3. $\mathcal{S}$ is finer than $\mathcal{P} \Leftrightarrow \sigma\{\mathcal{P}\} \subseteq \sigma\{\mathcal{S}\}$.
4. f is $\mathcal{P}$-measurable $\Leftrightarrow f$ is $\sigma\{\mathcal{P}\}$-measurable.

We interpret $\mathcal{F} \subseteq \mathcal{G}$ as $\mathcal{G}$ contains more information than $\mathcal{F}$.

Definition 3.24 $\sigma\{X\}$ is the smallest σ-algebra such that X is $\mathcal{F}$–measurable.

Definition 3.25 If a Wiener process W, is adapted to a filtration $\mathcal{F}$ and if $W(t) - W(s)$ is independent of $\mathcal{F}_s$ we call W a *$\mathcal{F}$-Wiener process.*

Consider the stochastic differential equation (SDE)

$$\begin{cases} dX(t) = \mu[t, X(t)]dt + \sigma[t, X(t)]dW(t) \\ X(0) = x, \end{cases}$$

where $\mu(t, x)$ and $\sigma(t, x)$ is given, continuous and *Lipschitz* in x. The Lipschitz condition says that there exists a constant L such as for all μ and σ:

$$\begin{aligned} |\mu(t,x) - \mu(t,y)| &\leq L|x - y| \quad \forall t, x, y \\ |\sigma(t,x) - \sigma(t,y)| &\leq L|x - y| \quad \forall t, x, y \end{aligned}$$

Then we are able to solve the SDE by integration

$$X(t) = x + \int_0^t \mu[s, X(s)]ds + \int_0^t \sigma[s, X(s)]dW(s)$$

Here the last integral is not a Reiman–Stjeltsin integral since W has infinite variation.

Definition 3.26 The *expectation value* (or mean value) of X given $(\Omega, \mathcal{F}, P)$ is in the discrete case given by

$$E[X] = \sum_{\omega \in \Omega} X(\omega)P\{\omega\}$$

and

$$E[X] = \int_\Omega X(\omega)dP\{\omega\}$$

in the continuous case.

For a finite set $X = \{x_1, x_2, \ldots, x_n\}$ we can partition Ω into subsets $\{X_i = x_1\}, \ldots, \{X_i = x_n\}$, and then write

$$E[X] = \sum_{k=1}^{n} \sum_{\omega \in \{X_k = x_k\}} X(\omega) \cdot P\{\omega\} = \sum_{k=1}^{n} x_k \sum_{\omega} P\{\omega\}$$
$$= \sum_{k=1}^{n} x_k P(X_k = x_k) = \sum_{k=1}^{n} x_k \mu_X\{x_k\}$$

Therefore we can sum over either Ω or $\boldsymbol{R}$.

Example 3.27

Calculate $E[S_3]$ in the binomial tree

$$\begin{aligned} E[S_3] &= S_2(uuu)P\{uuu\} + S_2(uud)P\{uud\} + S_2(udu)P\{udu\} \\ &+ S_2(udd)P\{udd\} + S_2(duu)P\{duu\} + S_2(ddu)P\{ddu\} \\ &+ S_2(dud)P\{dud\} + S_2(ddd)P\{ddd\} \\ &= 16 \cdot P(A_{uu}) + 4 \cdot P(A_{ud} \cup A_{du}) + P(A_{dd}) \\ &= 16 \cdot P\{S_2 = 16\} + 4 \cdot P\{S_2 = 4\} + P\{S_2 = 1\} \\ &= 16 \cdot \mu_{S_2}\{16\} + 4 \cdot \mu_{S_2}\{4\} + \mu_{S_2}\{1\} = 16 \cdot \frac{1}{9} + 4 \cdot \frac{4}{9} + \frac{4}{9} = \frac{36}{9} = 4 \end{aligned}$$

Definition 3.28 The *Tower property.* Let X be a random variable on $(\Omega, \mathcal{F}, P)$ and let $\mathcal{H} \subseteq \mathcal{G}$ be any two σ-algebras on $\mathcal{P}$, then

$$\mathrm{E}[\mathrm{E}[X|\mathcal{G}]|\mathcal{H}] = \mathrm{E}[X|\mathcal{H}].$$

Definition 3.29 The *Variance of X* is defined as

$$Var[X] = \int_{\Omega} [X(\omega) - E[X]]^2 dP\{\omega\}$$

$$Var[X] = \sum_{\omega=\Omega} \left(X(\omega) - E[X(\omega)]\right)^2 P\{\omega\} = \sum_{k=1}^{n} \left(x_k - E[X(\omega)]\right)^2 \mu_X(x_k) =$$
$$= E\left[\left(X(\omega) - E\left[X(\omega)\right]\right)^2\right] = E\left[X^2(\omega)\right] - \left(E\left[X(\omega)\right]\right)^2$$

Definition 3.30 An *indicator function* $I: R \rightarrow R$ is defined by

$$I_A(x) = \begin{cases} 0 & x \notin A \\ 1 & x \in A \end{cases}$$

where A is called *a set indicated by* I_A.

Definition 3.31 A function h is called *simple* if we can write

$$h(x) = \sum_{k=1}^{n} c_k I_k(x)$$

Definition 3.32 Let f: $X \rightarrow R$ be non-negative and measurable. Then, the integral of f is defined by

$$\int_X f(x) d\mu(x) = \sup_{\varphi} \int_X \varphi(x) d\mu(x),$$

where the supremum is taken over all simple functions $\varphi \leq f$.

Definition 3.33 A measurable function f is said to be *integrable* if

$$\int_X |f(x)| d\mu(x) < \infty$$

We write this as $f \in L^1(X, \mathcal{F}, \mu)$. For an integrable function f, the integral on f on X is defined by

$$\int_X f(x) d\mu(x) = \int_X f^+(x) d\mu(x) - \int_X f^-(x) d\mu(x).$$

If A is a measurable set, the integral of f on A is defined by

$$\int_A f(x) d\mu(x) = \int_X I_A(x) f(x) d\mu(x).$$

3.3.1 Introduction to Integration Theory

Definition 3.34 A *Borel algebra* $\mathbf{B}(\boldsymbol{R})$ is defined as the smallest σ-algebra that contains all open intervals on $\boldsymbol{R}$. The subsets in $\mathbf{B}$ are called *Borel sets*. All sets we can think of and all writable sets on $\boldsymbol{R} \in \mathbf{B}$, e.g.

$$(a,\infty) = \bigcup_{n=1}^{\infty} (a, a+n) \qquad (-\infty, a) = \bigcup_{n=1}^{\infty} (a-n, a)$$
$$[a,\infty) = \bigcup_{n=1}^{\infty} [a, a+n] \qquad (-\infty, a] = \bigcup_{n=1}^{\infty} [a-n, a]$$
$$(-\infty, a) \cup (b, \infty) \qquad [a,b] = ((-\infty,a) \cup (b,\infty))^c$$
$$(a,b] = (-\infty, b] \cap (a,\infty) \qquad \{a\} = \bigcap_{n=1}^{\infty} \left(a - \frac{1}{n}, a + \frac{1}{n}\right)$$

are Borel sets. This means e.g. that all sets with infinitely number of real numbers is a Borel set; e.g. $A = \{a_1, a_2, \ldots, a_n\}$ is a Borel set

$$A = \bigcup_{k=1}^{n} \{a_k\}.$$

Therefore, all irrational numbers are a Borel set since these are complement to all real numbers.

Definition 3.35 A measure on $(\boldsymbol{R}, \mathbf{B}(\boldsymbol{R}))$ is a function μ, which maps $\mathbf{B}$ on the interval $[0, \infty]$ with

$$\begin{cases} \mu(\varnothing) = 0 \\ \mu\left(\bigcup_{k=1}^{\infty} A_k\right) = \sum_{k=1}^{\infty} \mu(A_k). \end{cases}$$

Definition 3.36 A function f: $\boldsymbol{R} \to \mathbf{B}(\boldsymbol{R})$ is called *Borel measurable* if

$$\{x \in B | f(x) \in A\} \in \mathcal{B}(\boldsymbol{R}).$$

To define measures of un-countable sets we have to generalize the concept of integration and introduce Lebesgue-integrals (see the literature about integration theory). We will only need selected definitions, as given below.

Definition 3.37 Some *Lebesgue integrals*

$$\int_R I_A d\mu_0 = \mu_0(A).$$

For a simple function we have

$$h(x) = \sum_{k=1}^{n} c_k I_k(x) \Rightarrow$$
$$\int_R hd\,\mu_0 = \sum_{k=1}^{n} c_k \int_R I_A d\mu_0 = \sum_{k=1}^{n} c_k \mu_0(A).$$

For a simple function $h(x) \leq f(x)\ \forall\, x \in R \Rightarrow$

$$\int_R fd\mu_0 = \sup\left\{\int_R hd\mu_0\right\}.$$

If this is $\neq \infty$ it is called f-integrable.

$$\int_R fd\,\mu_0 = \int_R f^+ d\mu_0 - \int_R f^- d\mu_0,$$

where $f^+(x) = \max\{f(x), 0\}$ and $f^-(x) = \max\{-f(x), 0\}$.

$$\int_A fd\,\mu_0 = \int_R I_A f d\mu_0,$$

where I_A is an indicator function to A.

3.3.2 Probability Spaces

Definition 3.38 A *probability space* is defined by $(\Omega, \mathcal{F}, P)$, where

- Ω is a non-empty set, called sample space, which contains all possible outcomes of some random experiment.
- $\mathcal{F}$ is a σ-algebra of all subsets of Ω.
- P is a probability measure on $(\Omega, \mathcal{F})$ which assigns to each set $A \in \mathcal{F}$ a number $P(A) = [0, 1]$, which represent the probability that the outcome of the random experiment lies in A.

Definition 3.39 Given $(\Omega, \mathcal{F}, P)$ and a stochastic variable X. If X is an indicator function (e.g., $X(\omega) = I_A(\omega) = 1$ if $\omega \in A$ and 0 otherwise) then

$$\int_{\Omega} XdP = P(A).$$

If X is simple, we have

$$\int_{\Omega} XdP = \sum_{k=1}^{n} c_k \int_{\Omega} I_{A_k} dP = \sum_{k=1}^{n} c_k P(A_k)$$

$$\int_{A} XdP = \int_{\Omega} X \cdot I_A dP$$

Definition 3.40 The *expectation value* for a stochastic variable is given by

$$E[X] = \int_{\Omega} X \cdot I_A dP = \int_{\Omega} X(\omega) \cdot dP(\omega).$$

Definition 3.41 If X is a positive stochastic variable, then

$$E[X] = \int_{0}^{\infty} P(X \geq t) dt.$$

Definition 3.42 φ is a *density function* on $\boldsymbol{R}$ if $\varphi > 0$ and $\int_{\Omega} \varphi \cdot d\mu_0 = 1$. The associated probability measure is given by

$$P(A) = \int_{A} \varphi \cdot d\mu_0 \forall \mathrm{A} \in \mathcal{F}(\boldsymbol{R}).$$

Here φ is called the *Radon–Nikodym derivative* with respect to μ_0 and

$$\varphi = \frac{dP}{d\mu_0}.$$

3.3.3 Independence

Definition 3.43 $A \in \mathcal{F}$ and $B \in \mathcal{F}$ is *independent* if $P(A \cap B) = P(A)P(B)$.

Definition 3.44 $\mathcal{G}$ and $\mathcal{H}$ is *independent σ-algebras* if $P(A \cap B) = P(A)P(B)$ $\forall$ $A \in \mathcal{G}$ and $B \in \mathcal{H}$.

Definition 3.45 X and Y are *independent stochastic variables* if they generate independent σ-algebras.

Example 3.46

Let $P\{HH\} = p^2$, $P\{HT\} = P\{TH\} = qp$, $P\{TT\} = q^2$ and define $A = \{HH, HT\}$ and $B = \{HT, TH\} \Rightarrow A \cap B = \{HT\}$. A and B is independent if $P\{HT\} = P\{HH, HT\}P\{HT, TH\}$ $\Rightarrow qp = P(A)P(B) = (p^2 + qp)*2qp = p*2qp = 2qp^2 \Rightarrow p = q = ½$.

Example 3.47

If $\mathcal{G} = \{\emptyset, \Omega, \{HH, HT\}, \{TH, TT\}\}$ and $\mathcal{H} = \{\emptyset, \Omega, \{HH, TH\}, \{HT, TT\}\}$. If we now let $A = \{HH, HT\}$ and $B = \{HH, TH\}$ we have $P(A)P(B) = (p^2 + qp)(p^2 + qp) = p^2$, $P(A \cap B) = P\{HH\} = p^2$

3.3.4 Conditioned Expectations

Study the binomial model

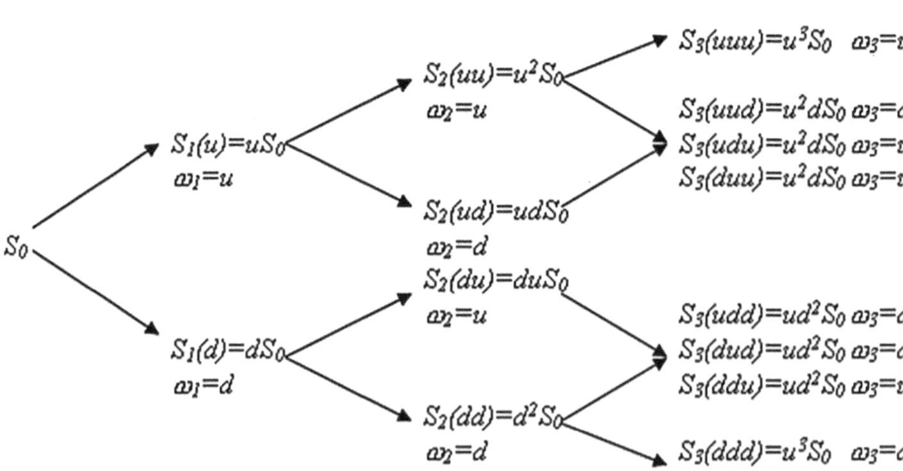

where every S_k is a stochastic process on $\Omega = \{uuu, uud, udu, duu, udd, dud, ddu, ddd\}$. $\mathcal{F} = P(\Omega)$ is a σ-algebra and $(\Omega, \mathcal{F})$ a measurable space. Every S_k is then a measurable function $\Omega \to R$.

Definition 3.48 The *conditioned expectation* of A given B is defined by

$$P(A|B) = \frac{P(A \cap B)}{P(B)}.$$

Definition 3.49 Suppose that we know the outcome $\omega \in B$ where B is measurable and $P(B) > 0$ then we define the expectation value of X conditioned B as

$$E[X|B] = \frac{1}{P(B)} \int_{\Omega} X(\omega) dP(\omega).$$

Example 3.50

Let us calculate S_1 given S_2: $E[S_1 \mid S_2]$. We know this is a stochastic variable

$$Y : Y(\omega) = E[S_1|S_2 = y] \text{ where } y = S_2(\omega).$$

Here $E[S_1 \mid S_2]$ has the following properties

* It is independent of ω.

* If S_2 is known then $E[S_1 \mid S_2]$ is known. Specifically, we have

– If $\omega = uuu$ or $\omega = uud \Rightarrow S_2(\omega) = u^2 S_0 \Rightarrow$ without knowing ω we know $S_1(\omega) = uS_0$

$$E[S_1|S_2](uuu) = E[S_1|S_2](uud) = uS_0.$$

Similarly, if $\omega = dd*$ we get

$$E[S_1|S_2](ddd) = E[S_1|S_2](ddu) = dS_0$$

– If $\omega = A = \{udu, udd, duu, dud\} \Rightarrow S_2(\omega) = u^d S_0 \Rightarrow$ but we don't know if $S_1(\omega) = uS_0$ or $S_1(\omega) = dS_0$. Therefore, we take the average value

$$P(A) = p^2 q + pq^2 + p^2 q + pq^2 = \{p + q = 1\} = 2pq.$$

Furthermore, we have

$$\int_A S_1 dP = p^2 q u S_0 + pq^2 u S_0 + p^2 q d S_0 + pq^2 d S_0 = pq(u + d) S_0.$$

For $\omega \in A$ we define

(continued)

Example 3.50 (continued)

$$E[S_1|S_2](\omega) = \frac{\int_A S_1 dP}{P(A)} = \frac{1}{2}(u+d)\cdot S_0.$$

Then

$$\int_A E[S_1|S_2]dP = \int_A S_1 dP$$

To summarize, we can write

$$E[S_1|S_2](\omega) = g(S_2(\omega)),$$

where

$$g(x) = \begin{cases} u\cdot S_0 & \text{if } x = u^2 S_0 \\ \frac{1}{2}(u+d)\cdot S_0 & \text{if } x = udS_0 \\ d\cdot S_0 & \text{if } x = d^2 S_0. \end{cases}$$

In other words $E[S_1 \mid S_2]$ is random only in the dependence of S_2. We can also write $E[S_1|S_2 = x] = g(x)$, where g is the function defined above. The random variable $E[S_1 \mid S_2]$ has two fundamental properties

* $E[S_1 \mid S_2]$ is $\sigma(S_2)$-measurable
* For all sets $A \in \sigma(S_2)$

$$\int_A E[S_1|S_2]dP = \int_A S_1 dP.$$

Some important properties of expectation values:

$E[E[X\|G]] = E[X]$	
$E[X\|G] = X$	if X is G-measurable
$E[X\|G] \geq 0$	if $X \geq 0$
$E[a_1X_1 + a_2X_2\|G] = a_1E[X_1\|G] + a_2E[X_2\|G]$	0
$E[\phi(X)\|G] \geq \phi(E[X\|G]) \quad \phi: R \to R, \quad E\|[\phi(X)]\| \leq \infty$	Jensen's unlikeness
$E[E[X\|G]\|H] = E[X\|H]$	if H is a sub-σ-algebra on G.
$E[ZX\|G] = ZE[X\|G]$	if Z is G-measurable.
$E[X\|G] = E[X]$	if X is independent of G.

3.3.5 Martingales

We have mentioned before that a martingale describes a fair game, where the profit in average will be zero even if the gambler is allowed to use previous results on a new stake. A stochastic process $\{X_t\}$ is a martingale if the conditional expectation value of X_t is given by: $E[X_t \mid X_u\,;\, u \le s] = X_s$ for all $s < t$.

For a general definition we need

- A probability space $(\Omega, \mathcal{F}, P)$.
- A filtration, i.e. a sequence of σ-algebras $\mathcal{F}_0 \subset \mathcal{F}_1 \subset \ldots \subset \mathcal{F}_n \subset \mathcal{F}$. (Finer and finer sets.)
- A stochastic process $X = \{x_k\}$ with random variables $x_0, x_1, \ldots$.

Definition 3.51 The process X is *martingale* (MG) if:

(i)	X is $\underline{\mathcal{F}}$–adapted	(X is generated by $\mathcal{F}$).
(ii)	$E[\lvert X(t)\rvert] < \infty\ \forall\, t \ge 0$.	
(iii)	$E[X(t)\lvert\, \mathcal{F}_s] = X(s)\ \forall\, s \le t$	(the martingale property)

The meaning that X is $\underline{\mathcal{F}}$–adapted is that all x_k are $\mathcal{F}_k$-measurable. In other words, if we know the information in $\mathcal{F}_k$ then we know the value of x_k. If we in (iii) use $\le$ or $\ge$ instead of $=$ we have a *super-martingale* and a *sub-martingale* respectively.

Lemma 3.52 If X is a martingale, then

$$E[\Delta X_n \lvert \mathcal{F}_{n-1}] = 0\ \forall\ n > 0,\ \ \Delta X_n = X_n - X_{n-1}.$$

Example 3.53

Let Y be an $\mathcal{F}$-measurable stochastic variable on $(\Omega, \mathcal{F}, P, \underline{\mathcal{F}})$ and define $X : X_t = E[Y\lvert \mathcal{F}_t],\ t \ge 0$. Then X is martingale because ($s < t$):

$$E[X_t\lvert\mathcal{F}_s] = E[E[Y\lvert\mathcal{F}_t]\lvert\mathcal{F}_s] = E[Y\lvert\mathcal{F}_s] = X_s.$$

Theorem 3.54 *Under the risk-neutral measure Q:* (p, q)*, the discounted stock price* $\left\{(1+r)^{-k}S_k, \mathcal{F}_k\right\}_{k=0}^{n}$ *from the binomial model is martingale.*

Proof.

$$E^Q\left[(1+r)^{-(k+1)}S_{k+1}|F_k\right] = (1+r)^{-(k+1)}(p\cdot u + q\cdot d)S_k$$
$$= \left(\tfrac{1}{1+r}\right)^{k+1}\left(\frac{u\cdot(1+r-d)}{u-d} + \frac{d\cdot(u-1-r)}{u-d}\right)S_k$$
$$= \left(\tfrac{1}{1+r}\right)^{k+1}\frac{(1+r)(u-d)}{u-d}S_k$$
$$= (1+r)^{-k}S_k$$

Definition 3.55 A martingale is said to be *quadratic integrable* if

$$\sup_{0\le t\le\infty} E\left[X^2(t)\right] < \infty.$$

The class of these martingales has the following notation: $M^2(\Omega, \mathcal{F}, P, \underline{\mathcal{F}})$.

3.3.6 Markov Processes

We start by studying a European lookback option with values $S_0 = 4$, $u = 2$, $d = ½$, $p = q = ½$ and $r = ¼$ with a strike price $K = 5$ with a two-period binomial model.

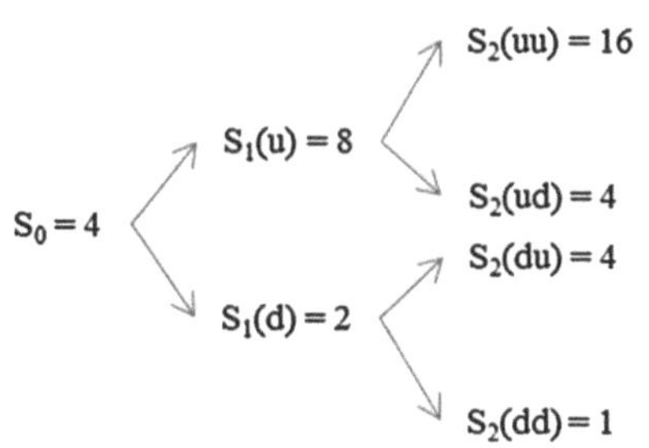

The value of the lookback option is given by:

$$V_2 = \max_{0\le t\le 2}(S_k - 5, 0).$$

We study the evolution backwards to calculate the value, thereby the name lookback. We have: $V_{uu} = (16-5) = 11$, $V_{ud} = (8-5) = 3$, $V_{du} = 0$ and $V_{dd} = 0$. (Remark $V_{ud} \neq V_{du}$). By travelling backwards in the tree we get

$$\begin{aligned}V_u &= \frac{1}{1+r}[pV_{uu} + qV_{ud}] = \frac{4}{5}\left\{\frac{1}{2}\cdot 11 + \frac{1}{2}\cdot 3\right\} = 5.60\\ V_d &= 0\\ V &= \frac{4}{5}\cdot\frac{1}{2}\cdot 5.60 = 2.24\end{aligned}$$

with

$$\Delta_{t-1} = \frac{V_t(u) - V_t(d)}{S_t(u) - S_t(d)}$$

we get $\Delta_0 = (5.6 - 0.0)/(8- 2) = 0.93$, $\Delta_1(u) = (11.0 - 3.0)/(16 - 4) = 0.67$ and $\Delta_1(d) = 0$. If we now sell one option at $X_0 = 2.24$ and hedge us with Δ_0 shares we get:

$$\begin{aligned}X_1(u) &= \Delta_0 S_1(u) + (1+r)(X_0 - \Delta_0 S_0)\\ &= 0.93\cdot 8 + (1+0.25)(2.24 - 0.93\cdot 4)\\ &= 5.60\end{aligned}$$

$$\begin{aligned}X_1(d) &= \Delta_0 S_1(d) + (1+r)(X_0 - \Delta_0 S_0)\\ &= 0.93\cdot 2 + (1+0.25)(2.24 - 0.93\cdot 4)\\ &= 0\end{aligned}$$

$$\begin{aligned}X_2(uu) &= \Delta_1(u)S_2(uu) + (1+r)(X_1(u) - \Delta_1(u)S_1(u))\\ &= 0.67\cdot 16 + (1+0.25)(5.60 - 0.67\cdot 8)\\ &= 11.0\end{aligned}$$

$$\begin{aligned}X_2(ud) &= \Delta_1(u)S_2(ud) + (1+r)(X_1(u) - \Delta_1(u)S_1(u))\\ &= 0.67\cdot 4 + (1+0.25)(5.60 - 0.67\cdot 8)\\ &= 3.0\end{aligned}$$

An ordinary European call option with the same data as above

$$V_2 = (S_k - 5)^+$$

gives $V_{uu} = 11$, $V_{ud} = V_{du} = 0$ and $V_{dd} = 0$. (**Remark** Since we cannot exercise before maturity, $V_{ud} = V_{du}$). Further

$$
\begin{aligned}
V_u &= \frac{1}{1+r}[pV_{uu} + qV_{ud}] = \frac{4}{5}\left[\frac{1}{2}\cdot 11 + \frac{1}{2}\cdot 0\right] = 4.40 \\
V_d &= 0 \\
V &= \frac{4}{5}\cdot\frac{1}{2}\cdot 4.40 = 1.76
\end{aligned}
$$

with

$$\Delta_{t-1} = \frac{V_t(u) - V_t(d)}{S_t(u) - S_t(d)}$$

we get $\Delta_0 = (4.4 - 0.0)/(8 - 2) = 0.733$, $\Delta_1(u) = (11.0 - 0.0)/(16 - 4) = 0.917$ and $\Delta_1(d) = 0$. If we now sell one option at $X_0 = 1.76$ and hedge us with Δ_0 shares we get

$$
\begin{aligned}
X_1(u) &= \Delta_0 S_1(u) + (1+r)(X_0 - \Delta_0 S_0) \\
&= 0.733\cdot 8 + (1+0.25)(1.76 - 0.733\cdot 4) \\
&= 4.40 \\
X_1(d) &= \Delta_0 S_1(d) + (1+r)(X_0 - \Delta_0 S_0) \\
&= 0.743\cdot 2 + (1+0.25)(1.76 - 0.733\cdot 4) \\
&= 0
\end{aligned}
$$

A general problem we have is that, for a model with n periods, we have Ω 2^n elements giving 2^n equations. For a three-month option we have 66 trading days and with a period length of one day we get $2^{66} \approx 7{*}10^{19}$ equations. The solution is, and we can solve it in three ways:

1. By simulations and averaging.
2. Approximate in continuous time. This gives a PDE-theory.
3. Using a Markov structure.

What we are doing in the binomial model is exactly 3.) above. Instead or four values at $n = 2$ (V_{uu}, V_{ud}, V_{du} and V_{dd}) we have three, because of $V_{ud} = V_{du}$. This gives us $n + 1$ equations instead of 2^n.

Definition 3.56 Let $(\Omega, \mathcal{F}, P)$ be a probability space with filtration $\{\mathscr{F}_t\}_{t\geq 0}$. An adapted process (X_t) is said to be a *Markov process* with respect to the filtration $(\mathscr{F}_t)$ if

$$E[f(X_t)|\mathcal{F}_s] = E[f(X_t)|X_s] \quad \text{for all } t \geq s \geq 0.$$

for every bounded real-valued Borel function f defined on R^d.

In other words this means that if we are studying a path, described by a geometrical Brownian motion (GBM) from 0 to t_0 and want to estimate the value of $f(X(t_1))$, the only relevant information is the value of $X(t_0)$.

Example 3.57

The stock price in the binomial model is a Markov process.

Theorem 3.58 *A Wiener process is a Markov process and*

$$P(W(t)\in B|W(s)) = \frac{1}{\sqrt{2\pi(t-s)}}\int_B \exp\left\{-\frac{(x-W(s))^2}{2(t-s)}\right\}dx.$$

3.3.7 Stopping Times and American Options

In a Markov model for a European contract with a value process $V_n = g(S_n)$ we define the backward recursion as

$$\begin{cases} V_n(x) = g(x) \\ V_k(x) = \dfrac{1}{1+r}(pV_{k+1}(ux) + qV_{k+1}(dx)). \end{cases}$$

Here $V_k(S_k)$ is the value of the option at time k and

$$\Delta_k = \frac{V_{k+1}(uS_k) - V_{k+1}(dS_k)}{(u-d)S_k}, \quad k = 0, 1, \ldots, n-1\,.$$

We will now study the binomial model for American contracts. In each node, k the holder can use his right to exercise the option and get $g(S_k)$. Therefore, the portfolio is given a value process satisfying $X_k \geq g(S_k)\ \forall\ k$. We then get

$$\begin{cases} V_n(x) = g(x) \\ V_k(x) = \max\left\{\dfrac{1}{1+r}(pV_{k+1}(ux) + qV_{k+1}(dx)), g(x)\right\}. \end{cases}$$

Let us study again, the two-period binomial tree, now for an American put option med with $S_0 = 4$, $u = 2$, $d = ½$, $p = q = ½$ and $r = ¼$ with a strike price $K = 5$.

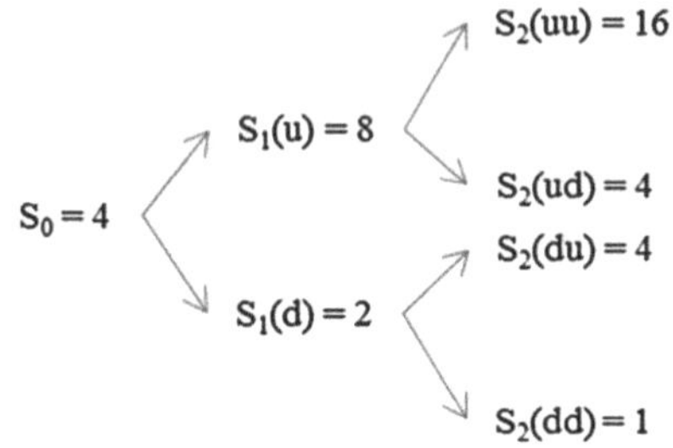

At maturity we have the value

$$V_2 = (5 - S_k)^+.$$

We now have $V_{uu} = 0$, $V_{ud} = V_{du} = 1$ and $V_{dd} = 4$ and the tree gives the values

$$V_u = \max\left\{\frac{1}{1+r}[pV_{uu} + qV_{ud}], (5-8)^+\right\} = \max\left\{\frac{4}{5}\left[\frac{1}{2}\cdot 0 + \frac{1}{2}\cdot 1\right], 0\right\} = 0.40$$

$$V_d = \max\left\{\frac{1}{1+r}[pV_{ud} + qV_{dd}], (5-2)^+\right\} = \max\left\{\frac{4}{5}\left[\frac{1}{2}\cdot 1 + \frac{1}{2}\cdot 4\right], 3\right\} = 3.00$$

$$V = \max\left\{\frac{1}{1+r}[pV_u + qV_d], (5-4)^+\right\} = \max\left\{\frac{4}{5}\left[\frac{1}{2}\cdot 0,4 + \frac{1}{2}\cdot 3\right], 1\right\} = 1.36.$$

Since

$$\Delta_{k-1} = \frac{V_k(u) - V_k(d)}{S_k(u) - S_k(d)}$$

we find $\Delta_0 = (0.40 - 3.00)/(8 - 2) = -0.43$. So we start to go short in the option and get 1.36. Then we hedge the position with Δ_0 stocks. For $k = 1$ we get ($X_1(d) = V_d$, $X_1(u) = V_u$)

$$\begin{aligned} 1 &= V_{du} = S_2(du)\Delta_1(d) + (1+r)(X_1(d) - \Delta_1(d)S_1(d)) \\ &= 4\cdot\Delta_1 + \frac{5}{4}(3 - 2\cdot\Delta_1) \Rightarrow \Delta_1(4 - 2.5) = 1 - 3.75 \\ &= -2.75 \Rightarrow \Delta_1 = -1.83 \end{aligned}$$

and

$$
\begin{aligned}
4 = V_{dd} &= S_2(dd)\Delta_1(d) + (1+r)(X_1(d) - \Delta_1(d)S_1(d)) \\
&= 1 \cdot \Delta_1 + \frac{5}{4}(3 - 2 \cdot \Delta_1) => \Delta_1(1 - 2.5) = 4 - 3.75 = 0.25 => \Delta_1 \\
&= -0.16
\end{aligned}
$$

We have bought Δ_1 stocks (if $\Delta_1 > 0$). The money we have left, that is, $(X_1 - \Delta_1 S_1)$ earns interest rate r, giving us

$$(1+r)(X_1 - \Delta_1 S_1).$$

If this was a European option $X_1(d) = S_1(d) = 2$ and Δ_1 were equal $(= -1)$

$$
\begin{aligned}
V_u &= \frac{1}{1+r}[pV_{uu} + qV_{ud}] = \frac{4}{5}\left[\frac{1}{2} \cdot 0 + \frac{1}{2} \cdot 1\right] = 0.40 \\
V_d &= \frac{1}{1+r}[pV_{ud} + qV_{dd}] = \frac{4}{5}\left[\frac{1}{2} \cdot 1 + \frac{1}{2} \cdot 4\right] = 2 \\
V &= \frac{1}{1+r}[pV_u + qV_d] = \frac{4}{5}\left[\frac{1}{2} \cdot 0,4 + \frac{1}{2} \cdot 2\right] = 0,96
\end{aligned}
$$

so

$$
\begin{aligned}
1 = V_{du} &= S_2(du)\Delta_1(d) + (1+r)(X_1(d) - \Delta_1(d)S_1(d)) \\
&= 4 \cdot \Delta_1 + \frac{5}{4}(2 - 2 \cdot \Delta_1) => \Delta_1(4 - 2.5) = 1 - 2.5 = -1.5 => \Delta_1 = -1.0
\end{aligned}
$$

The value of a hedged portfolio with an American option is given by

$$
\begin{aligned}
X_{k+1} &= S_{k+1}\Delta_k + (1+r)(X_k - \Delta_k S_k - C_k) \\
&= (1+r)X_k + \Delta_k\left(S_{k+1} - (1+r)S_k - (1+r)C_k\right),
\end{aligned}
$$

where C_k is the part that we can consume at time $t = k$.

Properties:

- The discounted portfolio value is a super-martingale.
- The value satisfy $X_k \geq g(S_k)$, $k = 0, 1, \ldots, n$.
- The value process is the process with the lowest value with these properties.

Question: When do we consume?
Answer: If

$$E\Big[(1+r)^{-(k+1)}V_{k+1}(S_{k+1})|\mathcal{F}_k\Big] < (1+r)^{-k}V_k(S_k)$$

We have

$$\frac{1}{1+r}E[V_{k+1}(S_{k+1})|\mathcal{F}_k] < V_k(S_k)$$

If the holder of the option doesn't exercise, then we can consume and close the gap between the values. In that case, when $X_k = V_k(S_k)$ for all values of k and where

$$\begin{cases} V_n(x) = g(x) \\ V_k(x) = \max\left\{\frac{1}{1+r}(pV_{k+1}(ux) + qV_{k+1}(dx)), g(x)\right\}. \end{cases}$$

In the previous example where, $V_1(S_1(u)) = 3$, $V_2(S_2(ud)) = 1$, $V_2(S_2(uu)) = 4$, we get

$$\frac{1}{1+r}E[V_2(S_2)|\mathcal{F}_1] = \frac{4}{5}\left[\frac{1}{2}\cdot 1 + \frac{1}{2}\cdot 4\right] = 2.$$

If the holder doesn't exercise at $t = 1$ we can consume one cash unit and hedge as

$$\Delta_k = \frac{V_{k+1}(uS_k) - V_{k+1}(dS_k)}{(u-d)S_k}.$$

As we can see, from the holder's point of view, it is optimal to exercise when $V_k(S_k) = g(S_k)$. I.e., at the intrinsic and not the discounted value.

Definition 3.59 Given the probability space $(\Omega, \mathcal{F}, P)$ and the filtration $\{\mathcal{F}_k\}_{k=0}^n$ of $\mathcal{F}$ we define the *stopping time* as a stochastic variable $\tau : \Omega \rightarrow \{0, 1, \ldots, n\} \cup \{\infty\}$ such as
$\{\omega \in \Omega; \tau(\omega) = k\} \in \mathcal{F}_k \,\forall\, k = 0, 1, \ldots, n, \infty$.

Example 3.60

We define (from the tree above) the stopping time

$$\tau(\omega) = \min\{k|V_k(S_k) = (5 - S_k)^+\}.$$

This stopping time is the time when the option value for the first time is equal to the instantaneous (intrinsic) value. This time is the optimal time to exercise the option. A stopping time is characterized by the fact that at every time $t < \tau$ we can decide if τ has occurred or not, based on the information we really have at time t. In our binomial model we have

$$\tau(\omega) = \begin{cases} 1 & \text{if } \omega = A_d \\ 2 & \text{if } \omega = A_u \end{cases} \quad \begin{array}{l} \{\omega : \tau(\omega) = 0\} = \varnothing \in \mathcal{F}_0 \\ \{\omega : \tau(\omega) = 1\} = A_d \in \mathcal{F}_1 \\ \{\omega : \tau(\omega) = 2\} = A_u \in \mathcal{F}_2. \end{array}$$

Definition 3.61 Let τ be a stopping time. We say that a set $A \subset \Omega$ is *determined by time* τ provided that

$$A \cap \{\omega|\tau(\omega) = k\} \in \mathcal{F}_k, \quad \forall k.$$

The collection of sets determined by τ is an σ-algebra, which we denote by $\mathcal{F}_\tau$.

Example 3.62

For the binominal model above, we have

$$\tau(\omega) = \min\{k \mid V_k(S_k) = (5 - S_k)^+\}$$

i.e.,

$$\tau(\omega) = \begin{cases} 1 & \text{if } \omega = A_d \\ 2 & \text{if } \omega = A_u \end{cases}$$

The set $\{ud\}$ is determined at the time τ, but the set $\{du\}$ is not. Indeed,

$$\begin{array}{l} \{ud\} \cap \{\omega|\tau(\omega) = 0\} = \varnothing \in \mathcal{F}_0 \\ \{ud\} \cap \{\omega : \tau(\omega) = 1\} = \varnothing \in \mathcal{F}_1 \\ \{ud\} \cap \{\omega : \tau(\omega) = 2\} = \{ud\} \in \mathcal{F}_2 \end{array}$$

but

$$\{du\} \cap \{\omega : \tau(\omega) = 1\} = \{du\} \notin \mathcal{F}_1$$

The atoms of $\mathcal{F}_\tau$ are $\{ud\}$, $\{uu\}$, $A_d = \{du, dd\}$.

3.3.8 The Radon–Nikodym Derivative

In mathematics, the Radon–Nikodym theorem below is a result in the functional analysis. In finance the Radon–Nikodym derivative is used to change measures. In Chap. 5 we will study this in detail.

Theorem 3.63 *Let P and Q being two probability measures on* $(\Omega, \mathcal{F})$. *Suppose that for each* $A \in \mathcal{F}$ *with* $P(A) = 0$, *and also* $Q(A) = 0$, *then we say that Q is absolute continuous with respect to P. Furthermore, then there exists a stochastic variable* $Z\ (\geq 0)$ *such that*

$$Q(A) = \int_{\Omega} Z dP(A).$$

We name Z as the Radon-Nikodym derivative of Q with respect to P.

It follows trivially from the definition of the derivative that, when P and Q are probability measures over the probability space Ω and X is a random variable. Then

$$E^{Q}[X] = \int_{\Omega} X dQ = \int_{\Omega} X \frac{dQ}{dP} dP = E^{P}\left[\frac{dQ}{dP} X\right].$$

If P at the same time is absolute continuous with respect to Q we say that P and Q are equivalent. I.e., if and only if $Q(A) = 0$ exactly when $P(A) = 0$ we have

$$\begin{aligned} E^{Q}[X] &= E^{P}[XZ] \quad \forall X \\ E^{P}[Y] &= E^{Q}\left[Y\frac{1}{Z}\right] \quad \forall Y. \end{aligned}$$

Example 3.64

Let $\Omega = \{uu, ud, du, dd\}$, $P(u) = 1/3$, $P(d) = 2/3$ and $Q(u) = Q(d) = 1/2$. Define $Z(\omega)$ as $Q(\omega)/P(\omega)$. Then

$$Z(uu) = (1/2)^2/(1/3)^2 = 9/4,\ Z(ud) = 9/8,\ Z(du) = 9/8 \text{ and } Z(dd) = 9/16.$$

As we have said before, in financial analysis, the Radon–Nikodym derivative is used to change measures. If we have a sample space Ω, with market probabilities P and if Q is the risk-neutral probability distribution, then we can find a transformation between P and Q with the help of the Radon–Nikodym derivative. If $P(\omega) > 0$ and $Q(\omega) > 0$ for all $\omega \in \Omega$, P and Q are equivalent. We write this as $Q \sim P$. If P and Q are absolute continuous we write this as $Q << P$.

Two measures are equivalent if they have the same sample space and the same set of "possibilities". Note the use of the word possibilities instead of probabilities. The two measures can have different probabilities for each outcome but must agree on what is possible.

Another way to formulate the Radon–Nikodym is using two different measures, μ and ν on (Ω, X). Absolute continuity means that $\mu << \nu$ and equivalence means that $\mu \sim \nu$. If $\nu \sim \mu$, i.e. they have exactly the same empty measure $\emptyset$, then we write the Radon–Nikodym derivative as

$$f = \frac{dv}{d\mu} \quad \Leftrightarrow \quad dv(x) = f(x) \cdot d\mu(x)$$

With this definition we can always find f, also on point sets (as we just did in Example 3.64).

$$f(n) = \begin{cases} \nu(n)/\mu(n) & \text{if } \mu(n) \neq 0 \\ 0 & \text{else} \end{cases}.$$

Remark If we make a σ-algebra finer and finer we may lose the absolute continuity. Suppose we have a given probability spaces $(\Omega, \mathcal{F}, P)$, with a filtration $\underline{\mathcal{F}} = \{\mathcal{F}\}$ on the interval $[0, T]$. Then, if $L_T \geq 0$ is a $\mathcal{F}$-measurable stochastic variable we can find a new measure Q on $(\Omega, \mathcal{F}_T)$ via

$$dQ = L_T dP$$

where Q will be a probability measure if $E^P[L_T] = 1$

$$\int_\Omega dQ = \int_\Omega L_T dP = E^P[L_T] = 1.$$

Definition 3.65 Z_k says to be a *P-martingale* if

$$\begin{aligned} Z_k &= E^P[Z|\mathcal{F}_k] \quad k = 0, 1, \ldots, n \\ E^P[Z_{k+1}|\mathcal{F}_k] &= E^P\left[E^P[Z|\mathcal{F}_{k+1}]|\mathcal{F}_k\right] = E^P[Z|\mathcal{F}_k] = Z_k. \end{aligned}$$

Lemma 3.66 If X is $\mathcal{F}_k$*-measurable* and $0 \leq j \leq k$, then

$$E^Q\left[X|\mathcal{F}_j\right] = \frac{1}{Z_j} E^P\left[XZ_k|\mathcal{F}_j\right].$$

Theorem 3.67 *L (as above) is a* $(\underline{\mathcal{F}}, P)$*-martingale.*

Proof: We have to show that $L_t = E^P[L_T \mid \mathcal{F}_t]$ for all $t \leq T$ or

$$\int_F L_t dP = \int_F L_T dP \quad \text{for all } \mathcal{F} \in \mathcal{F}_t$$

$$\int_F L_t dP = \{F \in \mathcal{F}_t\} = Q_t(F) = Q_T(F) = \{F \in \mathcal{F}_t \subseteq \mathcal{F}_T\} = \int_F L_T dP.$$

A risk-neutral measure is a probability measure Q equivalent to the real probabilities P (Q ~ P) under which all tradable assets are martingales after discounting.

Theorem 3.68 *Given a probability space* $(\Omega, \mathcal{F}, P)$, $X \in L^1(\Omega, \mathcal{F}, P)$ *and a probability measure Q on* $(\Omega,\mathcal{F})$ *where* $Q << P$ *and* $L = \frac{dQ}{dP}$. *If* $\mathcal{G}$ *is a* σ*-algebra such as* $\mathcal{G} \subseteq \mathcal{F}$ *then*

$$E^Q[X|\mathcal{G}] = \frac{E^P[LX|\mathcal{G}]}{E^P[L|\mathcal{G}]}$$

Proof: We will show that

$$E^Q[X|\mathcal{G}]E^P[L|\mathcal{G}] = E^P[LX|\mathcal{G}]$$

It is enough to show

$$\int_G E^Q[X|\mathcal{G}]E^P[L|\mathcal{G}]dP = \int_G LX\,dP$$
$$\Leftrightarrow$$
$$\int_G E^P\left[LE^Q[X|\mathcal{G}]|\mathcal{G}\right]dP = \int_G LX\,dP$$
$$\Leftrightarrow$$

$$\int_G LE^Q[X|\mathcal{G}]dP = \int_G LX\,dP$$
$$\Rightarrow \ \{LdP = dQ\} \ \Rightarrow$$
$$\int_G E^Q[X|\mathcal{G}]dQ = \int_G XdQ\int_G LX\,dP.$$

3.4 Properties of Normal and Log-Normal Distributions

When we will study the Black–Scholes model, which is a continuous time model, the normal and log-normal distributions will be used. Therefore, we will now give the most important properties of them.

If the density function φ is given by

$$\varphi(x) = \frac{1}{\sqrt{2\pi\sigma^2}}\exp\left\{-\frac{(x-m)^2}{2\sigma^2}\right\} = N'\left(\frac{x-m}{\sigma}\right)$$

we say that x has a Gaussian (or normal) distribution, with mean m and variance σ^2. In this case we say that x is an $N(m, \sigma^2)$ random variable.

If X is an $N(m, \sigma^2)$-random variable, then

$$E(X) = \frac{1}{\sqrt{2\pi\sigma^2}}\int_{-\infty}^{\infty} x\cdot\exp\left\{-\frac{(x-m)^2}{2\sigma^2}\right\}dx = m$$

which is called the 1st moment of the probability distribution. The second moment gives the variance

$$Var(X) = \frac{1}{\sqrt{2\pi\sigma^2}} \int_{-\infty}^{\infty} (x-m)^2 \cdot \exp\left\{-\frac{(x-m)^2}{2\sigma^2}\right\} dx = \sigma^2.$$

The 3rd and 4th moments are given by

$$\frac{1}{\sqrt{2\pi\sigma^2}} \int_{-\infty}^{\infty} (x-m)^3 \cdot \exp\left\{-\frac{(x-m)^2}{2\sigma^2}\right\} dx = 0$$

$$\frac{1}{\sqrt{2\pi\sigma^2}} \int_{-\infty}^{\infty} (x-m)^4 \cdot \exp\left\{-\frac{(x-m)^2}{2\sigma^2}\right\} dx = 3\sigma^4$$

They are called the skewness and the kurtosis (or flatness). We note that if x is $N(m, \sigma^2)$, then $x = m + \sqrt{\Delta t}\xi$ where ξ is a standard Gaussian variable with mean zero and variance 1; i.e., ξ is $N(0, 1)$.

If $x = \ln y$ the probability density function of y is called a log-normal distribution and is given by

$$\varphi(y) = \frac{1}{\sqrt{2\pi\sigma^2 y}} \exp\left\{-\frac{(\ln y - m)^2}{2\sigma^2}\right\} \quad y > 0.$$

This can be seen from $y = e^x$ and x $\sim N(m, \sigma^2)$ and

$$\begin{aligned} P_Y(y) &= P(e^x \le y) = P(x \le \ln(y)) = \frac{1}{\sqrt{2\pi\sigma^2}} \int_{-\infty}^{\ln(y)} \exp\left\{-\frac{(x-m)^2}{2\sigma^2}\right\} dx \\ &= \int_{-\infty}^{\ln(y)} \varphi(x) dx = \Phi(\ln(y)) - \Phi(-\infty). \end{aligned}$$

If we take the derivative with respect to y we get

$$\frac{\partial P_Y}{\partial y} = \frac{\partial \Phi(\ln(y))}{\partial y} = \frac{\partial \ln(y)}{\partial y} \frac{\partial \Phi(\ln(y))}{\partial (\ln(y))} = \frac{1}{y} \varphi(\ln(y)).$$

Theorem 3.69 *If X is a Gaussian (normal) process with mean m and variance σ^2, i.e., $X \sim N(m, \sigma^2)$ and $\gamma \in R$ we have*

$$E\left[e^{-\gamma X}\right] = \exp\left\{-\gamma m + \frac{1}{2}\gamma^2\sigma^2\right\}.$$

Proof: Per definition and manipulation with the exponent we have

$$\begin{aligned}
E[e^{-\gamma X}] &= \frac{1}{\sqrt{2\pi\sigma^2}}\int_{-\infty}^{\infty} e^{-\gamma x} e^{-\frac{(x-m)^2}{2\sigma^2}} dx = \frac{1}{\sqrt{2\pi\sigma^2}}\int_{-\infty}^{\infty} e^{-\frac{1}{2\sigma^2}[2\gamma x\sigma^2 + x^2 - 2xm + m^2]} dx \\
&= \frac{1}{\sqrt{2\pi\sigma^2}}\int_{-\infty}^{\infty} e^{-\frac{1}{2\sigma^2}[x^2 - 2(m-\gamma\sigma^2)x + m^2]} dx \\
&= \frac{1}{\sqrt{2\pi\sigma^2}}\int_{-\infty}^{\infty} e^{-\frac{1}{2\sigma^2}\left[(x-(m-\gamma\sigma^2))^2 + 2m\gamma\sigma^2 - \gamma^2\sigma^4\right]} dx \\
&= e^{-\gamma m + \frac{1}{2}\gamma^2\sigma^2}\frac{1}{\sqrt{2\pi\sigma^2}}\int_{-\infty}^{\infty} e^{-\frac{1}{2\sigma^2}\left[(x-[m-\gamma\sigma^2])^2\right]} dx \\
&= e^{-\gamma m + \frac{1}{2}\gamma^2\sigma^2}
\end{aligned}$$

where the last equality is due to the fact that the function

$$\frac{1}{\sqrt{2\pi\sigma^2}} e^{\frac{1}{2\sigma^2}\left[(x-[m-\gamma\sigma^2])^2\right]}$$

is a probability density function, namely the density function for an $N(m - \gamma\sigma^2, \sigma^2)$ distributed random variable.

Using Theorem 3.69., we can easily compute the mean and the variance of the lognormal distributed random variable $Y = e^X$. The mean is (let $\gamma = -1$)

$$E[Y] = E\left[e^X\right] = \exp\left\{m + \frac{1}{2}\sigma^2\right\}.$$

With $\gamma = -2$ we get

$$E\left[Y^2\right] = E\left[e^{2X}\right] = e^{2\left(m+\sigma^2\right)}$$

so that the variance of Y is

$$Var[Y] = E[Y^2] - (E[Y])^2 = e^{2(m+\sigma^2)} - e^{2m+\sigma^2} = e^{2m+\sigma^2}\left(e^{\sigma^2} - 1\right).$$

The next theorem provides an expression of the truncated mean of a lognormal distributed random variable, i.e. the mean of the part of the distribution that lies above some level. We define the indicator function $I_{\{Y>K\}}$ to be equal to 1 if the outcome of the random variable Y is greater than the constant K and equal to 0 otherwise.

Theorem 3.70 *If $X = \ln Y \sim N(m, \sigma^2)$ and $K > 0$, then we have*

$$E\left[Y \cdot I_{\{Y>K\}}\right] = e^{m+\frac{1}{2}\sigma^2} N\left(\frac{m - \ln K}{\sigma} + \sigma\right) = E[Y] N\left(\frac{m - \ln K}{\sigma} + \sigma\right)$$

Proof: Because $Y > K$, $X > \ln K$, it follows from the definition of the expectation of a random variable that

$$E\left[Y \cdot I_{\{Y>K\}}\right] = E\left[e^X \cdot I_{\{X>\ln K\}}\right] = \frac{1}{\sqrt{2\pi\sigma^2}} \int_{\ln K}^{\infty} e^x e^{-\frac{(x-m)^2}{2\sigma^2}} dx$$

$$= \frac{1}{\sqrt{2\pi\sigma^2}} \int_{\ln K}^{\infty} e^{-\frac{\left(x-\left(m+\sigma^2\right)\right)^2}{2\sigma^2}} e^{\frac{2m\sigma^2+\sigma^4}{2\sigma^2}} dx = e^{m+\frac{1}{2}\sigma^2} \int_{\ln K}^{\infty} f_X(x) dx$$

where

$$f_X(x) = \frac{1}{\sqrt{2\pi\sigma^2}} e^{-\frac{\left(x-\left(m+\sigma^2\right)\right)^2}{2\sigma^2}}$$

is a probability density function for an $N(m + \sigma^2, \sigma^2)$ distributed random variable. The calculation

$$\begin{aligned}
\int_{\ln K}^{\infty} f_X(x) dx &= P(X > \ln K) = P\left(\frac{X - [m + \sigma^2]}{\sigma} > \frac{\ln K - [m + \sigma^2]}{\sigma}\right) \\
&= P\left(-\frac{X - [m + \sigma^2]}{\sigma} < -\frac{\ln K - [m + \sigma^2]}{\sigma}\right) \\
&= N\left(-\frac{\ln K - [m + \sigma^2]}{\sigma}\right) = N\left(\frac{m - \ln K}{\sigma} + \sigma\right)
\end{aligned}$$

completes the proof.

Theorem 3.71 *If* $X = \ln Y \sim N(m, \sigma^2)$ *and* $K > 0$*, then we have*

$$\begin{aligned} E[\max\{Y-K,0\}] &= e^{m+\frac{1}{2}\sigma^2} N\left(\frac{m-\ln K}{\sigma}+\sigma\right) - KN\left(\frac{m-\ln K}{\sigma}\right) \\ &= E[Y]N\left(\frac{m-\ln K}{\sigma}+\sigma\right) - KN\left(\frac{m-\ln K}{\sigma}\right) \end{aligned}$$

Proof: Note that

$$E[\max\{Y-K,0\}] = E\left[(Y-K)I_{\{Y>K\}}\right] = E\left[Y \cdot I_{\{Y>K\}}\right] - K \cdot P(Y > K)$$

The first term is known from Theorem 3.70. The second term can be rewritten as

$$\begin{aligned} P(Y > K) &= P(X > \ln K) = P\left(\frac{X-m}{\sigma} > \frac{\ln K - m}{\sigma}\right) \\ &= P\left(-\frac{X-m}{\sigma} < -\frac{\ln K - m}{\sigma}\right) = N\left(-\frac{\ln K - m}{\sigma}\right) \\ &= N\left(\frac{m-\ln K}{\sigma}\right). \end{aligned}$$

The claim now follows immediately.

3.5 The Itô Lemma

One of the most important formulas in the financial analysis is the Lemma by Itô. The Lemma states how to differentiate functions of stochastic processes. To understand Itô's formula in its most simple form, we start with a Taylor expansion to the lowest orders for a function of two variables: $F(t, X)$

$$dF = \frac{\partial F}{\partial t}dt + \frac{\partial F}{\partial X}dX + \frac{1}{2}\frac{\partial^2 F}{\partial t^2}(dt)^2 + \frac{1}{2}\frac{\partial^2 F}{\partial X^2}(dX)^2 + \frac{\partial^2 F}{\partial t \partial X}dtdX + \ldots .$$

where X is described by the stochastic process given by

$$dX = \mu \cdot dt + \sigma \cdot dW.$$

Here μ represent a deterministic drift and σ the volatility. W is a Wiener process with the property $(dW)^2 = dt$. Thus, to the lowest order we get

$$(dX)^2 = \mu^2 \cdot (dt)^2 + \sigma^2 \cdot (dW)^2 + 2 \cdot \mu \cdot \sigma \cdot dt \cdot dW \to \sigma^2 \cdot dt.$$

In the lowest order, we ignore *dtdw* and *dtdt*. To the lowest order of *dF*, we then have

$$dF = \left(\frac{\partial F}{\partial t} + \mu \frac{\partial F}{\partial X} + \frac{1}{2}\sigma^2 \frac{\partial^2 F}{\partial X^2} \right) dt + \sigma \frac{\partial F}{\partial X} dW$$

which is the Itô's formula. In finance, this is the most useful expression of Itô. Sometimes the Itô's Lemma is expressed as

$$dF(X) = \frac{\partial F(X)}{\partial X} dX + \frac{1}{2} \frac{\partial^2 F(X)}{\partial X^2} dt.$$

But mathematically, a more meaningful form of Itô's Lemma is the integral version:

$$F(X(t)) = F(X(0)) + \int_0^t \frac{\partial F}{\partial X} \{X(\tau)\} dX(\tau) + \frac{1}{2} \int_0^t \frac{\partial^2 F}{\partial X^2} \{X(\tau)\} d\tau$$

This is because we have a solid definition of the integrals. The first integral is called the Itô integral and the second is a Riemann integral. This means that in stochastic calculus we have to use

$$df = \frac{\partial f}{\partial t} dt + \frac{\partial f}{\partial x} dx + \frac{1}{2} \frac{\partial^2 f}{\partial X^2} (dx)^2$$

to lowest order differentiation of $f(x(t), t)$ instead of

$$df = \frac{\partial f}{\partial t} dt + \frac{\partial f}{\partial x} dx$$

as in ordinary calculus. A generalization of Itô can be written as

$$dF(X_1, \ldots X_n, t) = \frac{\partial F}{\partial t} dt + \sum_{i=1}^{n} \frac{\partial F}{\partial X_i} dX_i + \frac{1}{2} \sum_{i,j=1}^{n} \frac{\partial^2 F}{\partial X_i \partial X_j} \sigma_i \sigma_j dt.$$

Example 3.72

Itô calculus is used in all calculations where we calculate differentials of stochastic processes where the uncertainty is modelled with a Wiener function. In particular we have

$$d(W^n) = \frac{\partial}{\partial W}(W^n)dW + \frac{1}{2}\frac{\partial^2}{\partial W^2}(W^n)(dW)^2$$
$$= nW^{n-1}dW + \frac{1}{2}n(n-1)W^{n-2}dt,$$

especially

$$d\left(W^2\right) = 2WdW + dt$$

so that

$$\int_s^t W(u)dW(u) = \frac{W^2(t) - W^2(s)}{2} - \frac{t-s}{2}.$$

Example 3.73

Let

$$F(t, W) = \exp\left\{\lambda W(t) - \frac{\lambda^2}{2}t\right\}.$$

Then

$$dF = \frac{\partial F}{\partial t}dt + \frac{\partial F}{\partial W}dW + \frac{1}{2}\frac{\partial^2 F}{\partial W^2}(dW)^2$$
$$= -\frac{\lambda^2}{2}\exp\lambda W(t) - \frac{\lambda^2}{2}t\}dt + \lambda\exp\lambda W(t) - \frac{\lambda^2}{2}t\}dW$$
$$+\frac{\lambda^2}{2}\exp\lambda W(t) - \frac{\lambda^2}{2}t\}dt$$
$$= \lambda\exp\lambda W(t) - \frac{\lambda^2}{2}t\}dW = \lambda F(t, W)dW$$

This shows that the solution to the stochastic differential equation

(continued)

Example 3.73 (continued)

$$\begin{cases} dX = \lambda X dW \\ X(0) = 1. \end{cases}$$

Is given by

$$X(t) = \exp\left\{\lambda X(t) - \frac{\lambda^2}{2}t\right\}.$$

3.5.1 Brownian Motion

When studying lattices-models of random processes, one naturally wonders whether these processes go to a limit as the step size is taken to be finer and finer. Or, more to the point, if we use lattices to model asset prices, does the model make sense in the limit that the step size goes to zero? To answer this question we now advance to continuous random processes.

We are aiming to develop models based on stochastic differential equations

$$dS = a(t, S)dt + b(t, S)dW$$

for the asset price $S(t)$, where the $a(t, S)dt$ term accounts for "deterministic motions", and the other term $b(t, S)dW$ accounts for "random motions".

The first step in developing such models is to decide what we should use for $W(t)$, the random part of the model. No matter how we sub-divide it, the curve $W(t)$ should still be random and composed of pieces with identical statistical properties, because stock prices appear random on even very fine time scales. In Fig. 3.2 we simulate for a year, 10 observations per day, a stock with initial price 30.00 with a volatility of 40 % and a risk-free interest rate 2 %.More formally, we wish $W(t)$ to have the following properties: First, $W(t)$ must have independent increments. For any date τ and for any $\Delta\tau > 0$, the value of $\Delta W = W(\tau + \Delta\tau) - W(\tau)$ is independent of $W(t)$ for all $t \leq \tau$. So increments of Brownian motion are independent of everything that has happened on or before the current date τ. In particular, $\Delta W_1 \equiv W(t_2) - W(t_1)$ and $\Delta W_2 \equiv W(t_3) - W(t_4)$ are independent whenever the two intervals $t_1 \leq t \leq t_2$ and $t_3 \leq t \leq t_4$ don't overlap. Said in another way, $W(t) - W(\tau)$ for $t > \tau$ does not depend on how one got to $W(\tau)$. As we shall see, this is a very powerful simplifying assumption.

Second, increments $\Delta W \equiv W(t_2) - W(t_1)$ is Gaussian random variables with mean 0 and variance $\Delta t \equiv t_2 - t_1$. We then have

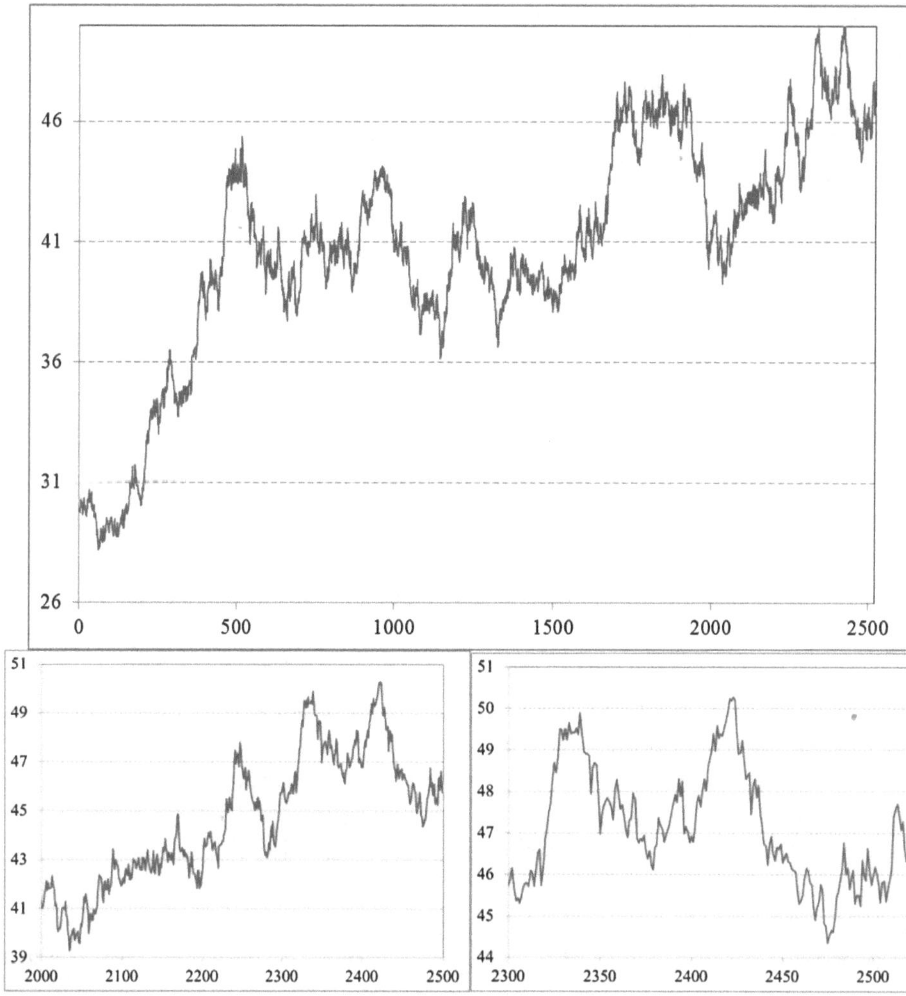

Fig. 3.2 A Brownian motion (Wiener process) illustrated as function of time on different time-scales

$$\Delta W \equiv W(t_2) - W(t_1) = \sqrt{t_2 - t_1}\xi$$

where ξ is $N(0, 1)$, that is, ξ is a Gaussian random variable with mean zero and variance 1. The reason we want ΔW to have mean zero is because we want it to represent the random part of the asset price movements. Any non-zero mean would represent a deterministic piece which we could put in the drift term $a(t, S)dt$.

The fact that ΔW is Gaussian with variance Δt follows directly from our desire to have $W(t)$ to be sub-dividable into finer and finer intervals, each with identical statistical properties. To show this, consider

$$\Delta W = W(t_1) - W(t_0) - W(t_0) \equiv \sum_{k=0}^{n-1}[W(\tau_{k+1}) - W(\tau_k)], \quad \tau_k = t_0 + \frac{k}{n}(t_1 - t_0),$$

where each $\delta W_k \equiv W(\tau_{k+1}) - W(\tau_k)$ are independent random variables (by the independent increment assumption) with identical distributions. Since the variables are independent, the variances can be written

$$Var[\Delta W] = \sum_{k=0}^{n-1} Var[W(\tau_{k+1}) - W(\tau_k)] = n \cdot Var[W(t_1) - W(t_0)]$$

Let $v(y) = Var[W(t + y) - W(t)]$. We have shown that

$$v(t_1 - t_0) = n \cdot v\left(\frac{t_1 - t_0}{n}\right)$$

for any t_1, t_0 and n, i.e., for any $\Delta t > 0$ and any n, we have $v(n\Delta t) = nv(\Delta t)$.

This is a functional equation, and it shouldn't be surprising that the only reasonable solutions are linear: $v(\Delta t) = \alpha\Delta t$ for some constant α. Brownian motion is normalized so that this constant is 1, i.e., $Var[\Delta W] \equiv Var[W(t_1) - W(t_0)] = t_1 - t_0$ for all t_0, t_1.

Thus ΔW is the sum of n independent, identically distributed variables with mean 0 and variance $(t_1 - t_0)/n$. As we take $n \to \infty$, the central limit theorem guarantees that ΔW is Gaussian with mean zero and variance $t_1 - t_0$.

Brownian motion has the following properties. Of these, the first two are part of the definition of Brownian motion, and the other three are derived below:

i. The increments ΔW are independent of the present and past values of $W(t)$. In particular, increments of non-overlapping intervals are independent

$\Delta W = W(t_2) - W(t_1)$ is independent of $W(t)$ for all $t \leq t_1$.

$\Delta W = W(t_2) - W(t_1)$ is independent of $\Delta W = W(t_4) - W(t_3)$ if $(t_1, t_2) \cap (t_3, t_4) = \varnothing$.

ii. $\Delta W = W(t + \Delta t) - W(t)$ is Gaussian with mean zero and variance $\sqrt{\Delta t}$. Said another way, $\Delta W = W(t + \Delta t) - W(t) = \sqrt{\Delta t}\xi$, where ξ is $N(0, 1)$, a Gaussian variable with mean 0 and variance 1.

iii. $W(t)$ is a continuous random process. This is easily proven, since for any $\delta > 0$,

$$\text{prob}\{|W(t+\Delta t) - W(t)| > \delta\} = \text{prob}\left\{ |\xi| > \frac{\delta}{\sqrt{\Delta t}} \right\} \underset{\Delta t \to 0}{\to} 0.$$

This is the definition of a continuous stochastic process.

iv. $W(t)$ is almost surely nowhere differentiable. This is again easily shown.

For any $K > 0$, we argue that

$$\text{prob}\left\{ \left| \frac{W(t+\Delta t) - W(t)}{\Delta t} \right| < K \right\} = \text{prob}\left\{ |\xi| > K\sqrt{\Delta t} \right\} \underset{\Delta t \to 0}{\to} 0.$$

So the probability that the slope is bounded is zero as $\Delta t \to 0$.

v. The continuity and non-differentiability followed directly from the scaling of ΔW. Since $\Delta W = \sqrt{\Delta t}\xi$, where ξ is $N(0, 1)$, we can write $\Delta W \sim O(\sqrt{\Delta t})$, or more succinctly $dW \sim O(\sqrt{\Delta t})$.

Later we shall prove a much more stunning result, the quadratic property of Brownian motion, $(dW)^2 = dt$. Note that the right side dt is not stochastic, which means that $(dW)^2$ is dt with certainty. This property is the key to deriving Itô's lemma, the backwards Kolmogorov equation, Feynman–Kač equation and many of the other day-to-day tools used in pricing. Before we can show this result, we need to define what we mean by differentials like dW and dt.

Black–Scholes

We have seen that the stock prices $S(t)$ follow a stochastic process, given as a Brownian motion described by the following stochastic differential equation

$$\begin{cases} dS(t) = \alpha \cdot S(t) \cdot dt + \sigma \cdot S(t) \cdot dW \\ S(0) = s \end{cases}.$$

This can easily be solved by letting $Z(t) = \ln\{S(t)\}$ and with use of the Itô lemma

$$
\begin{aligned}
dZ(t) &= \frac{1}{S(t)}dS(t) - \frac{1}{2}\frac{1}{S(t)^2}(dS(t))^2 \\
&= \frac{1}{S(t)}(\alpha \cdot S(t)dt + \sigma \cdot S(t)dW(t)) - \frac{1}{2}\frac{1}{S(t)^2}\sigma^2 S(t)^2 dt \\
&= \left(\alpha - \frac{1}{2}\sigma^2\right)dt + \sigma \cdot dW(t) \\
Z(0) &= \ln(s)
\end{aligned}
$$

Integration gives

$$Z(t) - Z(0) = \left(\alpha - \frac{1}{2}\sigma^2\right)t + \sigma \cdot W(t).$$

Thus

$$S(t) = s \cdot e^{\left\{\left(\alpha - \frac{1}{2}\sigma^2\right)t + \sigma \cdot W(t)\right\}}.$$

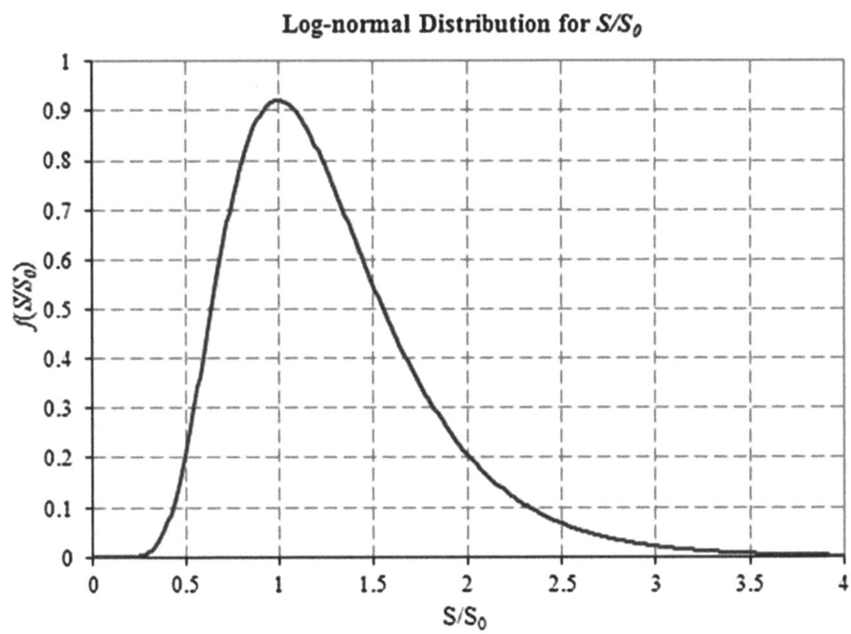

Fig. 3.3 The log-normal probability distribution with $\sigma^2 = 0.4$, $\mu = 0.16$ and $(t - t_0) = 1$

Since $W(t) - W(t_0)$ is normal distributed with mean zero and variance $(t - t_0)$, i.e.,

$N[0, (t - t_0)]$ we know that Z must be normal distributed as:

$$Z \sim N\left[\left(\alpha - \frac{\sigma^2}{2}\right)(t - t_0), \sigma^2(t - t_0)\right].$$

Therefore, $S(t)/S(t_0)$ follows a log-normal distribution ($\mu = \alpha$)

$$g(S(t)) = \frac{1}{\sigma S(t)\sqrt{2\pi(t - t_0)}} \exp\left\{-\frac{\left(\ln\{S(t)/S(t_0)\} - (\mu - \sigma^2/2)(t - t_0)\right)^2}{2\sigma^2(t - t_0)}\right\}.$$

A typical log-normal probability distribution for $S(t)/S(t_0)$ are shown in Fig. 3.3.

3.6 Stochastic Integration

To understand stochastic integration we will start by studying the integral $\int g(s)dW(s)$. We will do this in a few simple steps:

1. Split the interval $[0, t]$ into equal parts $0 = t_0 < t_1 \ldots < t_n = t$.
2. For each outcome ω define an integral

$$I_n(\omega) = \sum g(\xi_k, \omega)\left[W(t_{k+1}, \omega) - W(t_k, \omega)\right]$$

3. Let $n \to \infty$ and hope for $I_n \to I$.

Let $g = W$ and study the integral $\int W(s)dW(s)$ by defining A_n and B_n:

$$\begin{cases} A_n = \sum_{k=1}^{n} W(t_k)[W(t_{k+1}) - W(t_k)] & \xi_k = t_k \\ B_n = \sum_{k=1}^{n} W(t_{k+1})[W(t_{k+1}) - W(t_k)] & \xi_k = t_{k+1} \end{cases}$$

We then get

$$
\begin{aligned}
A_n &= \lim_{n\to\infty} \sum_{k=1}^{n} W(t_k)[W(t_{k+1}) - W(t_k)] \\
&= \frac{1}{2}\lim_{n\to\infty} \sum_{k=1}^{n} [W(t_{k+1}) + W(t_k) - (W(t_{k+1}) - W(t_k))] \\
&\times (W(t_{k+1}) - W(t_k)) \\
&= \frac{1}{2}\lim_{n\to\infty} \sum_{k=1}^{n} [(W(t_{k+1}) + W(t_k)) - (W(t_{k+1}) - W(t_k))] \\
&-\frac{1}{2}\lim_{n\to\infty} \sum_{k=1}^{n} (W(t_{k+1}) - W(t_k))^2 = \frac{1}{2}\lim_{n\to\infty} \sum_{k=1}^{n} \left[W^2(t_{k+1}) - W^2(t_k)\right] \\
&-\frac{1}{2}\lim_{n\to\infty} \sum_{k=1}^{n} (W(t_{k+1}) - W(t_k))^2 = \frac{1}{2}W^2(t) - \frac{1}{2}t
\end{aligned}
$$

$$
\begin{aligned}
B_n &= \lim_{n\to\infty} \sum_{k=1}^{n} W(t_{k+1})[W(t_{k+1}) - W(t_k)] \\
&= \frac{1}{2}\lim_{n\to\infty} \sum_{k=1}^{n} [W(t_{k+1}) + W(t_k) + (W(t_{k+1}) - W(t_k))] \\
&\times (W(t_{k+1}) - W(t_k)) \\
&= \frac{1}{2}\lim_{n\to\infty} \sum_{k=1}^{n} [(W(t_{k+1}) + W(t_k))(W(t_{k+1}) - W(t_k))] \\
&+\frac{1}{2}\lim_{n\to\infty} \sum_{k=1}^{n} (W(t_{k+1}) - W(t_k))^2 = \frac{1}{2}\lim_{n\to\infty} \sum_{k=1}^{n} \left[W^2(t_{k+1}) - W^2(t_k)\right] \\
&+\frac{1}{2}\lim_{n\to\infty} \sum_{k=1}^{n} (W(t_{k+1}) - W(t_k))^2 = \frac{1}{2}W^2(t) + \frac{1}{2}t
\end{aligned}
$$

I.e.

$$
\begin{cases}
A_n + B_n = W^2(t) \\
B_n - A_n = \sum_{k=0}^{n-1} (\Delta W_k)^2 = S_n
\end{cases}
$$

where $\lim_{n\to\infty} S_n(t) = t$. By letting $B_n - A_n \to t \Rightarrow A_n \to A$ and $B_n \to B$ where

$$
\begin{cases}
A = \dfrac{W^2(t)}{2} - \dfrac{t}{2} \\
B = \dfrac{W^2(t)}{2} + \dfrac{t}{2}
\end{cases}
$$

We then observe that ξ_k effects the integral concept and

$$\begin{cases} \int_0^t W(s)dW(s) = \frac{W^2(t)}{2} - \frac{t}{2} \text{ This is called (the forward – or) the} \\ \text{Itô – integral and} \\ \int_0^t W(s)dW(s) = \frac{W^2(t)}{2} + \frac{t}{2} \text{ this the backward integral.} \end{cases}$$

From this we learn a few things.

1. Since Wiener trajectories have unlimited variations, we cannot define integrals as

$$I_n(\omega) = \sum g(\xi_k, \omega)\big[W(t_{k+1}, \omega) - W\big(t_k, \omega\big)\big]$$

2. In any case, there seems to be a hope to define integrals as a limit value in L^2.
3. The choice of ξ_k will critical decide the value we get. Different choices of ξ_k will give us integral definitions with different properties.

We will use the Itô stochastic integral for an important reason; In all natural cases unknown future events cannot affect the present. This means that the value of a function $G(t)$ is non-anticipating in that it cannot be used to predict the future increment in dX. This is of course equivalent to saying that $G(t)$ is a martingale since what we mean is exactly that

$$E_s[G(t)] = G(s) \quad \text{för} \quad s \leq t.$$

We only know what is the present value of $G(t)$, which corresponds to that at the beginning-integrating interval. For this reason it is more appropriate to use the Itô integral. It is important also to note that integrating a non-anticipative function with respect to dt or dX is itself non-anticipating. So for $G(t) = X(t)$ the Itô integral becomes

$$\int_0^T X(s)dX(s) \approx \sum_{i=1}^{n} X_{i-1}(X_i - X_{i-1})$$

I.e.

$$I = \sum_{i=1}^{n} X_i(X_i - X_{i-1}) - \sum_{i=1}^{n} (X_i - X_{i-1})^2$$
$$= \sum_{i=1}^{n} (X_i^2 - X_{i-1}^2) - I - \sum_{i=1}^{n} (X_i - X_{i-1})^2.$$

So

$$2 \cdot I = X(T)^2 - T$$

or

$$\int_0^T X(s)dX(s) = \frac{X(T)^2 - T}{2}.$$

3.6.1 Proof of $(dW)^2 = dt$

We will now make a proof of the important property of the Wiener process $(dW)^2 = dt$. Like any other differential, this differential is defined in terms of its integral

$$\int_{t_0}^{t_1} (dW)^2 \equiv \lim_{n \to \infty} \sum_{k=0}^{n-1} [W(t_{k+1}) - W(t_k)]^2$$

where $t_k = t_0 + k(t_1 - t_0)/n$. Since

$$W(t_{k+1}) - W(t_k) = \sqrt{t_{k+1} - t_k}\xi_k = \sqrt{\frac{t_1 - t_0}{n}}\xi_k$$

we have

$$\int_{t_0}^{t_1} (dW)^2 \equiv \lim_{n \to \infty} \frac{t_1 - t_0}{n} \sum_{k=0}^{n-1} \xi_k^2$$

where $\xi_0, \xi_1, \ldots \xi_{n-1}$ are independent $N(0, 1)$ variables. Clearly the mean of the sum is

$$E\left[\frac{t_1 - t_0}{n}\sum_{k=0}^{n-1}\xi_k^2\right] = \frac{t_1 - t_0}{n}\sum_{k=0}^{n-1}E\left[\xi_k^2\right] = t_1 - t_0.$$

Since the ξ's are independent, the variance of the sum is

$$Var\left[\frac{t_1 - t_0}{n}\sum_{k=0}^{n-1}\xi_k^2\right] = \frac{(t_1 - t_0)^2}{n^2}\sum_{k=0}^{n-1}Var\left[\xi_k^2\right] = \frac{(t_1 - t_0)^2}{n^2}\sum_{k=0}^{n-1}E\left[\left(\xi_k^2 - 1\right)^2\right].$$

For unit Gaussian variables, $E[(\xi_k{}^2 - 1)^2] = 2$, so the variance of the sum works out to

$$Var\left[\frac{t_1 - t_0}{n}\sum_{k=0}^{n-1}\xi_k{}^2\right] = \frac{2}{n}(t_1 - t_0)^2.$$

Thus

$$\int_{t_0}^{t_1}(dW)^2 \equiv \lim_{n\to\infty} S_n$$

where the sum S_n has mean $t_1 - t_0$ and variance $O(1/n)$. We conclude that in the limit $n \to \infty$, this integral is $t_1 - t_0$ with certainty. Thus

$$\int_{t_0}^{t_1}(dW)^2 = t_1 - t_0$$

for any t_0 and t_1. Since differentials are defined only in terms of their integrals, we can re-write this as

$$(dW)^2 = dt.$$

The other quadratic differentials are zero: $(dt)^2 = 0$ and $dWdt = 0$. To show this, let us write out their integrals. First,

$$\int_{t_0}^{t_1}(dt)^2 = \lim_{n\to\infty}\sum_{k=0}^{n-1}[t_{k+1} - t_k]^2 = \lim_{n\to\infty}\frac{1}{n}[t_1 - t_0]^2 = 0$$

and

$$\int_{t_0}^{t_1} dtdW = \lim_{n\to\infty}\sum_{k=0}^{n-1}(t_{k+1}-t_k)[W(t_{k+1})-W(t_k)] = \lim_{n\to\infty}\sum_{k=0}^{n-1}(t_{k+1}-t_k)^{3/2}\xi_k$$

$$= \lim_{n\to\infty}\frac{1}{n^{3/2}}[t_1-t_0]^{3/2}\sum_{k=0}^{n-1}\xi_k = \lim_{n\to\infty}\frac{[t_1-t_0]^{3/2}}{n}\xi$$

Since means and variances of independent variables are additive, clearly the sum of the ξ_k gives a Gaussian variable with mean 0 and variance n. ξ is as usual a

$N(0, 1)$ variable. Clearly this is zero in the limit.

Putting this together with our preceding results gives the so-called box algebra

$$(dW)^2 = dt, \quad dWdt = 0, \quad (dt)^2 = 0.$$

Of course, all higher powers are also zero

$$\begin{aligned} (dW)^k &= 0 \quad \text{for} \quad k > 2, \\ (dW)^k dt &= 0 \quad \text{for} \quad k > 1, \\ (dt)^k &= 0 \quad \text{for} \quad k > 1. \end{aligned}$$

3.6.2 Monte Carlo Simulations

Suppose we have some variable, an asset price perhaps, which we model by an Itô process

$$dX = a(t,X)dt + b(t,X)dW.$$

Commonly the value of a financial instrument will turn out to be the expected value of some payoff at the expiry date T,

$$V = E[P(X(T)].$$

The Monte Carlo method is the most direct method of calculating such expected values. Recall that the Itô process above is equivalent to stating that $X(t)$ is the limit as $n \to \infty$ of $X(\tau_n)$, where

$$X(t_j) = X(t_0) + \lim_{n\to\infty}\left\{\sum_{k=0}^{j-1} a(t_k, X(t_k))(t_{k+1} - t_k) + \sum_{k=0}^{j-1} b(t_k, X(t_k))\sqrt{t_{k+1} - t_k}\xi_k\right\}$$

for $j = 0, 1, \ldots, n-1$. Here $\tau_k = t_0 + k(T - t_0)/n$ and $\xi_0, \xi_1, \ldots, \xi_{n-1}$ are independent $N[0, 1]$ variables. For the MC method, we first discretize in time, picking $\tau_0, \tau_1, \ldots, \tau_n = T$. We then pick the n independent $N[0, 1]$ variables $\xi_0, \xi_1, \ldots, \xi_{n-1}$. Substituting these above gives a possible path (also called a realization) or the asset price $X(t)$. Were the asset to follow this path, the financial instrument would yield $P(X(T)) = P_1$. Repeating this procedure for newly selected random variables $\xi_0, \xi_1, \ldots, \xi_{n-1}$ yields a second possible path and payoff P_2. Repeating this many times and averaging over the outcomes then gives the option price

$$V \equiv E[P(X(Y))] = \frac{1}{N}\sum_{k=1}^{N} P_k$$

The Monte Carlo (MC) method's key advantage is that it is very flexible. For example, suppose we have a path dependent financial instruments, whose payoffs depend on, say the maximum, minimum or average value of $X(t)$ between t_0 and T. Since MC simulates the entire path, the value of these options is no harder to determine that the value of a European option.

The weakness of MC is that it is slow and computationally expensive. In fact, let σ be the standard deviation of the payoff

$$\sigma^2 = \mathrm{Var}\{P(X(T)\} \approx \mathrm{Var}\{P_k\}.$$

The last equality is only an approximation due to our time discretization in the MC method. Since the random variables on each path are chosen independently, the variances on different paths add. So we have

$$Var\left\{\frac{1}{N}\sum_{k=1}^{N} P_k\right\} = \frac{1}{N^2} Var\left\{\sum_{k=1}^{N} P_k\right\} = \frac{1}{N} Var\{P_k\} = \frac{\sigma^2}{N}.$$

The typical error in the MC evaluation is the standard deviation $\sigma/\sqrt{N}$. So quite generally the error in the MC method goes down like $1/\sqrt{N}$. In other words, to reduce the error by a factor of 10, one needs to do 100 times as many paths.

3.6.3 Integration by Itô's Lemma

Itô's formula can also be used when we need to integrate stochastic processes. This integral is called the *Itô integral.* A Wiener trajectory is a continuous function of time, but not differentiable in any point. To find the integral we therefore try to do the following

1. Divide the interval $[0, t]$ in equal parts $0 = t_0 < t_1 < \ldots < t_n = t$.
2. Define for each outcomeω: $I_n(\omega) = \Sigma g(\xi_k)[W(t_{k+1}, \omega) - W(t_k, \omega)]$.
3. Sum and let $n \to \infty$ hoping that $I_n \to I$.

Example 3.74

Calculate the integral $\int_0^t W(s)dW(s)$ where W is a Wiener process. Let $Z(t) = W^2(t)$ and use Itô formula

$$
\begin{aligned}
dZ(t) &= \frac{\partial Z}{\partial W}dW + \frac{1}{2}\frac{\partial^2 Z}{\partial W^2}(dW)^2 \\
&= 2 \cdot W(t) \cdot dW(t) + \frac{1}{2} \cdot 2 \cdot (dW(t))^2 = 2 \cdot W(t) \cdot dW(t) + dt
\end{aligned}
$$

Integration gives: $W^2(t) = t + 2 \cdot \int_0^t W(s)dW(s)$, i.e.

$$\int_0^t W(s)dW(s) = \frac{1}{2}W^2(t) - \frac{t}{2}.$$

Therefore

$$\int_0^T X(t)dX(t) = \frac{X(T)^2 - T}{2}.$$

An alternative is to use Itô's integral formula

$$\int_0^T dF = \int_0^T \frac{\partial F}{\partial t}dt + \int_0^T \frac{\partial F}{\partial X}dX + \frac{1}{2}\int_0^T \frac{\partial^2 F}{\partial X}dt$$

and letting $\frac{\partial F}{\partial X} = X$, e.g. $F = X^2/2$

(continued)

Example 3.74 (continued)

$$\int_0^T \frac{\partial F}{\partial X} dX = \int_0^T X(t)dX(t) = \int_0^T dF - \int_0^T \frac{\partial F}{\partial t} dt - \frac{1}{2}\int_0^T \frac{\partial^2 F}{\partial X^2} dt =$$
$$= \frac{X(T)^2}{2} - 0 - \frac{T}{2} = \frac{X(T)^2 - T}{2}$$

Similarly, if we let $\frac{\partial F}{\partial X} = X^2$, e.g. $F = X^3/3$ we can show that

$$\int_0^T X(t)^2 dX(t) = \frac{X(T)^3}{3} - \int_0^T X(t)dt$$

and with $\frac{\partial F}{\partial X} = t$, e.g. $dF = d(X(t)t)$ we can show

$$TX(T) = \int_0^T t \cdot dX(t) + \int_0^T X(t)dt$$

since

$$TX(T) = \int_0^T \frac{\partial F}{\partial t} dt + \int_0^T \frac{\partial F}{\partial X} dX + \frac{1}{2}\int_0^T \frac{\partial^2 F}{\partial X^2} dt = \int_0^T X(t)dt + \int_0^T t \cdot dX(t) + 0.$$

Example 3.75

Calculate the expectation value $E[W^4(t)]$.

Let $Z(t) = W^4(t)$ and $X(t) = W(t)$ e.g. $dX(t) = dW(t)$ (we have no drift but only a diffusion = 1):

$$dZ = d(W^4) = 4 \cdot W^3 dW + \frac{1}{2} \cdot 12 \cdot W^2 (dW)^2 = 6 \cdot W^2 dt + 4 \cdot W^3 dW.$$

Integration gives

$$W^4(T) = 6\int_0^T W^2(s)ds + 4\int_0^T W^3(s)dW(s).$$

If we take the expectation value we get

(continued)

Example 3.75 (continued)

$$E[W^4(T)] = 6\int_0^T E[W^2(s)]ds + 4 \cdot E\left[\int_0^T W^3(s)dW(s)\right] = 6\int_0^T sds = 3 \cdot T^2$$

since the expectation value of a stochastic integral is zero by definition.

Definition 3.76 Let $L^2[a, b]$ represent the class of processes g satisfying:
g is $\underline{\mathcal{F}}$-adapted and

$$\int_0^t E\left\{[g(s)]^2\right\}ds < \infty.$$

For each choice of $a \leq b$ we will now define the integral

$$\int_a^b g(s)dW(s)$$

for any g in $L^2[a, b]$. For simplicity we suppose g is a simple function. Then, there exist values a and b such as $a = t_0 < t_1 < \ldots < t_n = b$ and

$$\begin{aligned} &g(s) = g(t_k)\forall s \in [t_k, t_{k+1}) \\ &g(t_k) \in F_{tk}\ k = 0, 1, \ldots, n \end{aligned}$$

Then

$$\int_a^b g(s)dW(s) = \sum_k g(t_k)[W(t_{k+1}) - W(t_k)].$$

Remark We use forward differences.

Theorem 3.77 *If g and h are simple $\mathcal{F}$-adapted processes, which have quadratic defined integral, also let $\alpha, \beta \in \boldsymbol{R}$. Then*

$$E\left[\int_a^b g(s)dW(s)\right] = 0$$

$$E\left[\left(\int_a^b g(s)dW(s)\right)^2\right] = \int_a^b E[g^2(s)]ds$$

$$E\left[\left(\int_a^b g(s)dW(s)\right)\left(\int_a^b h(s)dW(s)\right)\right] = \int_a^b E[g(s)h(s)]ds$$

$\int_a^b g(s)dW(s)$ is $\mathcal{F}_t$-measurable

$$E\left[\int_a^b g(s)dW(s)\middle|\mathcal{F}_a\right] = 0$$

$$\int_a^b [\alpha g(s) + \beta h(s)]dW(s) = \alpha\int_a^b g(s)dW(s) + \beta\int_a^b h(s)dW(s)$$

4

Continuous Time Models

4.1 Classifications of Partial Differential Equations

Before we begin the study of partial differential equations (PDEs) we will explain how to classify them. A general quadratic surface can be described by the expression

$$Ax^2 + Bxy + Cy^2 + Dx + Ey + F = 0.$$

Depending on the values of the constants (A, B, C, D, E and F), different geometrical objects will be represented:

$A = C$, $B = 0$:	$\Rightarrow$	a Circle,
$B^2 - 4AC < 0$:	$\Rightarrow$	an Ellipse,
$B^2 - 4AC = 0$:	$\Rightarrow$	a Parabola and
$B^2 - 4AC > 0$:	$\Rightarrow$	a Hyperbola

Similarly, we classify second order partial differential equations by the expression

$$A\frac{\partial^2 F(x,y)}{\partial x^2} + B\frac{\partial^2 F(x,y)}{\partial x \partial y} + C\frac{\partial^2 F(x,y)}{\partial y^2} + D\frac{\partial F(x,y)}{\partial x} + E\frac{\partial F(x,y)}{\partial y} + F(x,y) = 0.$$

This means that if $B^2 - 4AC = 0$ we call this a parabolic partial differential equation. As we will see, in the Black–Scholes PDE, x will represent the underlying (stock) price and y the time. Furthermore, $B = C = 0$, so this is a

J.R.M. Röman, *Analytical Finance: Volume I*,
DOI 10.1007/978-3-319-34027-2_4

parabolic PDE. If we let y becomes time and x a space variable, the PDE class is of great importance to how information evolves in time. A parabolic PDE is a so-called diffusion equation. In diffusion processes the information about the history is rubbed out. In physics, a typical diffusion equation is the heat equation. With a given temperature, we can estimate the equilibrium temperature in the future. But we cannot say anything about the previous temperature distribution.

Example 4.1

If we measure the temperature in each point in a room, you can use this measurement as an initial condition with a diffusion equation to solve the temperature distribution in a later time. But you can't solve the diffusion equation backwards in time to find the temperature distribution in the room an hour ago.

Example 4.2

If $B = D = E = 0$ we get a simple elliptical (circle) PDE describing a wave. The information will not be rubbed out in this case. This is because the information will move with a certain velocity given a growing circle in time, giving a growing cone, as time evolves. An equation as this can be solved backwards in time as well. Furthermore, the boundary condition only has to be given on a circle itself or even a point, since the information travels with a constant velocity.

4.2 Parabolic PDEs

We will now show how to solve partial differential equation of parabolic type by using stochastic processes. In physics this kind of PDEs are called diffusion equations. We therefore start by considering the following Cauchy problem on the interval $[0, T]$:

$$\begin{cases} \dfrac{\partial F}{\partial t}(t,x) + \mu(t,x)\dfrac{\partial F}{\partial x}(t,x) + \dfrac{1}{2}\sigma^2(t,x)\dfrac{\partial^2 F}{\partial x^2}(t,x) = 0 \\ F(T,x) = \phi(x) \end{cases}.$$

Instead of using traditional analytical methods, such as the Fourier method, we will find $F(t, x)$ in term of an associated diffusion process. Therefore, we suppose that there is a solution where we fix t and x as in Fig. 4.1 and let $X(t)$ solve the stochastic differential equation

$$\begin{cases} dX(t) = \mu(t,X(t))dt + \sigma(t,X(t))dW(t) \\ X(T) = x \end{cases}.$$

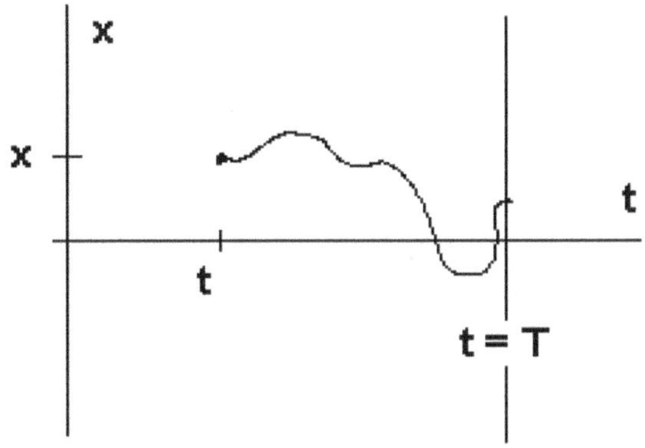

Fig. 4.1 An associated diffusion process used to solve a parabolic PDE

If we start by applying the Itô formula on the function $F(t, X)$, we get

$$dF = \left(\frac{\partial F}{\partial t} + \mu\frac{\partial F}{\partial x} + \frac{1}{2}\sigma^2\frac{\partial^2 F}{\partial x^2}\right)dt + \sigma\frac{\partial F}{\partial x}dW = \sigma\frac{\partial F}{\partial x}dW,$$

where the bracket itself is zero, due to the PDE. In other words, the bracket $(..)dt$ vanishes since this is by definition equal to zero. If we integrate this, we get

$$\phi(x) = F(T,X(T)) = F(t,X(t)) + \int_t^T \sigma(s,X(s))\frac{\partial F}{\partial x}(s,X(s))dW(s).$$

By taking the expectation value and let $X = x$, we get

$$F(t,x) = E^Q_{t,x}[\phi(X(T))].$$

This formula is called the *Feynman–Kač representation*. Observe that the expectation of the stochastic integral is zero by definition. This is one of the properties of the Wiener process, since it is a noise where the integration of the expected value is zero.

Example 4.3

Consider the heat equation

$$\begin{cases} F_t + \frac{1}{2}\sigma^2 F_{xx} = 0, \\ F(T,x) = x^2 \end{cases}$$

where the drift and volatility is given by $\mu(t,x) = 0$ and $\sigma(t,x) = \sigma$ and the boundary condition $\phi(x) = x^2$. The process is therefore given by

$$\begin{cases} dX(s) = \sigma \cdot dW(s) \\ X(t) = x \end{cases}.$$

We start by applying the Itô formula on the function $F(t, X)$:

$$dF = \left(\frac{\partial F}{\partial t} + \frac{1}{2}\sigma^2\frac{\partial^2 F}{\partial x^2}\right)dt + \sigma\frac{\partial F}{\partial x}dW = \sigma\frac{\partial F}{\partial x}dW.$$

If we integrate and take the expectation value, we end up with the Feynman–Kač representation

$$F(t,x) = E^Q_{t,x}\left[X_T^2\right].$$

Now, let $Z = X^2$ and use the Itô formula on Z

$$dZ = \frac{\partial Z}{\partial X}dX + \frac{1}{2}\frac{\partial^2 Z}{\partial X^2}(dX)^2 = 2 \cdot X_t dX + \frac{1}{2} \cdot 2 \cdot (dX)^2 = \sigma^2 dt + 2 \cdot \sigma \cdot XdW.$$

Integration gives

$$Z(T) - Z(t) = \int_t^T \sigma^2 ds + \int_t^T 2 \cdot \sigma \cdot XdW = \sigma^2(T-t) + 2 \cdot \sigma \int_t^T XdW.$$

Finally, we take the expectation value and get

$$E[Z(T)] = E[Z(t)] + \sigma^2(T,t) + 2\sigma\int_t^T E[X]dW = x^2 + \sigma^2(T-t).$$

We therefore have the following solution to our PDE

$$F(t,x) = x^2 + \sigma^2(T-t).$$

We can also find the solution by using $dX = \sigma \cdot dW$ and $X_T = x + \sigma[W_T - W_t]$

$$F(t,x) = E^Q_{t,x}\left[X_T^2\right] = Var[X_T] + \left\{E^Q_{t,x}[X_T]\right\}^2 = \sigma^2(T-t) + x^2.$$

In general, we can solve the following partial differential equation

$$\begin{cases} \frac{\partial F}{\partial t}(t,x) + \mu(t,x)\frac{\partial F}{\partial x}(t,x) + \frac{1}{2}\sigma^2(t,x)\frac{\partial^2 F}{\partial x^2}(t,x) = rF(t,x) \\ F(T,x) = \phi(x) \end{cases}$$

by starting with $dX(t) = \mu dt + \sigma dW$ and using the Itô formula on $F(t, X)$ we then gets (as always)

$$dF = \left(\frac{\partial F}{\partial t} + \mu\frac{\partial F}{\partial x} + \frac{1}{2}\sigma^2\frac{\partial^2 F}{\partial x^2}\right)dt + \sigma\frac{\partial F}{\partial x}dW$$

and therefore

$$dF = rFdt + \sigma\frac{\partial F}{\partial x}dW.$$

If we integrate this from t to T and taking the expectation value (so that the stochastic part vanishes) we get the Feynman–Kač formula

$$F(t,x) = e^{-r(T-t)}E^Q[\phi(x)].$$

For details of how to solve stochastic differential equations such as the one above, see example 4.5. We observe that the right-hand side of the PDE ($rF(t, x)$) gives a discount factor in front of the expectation value. This is due to the integrating factor we get when solving the stochastic differential equation.

Example 4.4

Solve the following partial differential equation

$$\begin{cases} \frac{\partial F}{\partial t}(t,x) + \frac{1}{2}\sigma^2 x^2\frac{\partial^2 F}{\partial x^2}(t,x) = 0 \\ F(T,x) = x^2 \end{cases}.$$

Suppose $F(t, X)$ solves the PDE, where $dX = \sigma XdW$ and $X(0) = x$. Using Itô we get

$$\begin{aligned} dF &= \frac{\partial F}{\partial t}dt + \frac{\partial F}{dX}dX + \frac{1}{2}\frac{\partial^2 F}{\partial X^2}(dX)^2 = \left(F_t + \frac{1}{2}x^2\sigma^2 F_{XX}\right)dt + \sigma XF_X dW \\ &= \sigma XF_X dW \end{aligned}$$

Integration gives

(continued)

Example 4.4 (continued)

$$X^2 = F(T,X(T)) = F(t,X(t)) + \sigma\int_t^T x\frac{\partial F}{\partial X}dW(s)$$

If we now take the expectation value we get the Feynman–Kač formula

$$F(t,x) = E^Q_{t,x}\left[X_T^2\right].$$

As always, to calculate such an expectation, we need the dynamics of $Z = X^2$ and use the Itô Lemma. We then get

$$\begin{cases} dZ = 2\cdot X_t dX + \dfrac{1}{2}\cdot 2\cdot (dX)^2\sigma^2 X^2 dt + 2\sigma X^2 dW = \sigma^2 Z dt + 2\sigma Z dW \\ Z(0) = X^2(0) = x^2 \end{cases}.$$

We now integrate

$$Z(T) = Z(t) + \sigma^2\int_t^T Zds + 2\sigma\int_t^T ZdW$$

and by taking the expectation value, we get the following integral equation.

$$E[Z] = x^2 + \sigma^2\int_t^T E[Z]ds.$$

This is an integral equation and the easiest way to solve this is to convert it to a differential equation. The standard technique to solve this equation is to is, first to define $m = E[Z]$ and then take the derivative with respect to time. We then get the following ordinary differential equation[1]

$$\begin{cases} \dfrac{dm}{dt} = -\sigma^2 m \\ m(T) = x^2 \end{cases}$$

This gives the solution to the partial differential equation

$$F(t,x) = m = x^2 e^{\sigma^2(T-t)}.$$

[1] **Remark:** $\frac{d}{dt}\left\{\sigma^2\int_t^T E[Z]ds\right\} = \frac{d}{dt}\left\{\sigma^2\int_t^T m(s)ds\right\} = \sigma^2\frac{d}{dt}\{M(T) - M(t)\} = -\sigma^2 m(t)$ where $M(t)$ is a primitive function to $m(t)$

The general solution to

$$\begin{cases} \frac{\partial F}{\partial t}(t,x) + \mu(t,x)\frac{\partial F}{\partial x}(t,x) + \frac{1}{2}\sigma^2(t,x)\frac{\partial^2 F}{\partial x^2}(t,x) - r(t)F(t,x) + k(t,x) = 0 \\ F(T,x) = \Phi(x) \end{cases}$$

can be found by letting $F(t, X)$ to be a solution where

$$\begin{cases} dX(s) = \mu(s,X(s))ds + \sigma(s,X(s))dW(s) \\ X(t) = x \end{cases}$$

If we use the Itô formula on $F(t, X)$ we get

$$\begin{aligned} dF &= \left(\frac{\partial F}{\partial t} + \mu\frac{\partial F}{\partial x} + \frac{1}{2}\sigma^2\frac{\partial^2 F}{\partial x^2}\right)dt + \sigma\frac{\partial F}{\partial x}dW \\ &= \{r(t)F(t,X) - k(t,X)\}dt + \sigma\frac{\partial F}{\partial x}(t,X)dW \end{aligned}.$$

By taking the expectation value and integrate we have

$$\begin{aligned} &F(t,x(t)) \\ &= E^Q_{t,x}\left[\Phi(X_T)\exp\left\{-\int_t^T r(s)ds\right\} + \int_t^T \exp\left\{-\int_t^T r(u)du\right\}k(s,X(s))ds\right]. \end{aligned}$$

Example 4.5

An SDE like

$$\begin{cases} dX = \mu X dt + \sigma dW \\ X(0) = x \end{cases}$$

can be solved as

$$X(t) = xe^{\mu t} + \sigma e^{\mu t}\int_0^t e^{-\mu s}dW(s) = xe^{\mu t} + \sigma\int_0^t e^{\mu(t-s)}dW(s)$$

by remembering the technique of solving the following ODE with integrating factor

(continued)

Example 4.5 (continued)

$$
\begin{aligned}
&\dot{x} + f(x)\cdot x = g(x)\\
&\Rightarrow\\
&e^{F(x)}\dot{x} + e^{F(x)}f(x)\cdot x = e^{F(x)}g(x)\\
&\Rightarrow\\
&\frac{d}{dt}\left(xe^{F(x)}\right) = e^{F(x)}g(x)\\
&\Rightarrow\\
&x = e^{-F(x)}\int_0^t e^{F(x)}g(x)dt
\end{aligned}
$$

where $F(x)$ is the primitive function to $f(x)$. From this we learn to use Itô on $Y = e^{-\mu t}X$

$$
\begin{aligned}
dY &= d(Xe^{-\mu t}) = \frac{\partial Y}{\partial t}dt + \frac{\partial Y}{\partial X}dX + \frac{1}{2}\frac{\partial^2 Y}{\partial X^2}(dX^2)\\
&= -\mu Xe^{-\mu t}dt + e^{-\mu t}\{\mu Xdt + \sigma dW\} + 0 = \sigma e^{-\mu t}dW
\end{aligned}.
$$

By integrating we get

$$
Y(t) = Y(0) = \sigma\int_0^t e^{-\mu s}dW(s) \;\Rightarrow\; e^{-\mu t}X(t) - x = \sigma\int_0^t e^{-\mu s}dW(s)
$$

Finally

$$
X(t) = xe^{\mu t} + \sigma\int_0^t e^{\mu(t-s)}dW(s).
$$

4.2.1 A Classical Result

We will now derive a classical result, the transition probabilities to a stochastic differential equation. The transition probability gives the probability to go from one state to another. Let X be the solution to

$$
dX(t) = \mu(t, X(t))dt + \sigma(t, X(t))dW(t).
$$

First, define A via

$$(Af)(s,y) = \mu(s,y)\frac{\partial f}{\partial y}(s,y) + \frac{1}{2}\sigma^2(s,y)\frac{\partial^2 f}{\partial y^2}(s,y)$$

and consider the boundary value problem given by

$$\begin{cases} \left(\frac{\partial u}{\partial s} + Au\right)(s,y) = 0 \quad (s,y)\in(0,T)\times R \\ u(T,y) = I_B(y) \end{cases}$$

where the indicator function $I_B(y)$ is defined such as it is one if $y \in B$ and zero otherwise. We then get

$$u(s,y) = E_{s,y}[I_B(X_T)] = P(X_T \in B|X_S = y).$$

We then have the following theorem.

Theorem 4.6 *The transition probabilities $P(s, y; t, B) = P(X_T \in B|\, X_s = y)$ is given by the solution to the Kolmogorov backward equation*

$$\begin{cases} \left(\frac{\partial u}{\partial s} + Au\right)(s,y) = 0 \quad (s,y)\in(0,T)\times R \\ u(T,y) = I_B(y) \end{cases}$$

Theorem 4.7 *Suppose that $P(s, y; t, dx)$ have the density $p(s, y; t, x)dx$. Then*

$$\begin{cases} \left(\frac{\partial}{\partial s} + A\right)p(s,y;t,x) = 0 \quad (s,y)\in(0,T)\times R \\ p(s,y;t,x) \to \delta_X \qquad \text{when } s \to t \end{cases}$$

The backward equation comes from the fact that A is acting on the backward variables (s, y). We will also derive a forward equation. Consider an arbitrary infinite differentiable "test-function" on $(s, T) \times R$ and use Itô

$$h(T,X_T) = h(s,X_s) + \int_s^T \left(\frac{\partial h}{\partial t} + Ah\right)(t,X_t)dt + \int_s^T \frac{\partial h}{\partial x}(t,X_t)dW_t.$$

Then take the expectation value and suppose $h(T, x) = h(s, x) = 0$

$$\int_{-\infty}^{\infty}\int_{s}^{T} p(s,y;t,x)\left(\frac{\partial}{\partial t}+A\right)h(t,x)dx\,dt = 0$$

Partial integration in x and t gives

$$\int_{-\infty}^{\infty}\int_{s}^{T} h(t,x)\left(\frac{\partial}{\partial t}-A^{*}\right)p(s,y;t,x)dx\,dt = 0$$

where

$$\left(A^{*}f\right)(t,x) = -\frac{\partial}{\partial x}\left[\mu(t,x)f(t,x)\right]+\frac{1}{2}\frac{\partial^2}{\partial x^2}\left[\sigma^2(t,x)f(t,x)\right]$$

This gives the *Fokker–Planck equation*

$$\begin{cases}\left(\dfrac{\partial}{\partial s}+A\right)p(s,y;t,x) = 0 & (s,y)\in(0,T)\times R\\ p(s,y;t,x)\rightarrow\delta_X & \text{when } s\rightarrow t\end{cases}$$

4.3 The Black–Scholes–Merton Model

We will now derive one of the most famous results in finance, the Black–Scholes partial differential equation. Myron Scholes and Robert C. Merton won the Nobel Prize in Economics in 1997. In 1973 Myron Scholes and Fischer Black († 1995) published their paper "The Pricing of Options and Corporate Liabilities" in the *Journal of Political Economy*. Robert C. Merton was the first to publish a paper expanding the mathematical understanding of the options pricing model, and coined the term "Black–Scholes options pricing model".

4.3.1 Modeling Asset Prices

The future price of most assets can be expected to have both deterministic and random components. A popular model for the deterministic piece is exponential growth,

$$dS(t) = \mu(t, S) \cdot S(t)dt.$$

In a small time interval, $S(t + dt) = S\{1 + \mu(t, S)dt\}$, i.e. the asset's rate of return is given by $\mu(t, S)dt$ for a short time interval dt. If $\mu(t, S)$ is constant, $\mu(t, S) = \mu_O$, then $S(t) = S(0)\exp\{\mu_O t\}$.

The random part of the price process is commonly modeled in terms of a Brownian motion,

$$dS(t) = \sigma(t, S) \cdot \mathrm{S}(t)dW(t).$$

Over a short time interval, $S(t + dt) = S(t) + \sigma(t, S)S\sqrt{dt}\xi$, where ξ is Gaussian with mean 0 and variance 1. The Brownian motion increases the variance of the price by $\sigma^2(t, S)S^2 dt$ in a short time interval dt. This means that the ratio of the standard deviation to the asset price itself is proportional to the volatility $\sigma(t, S)$

$$\frac{StdDev\{S(t + dt)\}}{S(t)} = \sigma(t, S)\sqrt{dt}.$$

The standard Black–Scholes's model uses constant drift $\mu(t, S) = \alpha$ and constant σ

$$dS(t) = \alpha \cdot S(t)dt + \sigma \cdot S(t)dW(t)$$

To derive the Black–Scholes PDE, we will study a market with two investment possibilities, a risk-free money-market account B that pays a constant interest rate r and a stock S. The price of the stock is characterized of a constant drift α and a stochastic term σSdW. The stochastic term is given by a geometric Brownian motion, (a Wiener process) where σ is called the volatility. The market is given by

$$\begin{cases} dB(t) = r \cdot B(t)dt \\ B(0) = 1 \end{cases} \Rightarrow \quad B(t) = e^{rt}$$

$$\begin{cases} dS(t) = \alpha \cdot S(t)dt + \sigma \cdot S(t)dW(t) \\ S(0) = s \end{cases}$$

The initial condition of the money-market account is 1 and the initial stock price is s. The Wiener process is normal distributed and $(dW(t))^2 = dt$. We will construct a portfolio h of the bond and the stock: $h = (h^0, h^1)$, where h holds the number of each instrument. h is then a stochastic process itself and the value process of the portfolio is defined as

$$V(t) = h^0(t) \cdot \mathrm{B}(t) + h^1(t) \cdot \mathrm{S}(t)$$

The portfolio is said to be *self-financed* if

$$\begin{aligned} dV(t) &= h^0(t) \cdot dB(t) + h^1(t) \cdot dS(t) \\ &= h^0(t) \cdot r \cdot B(t)dt + h^1(t) \cdot \alpha \cdot S(t)dt + h^1(t) \cdot \sigma \cdot S(t)dW(t) \\ &= \{h^0(t) \cdot r \cdot B(t) + h^1(t) \cdot \alpha \cdot S(t)\}dt + h^1(t) \cdot \sigma \cdot S(t)dW(t) \end{aligned}$$

We start by defining a relative portfolio $u = (u^0, u^1)$ by

$$u^0(t) = \frac{h^0(t) \cdot \mathrm{B}(t)}{V(t)}, u^1(t) = \frac{h^1(t) \cdot \mathrm{S}(t)}{V(t)}, u^0(t) + u^1(t) = 1.$$

The self-financed value process in terms of the relative portfolio is

$$dV(t) = V(t) \cdot \{r \cdot u^0(t) + \alpha \cdot u^1(t)\}dt + V(t) \cdot \sigma \cdot u^1(t)dW(t).$$

Suppose that $V(t) = V(t, S(t))$. The Itô lemma gives

$$\begin{aligned} dV &= \frac{\partial V}{\partial t}dt + \frac{\partial V}{\partial S}dS + \frac{1}{2}\frac{\partial^2 V}{\partial S^2}(dS)^2 \\ &= \frac{\partial V}{\partial t}dt + \frac{\partial V}{\partial S}(\alpha \cdot S(t)dt + \sigma \cdot S(t)dW(t)) + \frac{1}{2}\sigma^2 \cdot S^2(t)\frac{\partial^2 V}{\partial S^2}dt \\ &= \left\{\frac{\partial V}{\partial t} + \alpha \cdot S(t) \cdot \frac{\partial V}{\partial S} + \frac{1}{2}\sigma^2 \cdot S^2(t)\frac{\partial^2 V}{\partial S^2}\right\}dt + \sigma \cdot S(t) \cdot \frac{\partial V}{\partial S}dW(t) \end{aligned}$$

To make this similar to $dV(t)$ in our first expression, we multiply with $V(t)$ and use the notation $\frac{\partial V}{\partial t} = V_t$, $\frac{\partial V}{\partial S} = V_s$ etc. Then

$$dV(t) = V\left\{\frac{V_t + \alpha \cdot S \cdot V_s + \frac{1}{2}\sigma^2 \cdot S^2 \cdot V_{ss}}{V}\right\}dt + V\frac{\sigma \cdot S \cdot V_s}{V}dW.$$

We now compare the terms end immediately see that

$$u^1 = \frac{S \cdot V_s}{V}$$

so we get

$$dV(t) = V\left\{\frac{V_t + \frac{1}{2}\sigma^2 \cdot S^2 \cdot V_{ss}}{V \cdot r} \cdot r + \alpha \cdot u^1\right\}dt + V \cdot \sigma \cdot u^1 dW.$$

By studying the remaining terms, we also see that

$$u^0 = \frac{V_t + \frac{1}{2}\sigma^2 \cdot S^2 \cdot V_{ss}}{V \cdot r}.$$

Since $u^0(t) + u^1(t) = 1$ we finally get the result

$$\frac{\partial V}{\partial t} + rS\frac{\partial V}{\partial S} + \frac{1}{2}\sigma^2 S^2 \frac{\partial^2 V}{\partial S^2} = rV.$$

This is the Black–Scholes partial differential equation.

Remark The equation is *independent of* α. In a risk-neutral world we can explain the terms in the PDE as

$\frac{\partial V}{\partial t} = \Theta$	The change of value with respect to time.
$rS\frac{\partial V}{\partial S} = rS \cdot \Delta$	The change of value with respect to the underlying.
$\frac{1}{2}\sigma^2 S^2 \frac{\partial^2 V}{\partial S^2} = \frac{1}{2}\sigma^2 S^2 \Gamma$	The change of value with respect to volatility.
rV	The expected change of value of the derivative security.

4.3.2 An Alternative Approach to Black–Scholes

The Black–Scholes PDE can also be derived as follows. We now start with a capital of X_0 and receive at each time t, $\Delta(t)$ shares in a stock, modelled by a Brownian motion:

$$dS(t) = \alpha \cdot S(t)dt + \sigma \cdot S(t)dW(t).$$

The investor is financing the investment with a loan with interest rate r. The capital at time t is given by

$$\begin{aligned} dX(t) &= \Delta(t)dS(t) + r[X(t) - \Delta(t)S(t)]dt \\ &= \Delta(t)[\alpha S(t)dt + \sigma S(t)dW(t)] + r[X(t) - \Delta(t)S(t)]dt \\ &= [rX(t) + (\alpha - r)\Delta(t)S(t)]dt + \sigma\Delta(t)S(t)dW(t) \end{aligned}$$

where we have invested in $\Delta(t)$ shares and earning interest rate on the remaining capital. The factor $(\alpha - r)$ is called the risk-premium. Consider now, also a European option paying an amount of $g(S(T))$ at time T and let $F(t, S(t))$ represent the value at time t. Then

$$dF = \frac{\partial F}{\partial t}dt + \frac{\partial F}{\partial S}dS + \frac{1}{2}\frac{\partial^2 F}{\partial S^2}(dS)^2$$
$$= \left\{F_t + \mu \cdot S \cdot F_s + \frac{1}{2}\sigma^2 \cdot S^2 F_{ss}\right\}dt + \sigma \cdot S \cdot F_s dW$$

If $X(t) = F(t, S(t))$ we get a delta-hedge $\Delta(t) = F_s(t, S(t))$ and

$$dF = \left\{F_t + \alpha \cdot S \cdot F_s + \frac{1}{2}\sigma^2 \cdot S^2 F_{ss}\right\}dt + \sigma \cdot S \cdot \Delta \cdot dW$$
$$= \left\{F_t + \alpha \cdot S \cdot F_s + \frac{1}{2}\sigma^2 \cdot S^2 F_{ss}\right\}dt - [rF + (\alpha - r)\Delta S]dt$$

$$F_t + \alpha S\Delta + \frac{1}{2}\sigma^2 S^2 F_{ss} = rF + \Delta(\alpha - r)S$$

Then α vanish and

$$\begin{cases} F_t + rSF_s + \frac{1}{2}\sigma^2 S^2 F_{ss} - rF = 0 \\ F_T = g(S(T)) \end{cases}$$

4.3.3 Alternative Approach Using Risk Neutrality

The following alternative using risk neutrality argument to deriving the Black–Scholes equation was put forward by *Cox* and *Ross* in 1976, which does not involve delta hedging. Here we shall explore exactly this argument and make a direct comparison to the delta-hedging technique. The concept of risk neutrality is one associated with an investment that has zero risk to asset price movement which must therefore, due to arbitrage consideration, earn the same rate as the risk-free return (e.g. as the money-market account).

The pricing dynamics of the underlying asset can be described by a geometric random walk of the form

$$dS = \alpha S dt + \sigma S dX.$$

Further to this, we know that our call option C satisfies Itô's lemma

$$dC = \left(\frac{\partial C}{\partial t} + \alpha S \frac{\partial C}{\partial S} + \frac{1}{2}\sigma^2 S^2 \frac{\partial^2 C}{\partial S^2} \right) dt + \sigma S \frac{\partial C}{\partial S} dX$$

and we wish to express this in terms of a geometric random walk for the option as

$$dC = \alpha_c C dt + \sigma_c C dX.$$

Therefore

$$\alpha_c = \frac{1}{C} \left(\frac{\partial C}{\partial t} + \alpha S \frac{\partial C}{\partial S} + \frac{1}{2}\sigma^2 S^2 \frac{\partial^2 C}{\partial S^2} \right)$$

and

$$\sigma_c = \frac{\sigma S}{C} \frac{\partial C}{\partial S}$$

in order to satisfy this requirement. Rearranging our expression gives us

$$\frac{\partial C}{\partial t} + \alpha S \frac{\partial C}{\partial S} + \frac{1}{2}\sigma^2 S^2 \frac{\partial^2 C}{\partial S^2} + \sigma S \frac{\partial C}{\partial S} - \alpha_c C = 0.$$

This resembles precisely the Black–Scholes equation *if we let* $\alpha_c = r$. In many literatures you will find something along the lines of "we replace μ by r to take a risk neutral preference". This is not as straightforward as it is made to sound; in fact, the process of assuming the growth parameters to be equivalent to a risk-free investment is a subtle point and needs to be further expanded upon.

You can construct a portfolio consisting of options and assets that is instantaneously risk-less by holding $\sigma_c C$ units of asset and short selling σS units of options with a value

$$\Pi = \sigma_c CS - \sigma CS = (\sigma_c - \sigma)CS.$$

[Notice that this is different to delta hedging when one owns an option short and Δ units of the asset.] Π is now written as a function of 4 variables, 3 stochastic S, C, σ_c and time t. Therefore, $\Pi \sim \Pi(S, C, \sigma_c, t)$. To differentiate this we require Itô's lemma for many variables. All cross-terms in involving 't'

vanish to slightly simplify matters, and we expand only to the second order for all other variables except t. Therefore,

$$d\Pi = \frac{\partial \Pi}{\partial t}dt + \frac{\partial \Pi}{\partial S}dS + \frac{\partial \Pi}{\partial C}dC + \frac{\partial \Pi}{\partial \sigma_c}d\sigma_c + \frac{1}{2}\frac{\partial^2 \Pi}{\partial S^2}dS^2 + \frac{1}{2}\frac{\partial^2 \Pi}{\partial C^2}dC^2 + \frac{1}{2}\frac{\partial^2 \Pi}{\partial {\sigma_c}^2}d{\sigma_c}^2 + \frac{\partial^2 \Pi}{\partial S \partial C}dSdC + \frac{\partial^2 \Pi}{\partial S \partial \sigma_c}dSd\sigma_c + \frac{\partial^2 \Pi}{\partial C \partial \sigma_c}dCd\sigma_c.$$

After some very lengthy and tedious algebra this reduces to the much shorter expression

$$d\Pi = (\alpha_c \sigma_c - \alpha \sigma)CSdt.$$

From simple arbitrage consideration this must earn the same as a risk-less interest rate $d\Pi = r\Pi dt$ since the structure of the portfolio is such that the risk is eliminated. Using this and our expressions above, we arrive at the expression

$$\alpha_c \sigma_c - \alpha \sigma = r(\sigma_c - \sigma)$$

or

$$\frac{\alpha_c - r}{\sigma_c} = \frac{\alpha - r}{\sigma}.$$

The interpretation of this equation has great financial significance. It says that the ratio extra rate return over a risk free investment of option and asset with their respective volatilities is fixed. This ratio is often termed the *market price of risk,* where the risk is measured in the volatility. Here we have shown that an option and the underlying asset have the same ratio within a *risk-neutral world* framework. From this equation we can interpret what we already know—the bigger the returns, the greater the risk. By using the expression of market price of risk and substituting it into our expression for α_c and σ_c above we recover the Black–Scholes equation

$$\frac{\partial C}{\partial t} + r \cdot S \cdot \frac{\partial C}{\partial S} + \frac{1}{2}\sigma^2 S^2 \cdot \frac{\partial^2 C}{\partial S^2} - r \cdot C = 0.$$

So we have not simply taken the equation above with $\alpha_c = \alpha = r$, but done something more subtle. This choice means that this ratio of market price of risk can be satisfied for any set of values for σ_c and σ. This is what makes the Black–Scholes model attractive. Furthermore, it sets a simple yet well-defined universal standard as desired as in any field. By letting $\alpha = r$ is different to saying that in reality no investment can grow faster than the rate r, but simply to set a fair price for our derivative we must let the two be equivalent.

Here we have derived the Black–Scholes equation without performing a delta hedge as is most often presented in common literature. Delta-hedging is a more refined and sophisticated extension to the risk neutrality argument.

4.3.4 Forwards and Futures

Forwards and futures are traded on equities, equity indices, bonds, currencies and commodities. No initial payments are made and both of the parties are obligated to fulfil the contract at maturity. But few contracts will reach delivery. Most of the contracts are closed out by buying or selling the opposite before maturity. Futures are exchange traded and usually daily settled. Both parties take an equally risk by taking their position and margin requirements is claimed by the clearinghouse or the exchange.

The price, F is given by

$$F = (1 + R) \cdot S = S \cdot e^{r \cdot t}$$

Here R is the simple interest rate and r the continuous compounding rate. If the underlying pays a known income (cash flow) with present value D, the price is given by

$$F = (1 + R) \cdot (S - D) = (S - D) \cdot e^{r \cdot t}$$

If there are future incomes in yield q (interest rate) the price is given by

$$F = (1 + R - q) \cdot S = S \cdot e^{(r - q) \cdot t}$$

A forward in another currency r_f the price is

$$F = \left(1 + R - r_f\right) \cdot S = S \cdot e^{\left(r - r_f\right) \cdot t}.$$

Example 4.8

Futures are very common contracts in the interest rate market and Treasury bills are short-term instruments issued by the government. Bills are usually quoted as

(continued)

Example 4.8 (continued)

percentage of a nominal amount, so at par they have the price as 100. Suppose we have a 6-month Treasury bill. The value is then given by

$$V = 100 \cdot e^{f \cdot t}$$

where the price is given as percentage of the nominal amount, f is the forward rate. If we are three months from delivery of a 90 days' future contract with interest rate 5 %, the price is given by

$$F = 100 \cdot e^{-0.05 \cdot 90/365} = 98.7578$$

The value of the Treasury bill is, with the same interest rate

$$V_6 = 100 \cdot e^{-0.05 \cdot 182/365} = 97.5377$$

This is a corresponding return on yearly basis of

$$[F/V_6 - 1] \cdot 4 = 5.0 \ \%.$$

Suppose there is another Treasury bill T_3 with 3 months to maturity and with a rate of 4 % and a value of

$$T_3 = 100 \cdot e^{-0.04 \cdot 91/365} = 99.0077.$$

The corresponding return on yearly basis is

$$[V_3/V_6 - 1] \cdot 4 = 6.02 \%.$$

Therefore, if the interest rate of the 6 months Treasury bill is above 5 % the last 90 days, it is better to hold the bill instead of the future.

Example 4.9

How can we make arbitrage using the instruments above?

Answer-1: Arbitrage with only Treasury bills

1. Sell short the 3-month bill at 99.0077, with the value 100 after 3 months.
2. Buy 99.0077/97.5377 = 1.01507 of the six the month's bill at 97.5377. This gives a net investment of 0.
3. After 3 months we pay back 100 for the 3-month bill.
4. Receive (9.0077/97.5377)*100 for the 6-month bill.
5. This corresponds to 101.507 or a yearly interest rate of 6.028 %.

Answer-2: Arbitrage with the future

(continued)

Example 4.9 (continued)

1. Sell short the future with the value 98.7578 after 3 months.
2. Buy the 6-month bill at 97.5377. This gives a net investment of 97.5377.
3. Deliver the bill to the buyer after 6 months and receive 98.7578.
4. The result corresponds to (98.7578/97.5377) * 100 * 101.25 or 5.098 %.

As we have shown, we can earn 5 % when the interest rate is 4 %. To do this without taking a risk, we can use a repo (repurchase agreement) on the Treasury bill at 4 %.

Black–Scholes with a Forward as Underlying

If we have a forward or future as underlying instead of a stock we can transform the Black–Scholes PDE to get a new PDE with the Black-76 formula as solution. The relation between the stock, S and the forward, F is given by

$$F(t) = S(t)e^{r(T-t)}.$$

We start with the Black–Scholes PDE

$$\frac{\partial V(S,t)}{\partial t} + r \cdot S(S,t) \cdot \frac{\partial V(S,t)}{\partial S} + \frac{1}{2}\sigma^2 S^2(S,t) \cdot \frac{\partial^2 V(S,t)}{\partial S^2} - r \cdot V(S,t) = 0$$

and use the following substitutions

$$\begin{aligned}
\frac{\partial}{\partial S} &= \frac{\partial F}{\partial S}\frac{\partial}{\partial F} = e^{r(T-t)}\frac{\partial}{\partial F} \\
\frac{\partial^2}{\partial S^2} &= \frac{\partial}{\partial S}\left(\frac{\partial F}{\partial S}\frac{\partial}{\partial F}\right) = \frac{\partial^2 F}{\partial S^2}\frac{\partial}{\partial F} + \frac{\partial F}{\partial S}\frac{\partial}{\partial S}\left(\frac{\partial}{\partial F}\right) = 0 + e^{r(T-t)}\frac{\partial F}{\partial S}\left(\frac{\partial}{\partial F}\right)^2 \\
&= e^{2r(T-t)}\frac{\partial^2}{\partial F^2} \\
\frac{\partial}{\partial t} &= \frac{\partial}{\partial t} + \frac{\partial F}{\partial t}\frac{\partial}{\partial F} = \frac{\partial}{\partial t} - rF\frac{\partial}{\partial F}
\end{aligned}$$

The result is

$$\frac{\partial V(F,t)}{\partial t} + \frac{1}{2}\sigma^2 F^2 \cdot \frac{\partial^2 V(F,t)}{\partial F^2} - r \cdot V(F,t) = 0.$$

4.3.5 Simplified Expression of Black–Scholes

Sometimes, the Black–Scholes equation (and also the Term-Structure equation in interest rate theory where S is replaced by r) is written as

$$\mathcal{L}(t)V(t) = r \cdot V(t),$$

where the evolution operator L(t) is given by

$$\mathcal{L}(t) = \frac{\partial}{\partial t} + \mu(t) \cdot \frac{\partial}{\partial S} + \frac{1}{2}\sigma^2(t) \cdot \frac{\partial^2}{\partial S^2}.$$

4.3.6 The Solution to Black–Scholes

Now, we will solve the Black–Scholes PDE for a European call option with strike price K. We write the Black–Scholes partial differential equation with a call option as our contingent claim (boundary condition) as

$$\begin{cases} F_t + rSF_s + \frac{1}{2}\sigma^2 S^2 F_{ss} - rF = 0 \\ F_T = \max(S_T - K, 0) \end{cases}.$$

As usually, we suppose that $F(t, S_t)$ is a solution to the PDE above where

$$\begin{cases} dS(t) = \alpha \cdot S(t) \cdot dt + \sigma \cdot S(t) \cdot dW(t) \\ S(t) = S_t \end{cases}$$

Itô gives with $\alpha = r$

$$\begin{aligned} dF &= \frac{\partial F}{\partial t} dt + \frac{\partial F}{\partial S} dS + \frac{1}{2}\frac{\partial^2 F}{\partial S^2}(dS)^2 \\ &= F_t dt + Fs(r \cdot Sdt + \sigma \cdot SdW) + \frac{1}{2}\sigma^2 \cdot S^2 F_{ss}(dW)^2 \\ &= \left\{ F_t + r \cdot S(t) \cdot F_s + \frac{1}{2}\sigma^2 \cdot S^2 F_{ss} \right\} dt + \sigma \cdot S \cdot F_s dW \\ &= \mathrm{rFdt} + \sigma \cdot S \cdot F_s dW. \end{aligned}$$

If we integrate and take the expectation value, the stochastic part vanishes and we get the Feynmann–Kač formula

$$F(t,S) = e^{-r(T-t)} E^Q_{t,s}[\max\{S_T - K, 0\}]$$

We here need S_T, given by the stochastic differential equation

$$\begin{cases} dS(t) = r \cdot S(t) \cdot dt + \sigma \cdot S(t) \cdot dW(t) \\ S(t) = S_t \end{cases}.$$

To find a solution we set $Z(t) = \ln\{S(t)\}$. From Itô Lemma we then get

$$\begin{aligned} dZ &= \frac{\partial Z}{\partial t} dt + \frac{\partial Z}{\partial S} dS + \frac{1}{2} \frac{\partial^2 Z}{\partial S^2} (dS)^2 = \\ &= \frac{1}{S}(r \cdot S dt + \sigma \cdot S dW) - \frac{1}{2} \frac{1}{S^2} \sigma^2 \cdot S^2 dt = \left(r - \frac{1}{2}\sigma^2\right) dt + \sigma \cdot dW. \end{aligned}$$

Integration from t to T gives

$$Z_T = \ln S_t + \left(r - \frac{1}{2}\sigma^2\right)(T-t) + \sigma(W_T - W_t)$$

I.e.

$$S_T = S_t \cdot \exp\left\{ \left(r - \frac{1}{2}\sigma^2\right)(T-t) + \sigma(W_T - W_t) \right\} = S_t \cdot e^y.$$

The probability distribution of Z is therefore a $N[(r - \sigma^2/2)(T-t), \sigma^2(T-t)]$-distribution, where the probability density function $g(S)$ is given by

$$g(S_T) = \frac{1}{\sigma S_T \sqrt{2\pi(T-t)}} \exp\left\{ -\frac{\left(\ln\{S_T/S_t\} - (r - \sigma^2/2)(T-t)\right)^2}{2\sigma^2(T-t)} \right\}.$$

The price of the call option is now given by

$$\Pi[X|\mathcal{F}] = e^{-r(T-t)} E^Q_{t,s}[\max\{S_T - K, 0\}].$$

First, we define the following variables, to simplify the calculations

$$\widetilde{r} = r - \frac{1}{2}\sigma^2, \qquad \tau = T - t \qquad v = W_T - W_t = \sqrt{\tau} z.$$

Then

$$S_T = S_t \cdot \exp\{\widetilde{r}\tau + \sigma\sqrt{\tau} z\} = S_t \cdot e^y$$

and

$$g(S_T) = \frac{1}{\sigma S_T\sqrt{2\pi\tau}}\exp\left\{-\frac{(\ln\{S_T/S_t\} - \tilde{r}\tau)^2}{2\sigma^2\tau}\right\} = \frac{1}{\sigma S_T\sqrt{2\pi\tau}}\exp\left\{-\frac{(y - \tilde{r}\tau)^2}{2\sigma^2\tau}\right\}.$$

By the above definitions we have

$$y = \tilde{r}\cdot\tau + \sigma\cdot\sqrt{\tau}z \quad \Rightarrow \quad z = \frac{y - \tilde{r}\cdot\tau}{\sigma\cdot\sqrt{\tau}}$$

and

$$g(S_T) = \frac{1}{\sigma S_T\sqrt{2\pi\tau}}\exp\left\{-\frac{z^2}{2}\right\} = \frac{1}{\sigma S_T\sqrt{\tau}}N'(z).$$

For the call option we have

$$\begin{aligned}
&\Phi = \max\{S_t\cdot e^y - K, 0\} \\
&S_t\cdot e^{y_0} - K = 0 \quad \Rightarrow \quad y_0 = \ln\left\{\frac{K}{S_t}\right\} \\
&z_0 = \frac{\ln\{K/S_t\} - \tilde{r}\cdot\tau}{\sigma\cdot\sqrt{\tau}}.
\end{aligned}$$

Here we have defined y_0 as the value of y where $S_t\cdot e^y - K = 0$. This also gives the value z_0 and we can now start to integrate to get the price of the call option

$$\begin{aligned}
\Pi &= e^{-r\cdot\tau}\int_{-\infty}^{\infty}\Phi(S)\cdot g(S)dS = \left\{dS = \frac{\partial S}{\partial y}dy = S_t dy\right\} \\
&= e^{-r\cdot\tau}\int_{-\infty}^{\infty}S_t\cdot\Phi(y)\cdot g(y)dy = \left\{dy = \frac{\partial y}{\partial z}dz = \sigma\sqrt{\tau}dz\right\} \\
&= e^{-r\cdot\tau}\int_{-\infty}^{\infty}\Phi(z)\cdot\varphi(z)dz \\
&= e^{-r\cdot\tau}\int_{-\infty}^{\infty}\max\{S_t\cdot e^{\tilde{r}\cdot\tau+\sigma\cdot\sqrt{\tau}z} - K, 0\}\varphi(z)dz
\end{aligned}$$

where

$$\varphi(z) = \frac{1}{\sqrt{2\pi}}e^{-z^2/2}.$$

A normally distributed density function $N(\mu, \sigma)$ is given by

$$\varphi(z) = \frac{1}{\sqrt{2\pi\sigma}} \exp\left\{-\frac{1}{2}\left(\frac{z-\mu}{\sigma}\right)^2\right\}$$

and a log-normal distribution by

$$\varphi(z) = \frac{1}{z\sqrt{2\pi\sigma}} \exp\left\{-\frac{1}{2}\left(\frac{\ln(z)-\mu}{\sigma}\right)^2\right\}$$

Now we can continue to integrate

$$\begin{aligned}
\Pi &= e^{-r\cdot\tau}\int_{z_0}^{\infty}\left(S_t\cdot e^{\bar{r}\cdot\tau+\sigma\cdot\sqrt{\tau}z} - K\right)\varphi(z)dz - \\
&= S_t\cdot e^{-r\cdot\tau}\int_{z_0}^{\infty} e^{\bar{r}\cdot\tau+\sigma\cdot\sqrt{\tau}z}\varphi(z)dz - K\cdot e^{-r\cdot\tau}\int_{z_0}^{\infty}\varphi(z)dz = A - B \\
B &= K\cdot e^{-r\cdot\tau}N(-z_0) \\
A &= \frac{S_t\cdot e^{-r\cdot\tau}}{\sqrt{2\pi}}\cdot e^{r\cdot\tau}\int_{z_0}^{\infty} e^{-\frac{1}{2}\sigma^2\tau + \sigma\cdot\sqrt{\tau}z - z^2/2}dz \\
&= \frac{S_t}{\sqrt{2\pi}}\int_{z_0}^{\infty} e^{-(z-\sigma\cdot\sqrt{\tau})^2/2}dz = S_t\cdot N\left(-z_0+\sigma\cdot\sqrt{\tau}\right).
\end{aligned}$$

Finally, we can write this as

$$\begin{aligned}
\Pi &= S_t\cdot N\left(-z_0+\sigma\cdot\sqrt{T-t}\right) - K\cdot e^{-r\cdot(T-t)}N(-z_0) = \\
&= S_t\cdot N(d_1) - K\cdot e^{-r\cdot(T-t)}N(d_2)
\end{aligned}$$

where

$$d_1 = \frac{\ln\{S_t/K\} + (r+\sigma^2/2)(T-t)}{\sigma\cdot\sqrt{T-t}}, \quad d_2 = d_1 - \sigma\cdot\sqrt{T-t}.$$

Here we have used the symmetry $N(-x) = 1 - N(x)$. Then by integration from x to ∞ gives the same result as integrating from $-\infty$ to $-x$. To see that

$$\frac{1}{\sqrt{2\pi}}\int_{z_0}^{\infty} e^{-(z-\sigma\cdot\sqrt{\tau})^2/2}dz = N\left(-z_0+\sigma\cdot\sqrt{\tau}\right)$$

we also use that

$$\frac{1}{\sqrt{2\pi}}e^{-(z-\sigma\cdot\sqrt{\tau})^2/2}$$

is a Gaussian function centred at $z=\sigma\cdot\sqrt{\tau}$. If we like to have this centred at $z=0$ we translate the curve and the integral becomes

$$\frac{1}{\sqrt{2\pi}}\int_{z_0}^{\infty} e^{-(z-\sigma\cdot\sqrt{\tau})^2/2}dz = \frac{1}{\sqrt{2\pi}}\int_{z_0-\sigma\sqrt{\tau}}^{\infty} e^{-z^2/2}dz = \frac{1}{\sqrt{2\pi}}\int_{-\infty}^{-z_0+\sigma\sqrt{\tau}} e^{-z^2/2}dz$$
$$= N\left(-z_0+\sigma\cdot\sqrt{\tau}\right).$$

4.3.7 The Solution to Black–Scholes

There are many other European options with different payoffs at maturity. By repeating the above arguments we can show that also these options are solutions of the same PDE, the Black–Scholes partial differential equation, with their specific boundary conditions for the payoff at maturity. A European put option gives the holder the right to sell the asset at the strike price K at the maturity date. The appropriate boundary condition for such put option is

$$V(T,s) = [K-s]^{+} \text{at}\, t=T \quad \text{(European put)}.$$

A digit call option gives the holder a payout of 1, if the asset price is above the strike K at maturity date. A digital put option gives the holder a payout of 1, if the asset price is below the strike. For these options

$$V(T,s) = \begin{cases} 0 & \text{if } s<K \\ 1 & \text{if } s\geq K \end{cases} \quad \text{at } t=T \qquad \text{(digital call)}$$

$$V(T,s) = \begin{cases} 1 & \text{if } s<K \\ 0 & \text{if } s\geq K \end{cases} \quad \text{at } t=T \qquad \text{(digital put)}.$$

Similarly, power calls and power puts gives the holder a payout given by

$$V(T,s) = \left\{[s-K]^+\right\}^2 \quad \text{at } t=T \quad \text{(power call)}$$
$$V(T,s) = \left\{[K-s]^+\right\}^2 \quad \text{at } t=T \quad \text{(power put)}\,.$$

A convexity options gives the holder the payoff

$$V(T,s) = \{s-K\}^2 \quad \text{at } t=T \qquad \text{(convexity option)}.$$

4.3.8 A Green's Function Approach

A Green's function, $G(x, s)$, of a linear differential operator $L = L(x)$ acting on distributions over a subset of the Euclidean space $\mathbf{R}^n$, at a point s, is any solution of

$$LG(x,s) = \delta(x-s) \tag{4.1}$$

where δ is the Dirac delta function. This property of a Green's function can be exploited to solve differential equations of the form

$$Lu(x) = f(x) \tag{4.2}$$

If the kernel of L is non-trivial, then the Green's function is not unique. However, in practice, some combination of symmetry, boundary conditions and/or other externally imposed criteria will give a unique Green's function. Moreover, Green's functions in general are distributions, not necessarily proper functions.

Loosely speaking, if such a function G can be found for the operator L, then if we multiply the equation (4.1) for the Green's function by $f(s)$, and then perform integration in the s variable, we obtain

$$\int LG(x,s)f(s)ds = \int \delta(x-s)f(s)ds = f(x).$$

The right-hand side is now given by the equation (4.2) above to be equal to $Lu(x)$, thus

$$Lu(x) = \int LG(x,s)f(s)ds$$

Because the operator $L = L(x)$ is linear and acts on the variable x alone (not on the variable of integration s), we can take the operator L outside of the integration on the right-hand side, obtaining;

$$Lu(x) = L\left(\int G(x,s)f(s)ds\right).$$

This suggests;

$$u(x) = \int G(x,s)f(s)ds \tag{4.3}$$

Thus, we can obtain the function $u(x)$ through knowledge of the Green's function in equation (4.1), and the source term on the right-hand side in equation (4.2). This process has resulted from the linearity of the operator L.

In other words, the solution of equation (4.2), $u(x)$, can be determined by the integration given in equation (4.3). Although $f(x)$ is known, this integration cannot be performed unless G is also known. The problem now lies in finding the Green's function G that satisfies equation (4.1). For this reason, the Green's function is also sometimes called the fundamental solution associated to the operator L.

Green Functions and Black–Scholes

One solution to Black–Scholes equation is

$$V'(S,t) = \frac{e^{-r(T-t)}}{\sigma S'\sqrt{2\pi(T-t)}}\exp\left\{-\frac{\left(\ln\{S/S'\} + (r-\sigma^2/2)(T-t)\right)^2}{2\sigma^2(T-t)}\right\}$$

for any S'. You can verify this by substituting back into the equation. This solution is special because as $t \to T$ it becomes zero everywhere except at $S = S'$. In this limit the function becomes what is known as a *Dirac delta function*. This function is as we see zero everywhere, except in one point where it is infinite in such a way that the integral is one. Since Black–Scholes is a linear function we can multiply with a constant to get another solution. We can also add functions with different S', as the one above and still have a solution. We therefore have that also

$$\frac{e^{-r(T-t)}}{\sigma\sqrt{2\pi(T-t)}}\int_0^\infty \exp\left\{-\frac{\left(\ln\{S/S'\} + (r-\sigma^2/2)(T-t)\right)^2}{2\sigma^2(T-t)}\right\}f\left(S'\right)\frac{dS}{S'}$$

is a solution for any function $f(S')$. If we choose the arbitrary function $f(S')$ to be the payoff function then this expression becomes the solution of the problem

$$V(S,t) = \frac{e^{-r(T-t)}}{\sigma\sqrt{2\pi(T-t)}} \int_0^\infty \exp\left\{ -\frac{\left(\ln\{S/S'\} + (r-\sigma^2/2)(T-t)\right)^2}{2\sigma^2(T-t)} \right\} f\left(S'\right) \frac{dS}{S'}.$$

The function $V'(S, t)$ is called the *Green's function.*

4.3.9 Transformation of Black–Scholes

Thc Black–Scholes equation can be transformed to a simple diffusion equation. This is done by a change of variables. If we write

$$V(S,t) = U(x,\tau) \cdot \exp\{\alpha x + \beta\tau\}$$

where

$$\alpha = -\frac{1}{2}\left(\frac{2r}{\sigma^2} - 1\right), \quad \beta = -\frac{1}{4}\left(\frac{2r}{\sigma^2} + 1\right)^2, \quad S = e^x, \quad t = T - \frac{2\tau}{\sigma^2}$$

Then U satisfies the following equation

$$\frac{\partial U}{\partial \tau} = \frac{\partial U^2}{\partial x^2}$$

This can be useful when we want to find simple numerical schemes.

4.3.10 A Martingale Approach

In a martingale approach to find the price for a European call option we start with the terminal value

$$E[C_T] = (S_T - K)^+ = \max(S_T - K, 0).$$

In a risk-neutral world, the expected value at maturity is this terminal value discounted with the risk free rate r, i.e.

$$C_t = e^{r(T-t)}C_T = e^{r(T-t)}E[\max(S_T - K, 0)].$$

Therefore, we have to consider the expectation value. For any random variable X that can lie in the interval $[0, \infty]$ with a probability density function $p(X)$, its expected value is given by

$$E[X] = \int_0^\infty xp(x)dx.$$

As we know, $(\ln(S_T) \mid S_t)$ is a normal random variable with mean $\alpha(T-t)$ and variance $\sigma^2(T-t)$ and S_T a lognormal random variable where $E[S_T] = E\left[e^{\ln(S_T)}\right]$. Let $X_T = \ln\{S_T\}$, then $E[e^{X_T}|S_t]$ is given by

$$E\left[e^{X_T}|S_t\right] = S_t \int_0^\infty e^x p(x)dx.$$

We can also take the expectation of X with respect to another probability density function, say $q(x)$. Call this expectation $E^Q[X]$ where

$$E^Q[X] = \int_0^\infty xq(x)dx.$$

Then $E[S_T]$ is given by

$$E^Q[S_T] = \int_0^\infty e^x q(x)dx.$$

Now, if $X = (X-K)^+$

$$E^Q[X] = \int_K^\infty xq(x)dx$$

so that

$$E^Q[\max(S_T - K, 0)] = \int_{\ln(K)}^{\infty} \left[e^{\ln(x)} - K\right] q(x)dx$$
$$= \int_{\ln(K)}^{\infty} e^{\ln(x)} q(x)dx - K \int_{\ln(K)}^{\infty} q(x)dx$$

where $q(x) \sim N\left[r(T,t) - \sigma^2/2, \sigma\sqrt{T-t}\right]$. By calculating this integral we get the same result as above, the Black–Scholes formula.

4.3.11 Delta for a European Call Option

To calculate the Greeks (the hedge parameters) we have to take derivatives of integrals. The easiest way is to think like

$$\frac{d}{dx} \int_{o}^{f(x)} g(y)dy = \frac{d}{dx}[G(y)]_0^{f(x)} = \frac{df(x)}{dx} \cdot g(f(x)).$$

Using Black–Scholes formula for a call option is given by

$$C = S \cdot N(d_1) - X \cdot e^{-rT} \cdot N(d_2)$$

where

$$d_1 = \frac{\ln(S/X) + (r + \sigma^2/2) \cdot T}{\sigma \cdot \sqrt{T}}, \qquad d_2 = d_1 - \sigma \cdot \sqrt{T}$$
$$\frac{\partial d_1}{\partial S} = \frac{\partial d_2}{\partial S} = \frac{1}{S \cdot \sigma \cdot \sqrt{T}}$$

i.e.

$$\begin{aligned}
\Delta = \frac{\partial C}{\partial S} &= \frac{\partial}{\partial S}[S \cdot N(d_1)] - X \cdot e^{-rT} \cdot \frac{\partial}{\partial S} N(d_2) \\
&= N(d_1) + S \cdot N'(d_1) \cdot \frac{\partial d_1}{\partial S} - X \cdot e^{-rT} \cdot N'(d_2) \cdot \frac{\partial d_2}{\partial S} \\
&= N(d_1) + S \cdot N'(d_1) \cdot \frac{1}{S \cdot \sigma \cdot \sqrt{T}} - X \cdot e^{-rT} \cdot N'(d_2) \cdot \frac{1}{S \cdot \sigma \cdot \sqrt{T}} \\
&= N(d_1) + \frac{1}{S \cdot \sigma \cdot \sqrt{T}} \left[S \cdot N'(d_1) - X \cdot e^{-rT} \cdot N'(d_2) \right]
\end{aligned}$$

But

$$\begin{aligned}
X \cdot e^{-rT} \cdot N'(d_2) &= X \cdot e^{-rT} \cdot N'\left(d_1 - \sigma \cdot \sqrt{T}\right) \\
&= X \cdot e^{-rT} \cdot N'(d_1) \cdot e^{d_1 \cdot \sigma \cdot \sqrt{T}} \cdot e^{-\sigma^2 \cdot T/2} \\
&= X \cdot e^{-rT} \cdot N'(d_1) \cdot e^{-\sigma^2 \cdot T/2} \cdot \frac{S}{X} \cdot e^{rT} \cdot e^{\sigma^2 \cdot T/2} \\
&= S \cdot N'(d_1)
\end{aligned}$$

Finally, we get

$$\Delta = N(d_1).$$

This is also what we expect if we look at the formula

$$C = S \cdot N(d_1) - X \cdot e^{-rT} \cdot N(d_2).$$

Other hedge parameters can be calculated similarly.

4.3.12 Black–Scholes and Time-Dependent Parameters

We can generalize Black–Scholes by letting the parameters become time-dependent. Suppose that the interest rate and volatility are time dependent. We then replace; $r \to r(t)$ and $\sigma \to \sigma(t)$. We can also introduce a time dependent dividend yield $q \to q(t)$. The Black–Scholes equation then takes the form

$$\frac{\partial F(t,S)}{\partial t} + (r(t) - q(t))S(t)\frac{\partial F(t,S)}{\partial S} + \frac{1}{2}\sigma^2(t)S^2(t)\frac{\partial^2 F(t,S)}{\partial S^2} - r(t)F(t,S) = 0.$$

If we now introduce new variables

$$S(t) = S(t)e^{\alpha(t)}, \quad F(t, S) = F(t, S)e^{\beta(t)}, \quad \tau = \gamma(t)$$

to eliminate the time dependence we get the new equation

$$\dot{\gamma}(t)\frac{\partial \overline{F}}{\partial \tau} + (r(t) - q(t) + \dot{\alpha}(t))\overline{S}(t)\frac{\partial \overline{F}}{\partial \overline{S}} + \frac{1}{2}\sigma^2(t)\overline{S}^2(t)\frac{\partial^2 \overline{F}}{\partial \overline{S}^2} - \left(r(t) + \dot{\beta}(t)\right)\overline{F} = 0$$

where the dots denotes time derivatives. To eliminate the time-dependent terms we now choose

$$\alpha(t) = \int_t^T (r(\tau) - q(\tau))d\tau,$$

$$\beta(t) = \int_t^T r(\tau)d\tau$$

and

$$\gamma(t) = \int_t^T \sigma^2(\tau)d\tau.$$

Now, we get the following equation

$$\frac{\partial \overline{F}}{\partial \tau} + \frac{1}{2}\overline{S}^2(t)\frac{\partial^2 \overline{F}}{\partial \overline{S}^2} = 0$$

This means that a solution to the PDE in the original variables is

$$F(t, S) = e^{-\beta(t)}\overline{F}\left(\gamma(t), S(t)e^{\alpha(t)}\right)$$

If we let F_{BS} represents any solution to Black–Scholes with time-independent parameters we have

$$F_{BS} = e^{-r_c(T-t)}\overline{F}_{BS}\left(\sigma_c^2(T-t), Se^{-(r_c+q_c)(T-t)}\right)$$

where

$$r_c = \frac{1}{T-t}\int_t^T r(\tau)d\tau,$$

$$q_c = \frac{1}{T-t}\int_t^T q(\tau)d\tau,$$

$$\sigma_c^2 = \frac{1}{T-t}\int_t^T \sigma^2(\tau)d\tau.$$

To make everything clear, her`e is the formula for a European call option with time-dependent parameters:

$$C = S\exp\left\{-\int_t^T q(\tau)d\tau\right\}N(d_1) - K\exp\left\{-\int_t^T r(\tau)d\tau\right\}N(d_2)$$

where

$$d_1 = \frac{\ln\{\frac{S}{K}\} + \int_t^T (r(\tau) - q(\tau))d\tau + \frac{1}{2}\int_t^T \sigma^2(\tau)d\tau}{\sqrt{\int_t^T \sigma^2(\tau)d\tau}}$$

and

$$d_2 = \frac{\ln\{\frac{S}{K}\} + \int_t^T (r(\tau) - q(\tau))d\tau - \frac{1}{2}\int_t^T \sigma^2(\tau)d\tau}{\sqrt{\int_t^T \sigma^2(\tau)d\tau}}.$$

4.4 Volatility

In the Black–Scholes model, the volatility of the underlying asset is the only non-directly observable variable. For this reason, it is necessary to devise some method where by one can estimate (efficiently) and possibly anticipate the volatility. By using market prices, we can implicitly calculate the volatilities when we know the market prices. This method has an advantage over direct estimations based on historical price changes, since it reflects how much volatility the market currently assumes within the Black–Scholes framework. Quite often, the implied volatility is found to give rise to a *skew*, *smile* or *frown*, see Fig. 4.2 depending upon the asset or market.

One problem with using the implied volatility is that whilst it takes into account the current view of the market, it does not give us any insight into possible future changes in volatility. Given that the value of an option is primarily driven by the volatility, making predictions is a valuable tool from a practitioner's perspective. To achieve this, we turn back our attention to historical price movements. The historical volatility is defined via the standard deviation of the movements in price. Suppose we have n observations: a_i: a_0, a_1, ..., a_{n-1}. If we define $u_i = \ln(a_i/a_{i-1})$ we can calculate the standard deviation:

$$s = \sqrt{\frac{1}{n-1} \cdot \sum_{i=1}^{n} u_i^2 - \frac{1}{n \cdot (n-1)} \left(\sum_{i=1}^{n} u_i \right)^2}$$

The volatility is then given as $s\sqrt{d}$, where d is the number of trading days in a year (≈ 250).

4.4.1 The Volatility Surface

In many situations, we need to use a volatility surface. We can find such a surface by a least square method from implied volatilities. We need one such surface for the call options and another for the put options. By using a bid/ask volatility spread and the mid volatilities, we can also find a bid surface and an ask surface.

The calibration process is then:

1. Calculates the implied mid volatilities from option prices.
2. Calibrates the two volatility surfaces (call and put) to the implied mid volatilities.

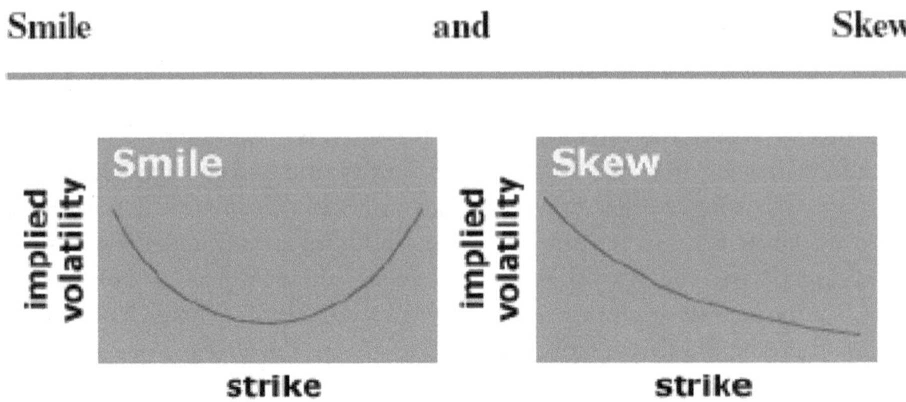

Fig. 4.2 The volatility smile and skew as function of the strike price

3. Applies a spread to extract bid and ask volatility surfaces. (Use larger spread for less liquid option series.)

The prices to use are either last paid, a bid/ask pair or a parity price, in that order of priority. We have to check that the implied volatility is not too low or not too high. Typical market data are shown in Table 4.1.

For option series with market prices, the mid volatility is calculated as the average of the implied bid and ask volatilities. The mid volatility for OMXS309V400 is for example (47.1 + 51.1)/2 = 49.1 %.

For option series with parity prices the mid volatility is given as the mid volatility of the option series with opposite option type adjusted with the average difference of the call and put mid volatilities of the option series that has market prices in both the call and the put options.

In this example the option series with strike prices 700, 710 and 720 have market prices in both the call and the put options. These option series show a slightly higher mid volatility for the put options compared to the call options and the average of the difference in mid volatility equals 1/3 (1.7 % + 1.0 % + 1.2 %) = 1.28 %.

The mid volatility for OMXS309V850 is therefore given by (30.1 + 32.0)/2 + 1.28 = 32.22 % and the mid volatility for OMXS309J400 49.1−1.28 = 47.82 %.

Option series with no prices will get a mid-volatility from the calibrated volatility surfaces.

Table 4.1 Option market data showing bid and ask prices and their implied volatilities for different strike

Call						Put				OMXS30, Oct 2009
Price type	Imp. bid	Bid	Ask	Imp. ask	Strike	Imp. bid	Bid	Ask	Imp. ask	Price type
None					380					None
Parity					400	47.1 %	2.1	3.3	51.1 %	Market
Parity					420	46.9 %	3.1	4.5	50.5 %	Market
...	...	...	...	...	...	...	...	...	...	...
Market	35.3 %	91.5	96.5	37.1 %	700	37.4 %	54.0	58.5	38.4 %	Market
Market	35.2 %	86.0	90.3	36.6 %	710	36.0 %	57.8	62.3	37.8 %	Market
Market	34.8 %	80.3	84.5	36.3 %	720	35.8 %	62.3	66.8	37.6 %	Market
...	...	...	...	...	...	...	...	...	...	...
Market	30.1 %	25.3	29.3	32.0 %	850					Parity
Market	29.6 %	22.3	26.3	31.6 %	860					Parity
None					870					None

The Method of Least Square

The concept of volatility surfaces implies that the mid volatility can be seen as a function of time to maturity T and strike price, X, i.e. $\sigma = \sigma(T, X)$. If we assume that the volatility surfaces are given by a 3rd degree polynomial surface we have

$$\sigma(T,X) = c_0 + c_1T + c_2T^2 + c_3T^3 + c_4X + c_5X^2 + c_6X^3 + c_7TX + c_8T^2X + c_9TX^2.$$

The problem is to find the coefficients c_k. If they are found, then the implied volatility can be calculated for any given time to expiration and strike price. The mid volatilities from the option series with market prices and parity prices provide several points. There are totally 10 unknown coefficients, c_k, for a 3rd degree polynomial surface. The minimum number of data points in order to calculate the coefficients are therefore 10. If there are market prices and parity prices corresponding to more than 10 data points these add up to an over determined system of linear equations. In this case there is no exact solution but there is a way to mathematically estimate the best approximation to the coefficients. This is called the method of least squares.

The set of data points can be converted into a linear system of equations using the equation above. The linear system of equations can be expressed as a matrix multiplication $A\cdot c = z$ where A is a matrix containing the times to expiration and the strike prices, c is a vector containing the unknown coefficients and z is a vector containing the mid volatilities.

$$A = \begin{bmatrix} 1 & T_1 & T_1^2 & T_1^3 & X_1 & X_1^2 & X_1^3 & T_1X_1 & T_1^2X_1 & T_1X_1^2 \\ \ddots & \ddots & \ddots & \ddots & \ddots & \ddots & \ddots & \ddots & \ddots & \ddots \\ 1 & T_k & T_k^2 & T_k^3 & X_k & X_k^2 & X_k^3 & T_kX_k & T_k^2X_k & T_kX_k^2 \end{bmatrix}$$

$$c = \begin{bmatrix} c_0 \\ \vdots \\ c_9 \end{bmatrix} \qquad z = \begin{bmatrix} \sigma_0 \\ \vdots \\ \sigma_9 \end{bmatrix}$$

We then do the simple algebra

$$A \cdot c = z \quad \Rightarrow \quad A^TA \cdot c = A^T \cdot z \quad \Rightarrow \quad c = \left(A^TA\right)^{-1} \cdot A^T \cdot z.$$

In Fig. 4.3 we see a typical volatility surface for European options on a stock index.

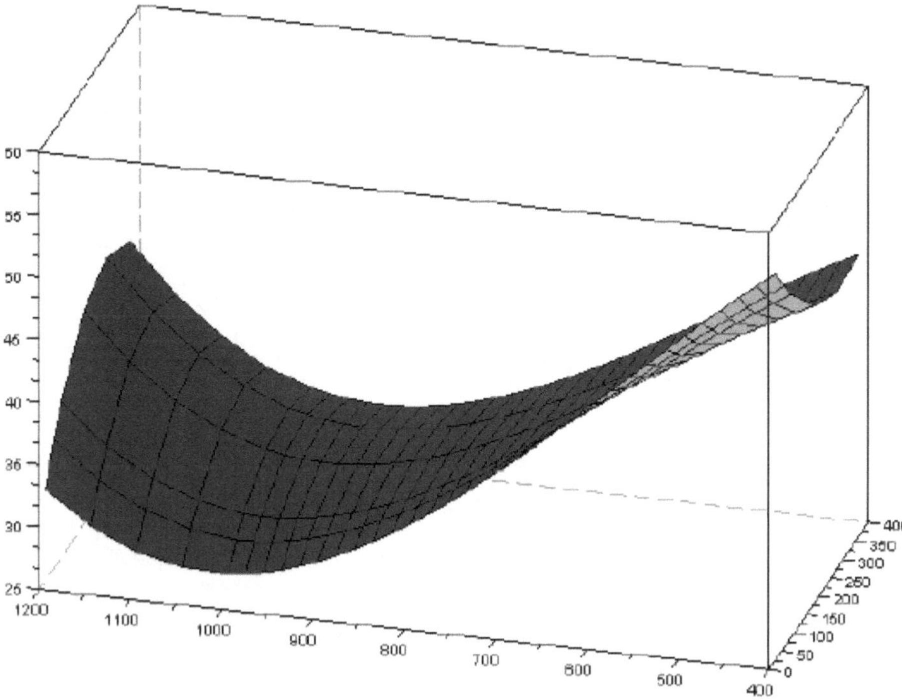

Fig. 4.3 A typical volatility surface. On the axis we have the index level and the strike price

The volatility for the options with no prices are calculated from the surface. Volatility for series outside the surface can be found by (flat) interpolation.

Example 4.10

We have the following points (x, y): (1, 1), (2, 3) and (3, 2) and we would like to find the equation of the line that fits these points with the least square.

We have then the following system of equations

$$\begin{cases} 1 = k \cdot 1 + m \\ 2 = k \cdot 3 + m \\ 3 = k \cdot 2 + m \end{cases} \Leftrightarrow \begin{pmatrix} 1 & 1 \\ 3 & 1 \\ 2 & 1 \end{pmatrix} \begin{pmatrix} k \\ m \end{pmatrix} = \begin{pmatrix} 1 \\ 2 \\ 3 \end{pmatrix} \Leftrightarrow Ac = y.$$

By using

$$A \cdot c = y \Rightarrow A^T A \cdot c = A^T \cdot y \Rightarrow c = (A^T A)^{-1} \cdot A^T \cdot y$$

(continued)

Example 4.10 (continued)

we have

$$\begin{pmatrix} k \\ m \end{pmatrix} = \left[\begin{pmatrix} 1 & 3 & 2 \\ 1 & 1 & 1 \end{pmatrix}\begin{pmatrix} 1 & 1 \\ 3 & 1 \\ 2 & 1 \end{pmatrix}\right]^{-1}\begin{pmatrix} 1 & 3 & 2 \\ 1 & 1 & 1 \end{pmatrix}\begin{pmatrix} 1 \\ 2 \\ 3 \end{pmatrix}$$

$$= \begin{pmatrix} 14 & 6 \\ 6 & 3 \end{pmatrix}^{-1}\begin{pmatrix} 13 \\ 6 \end{pmatrix} = \frac{1}{14 \cdot 3 - 6 \cdot 6}\begin{pmatrix} 3 & -6 \\ -6 & 14 \end{pmatrix}\begin{pmatrix} 13 \\ 6 \end{pmatrix}$$

$$= \frac{1}{6}\begin{pmatrix} 3 \\ 6 \end{pmatrix} = \begin{pmatrix} 1/2 \\ 1 \end{pmatrix}$$

i.e.,

$$y = \tfrac{1}{2}x + 1.$$

4.4.2 Volatility Models

According to the classical Black–Scholes options pricing model, all options based on the same underlying sharing a constant implied volatility under the assumption of a geometric Brownian motion process. But if this model is used to back-test the market-traded option, we can observe that different contracts produce significantly different implied volatilities. Options' implied volatilities actually vary with the different time to maturity. This is the term structure of implied volatility. For a given time to maturity, implied volatilities for different strikes are not the same either. This is the implied volatility skew and is often referred as the volatility smile. All these market evidences imply that the option market expects the future volatility of the underlying asset will not be a constant.

In Fig. 4.4 we show the S&P 500 and the VIX volatility index over several years (data provided by the Chicago Board of Option Exchange, www.cboe.com). S&P (Standard & Poor's) 500 is an American stock market index based on the market capitalizations of 500 large companies having common stock listed on the NYSE or NASDAQ. The S&P 500 index components and their weightings are determined by S&P Dow Jones Indices. It differs from other US stock market indices, such as the Dow Jones Industrial Average or the Nasdaq Composite index, because of its diverse constituency and weighting methodology. It is one of the most commonly followed equity indices, and many consider it one of the best representations of the US stock market, and a bellwether

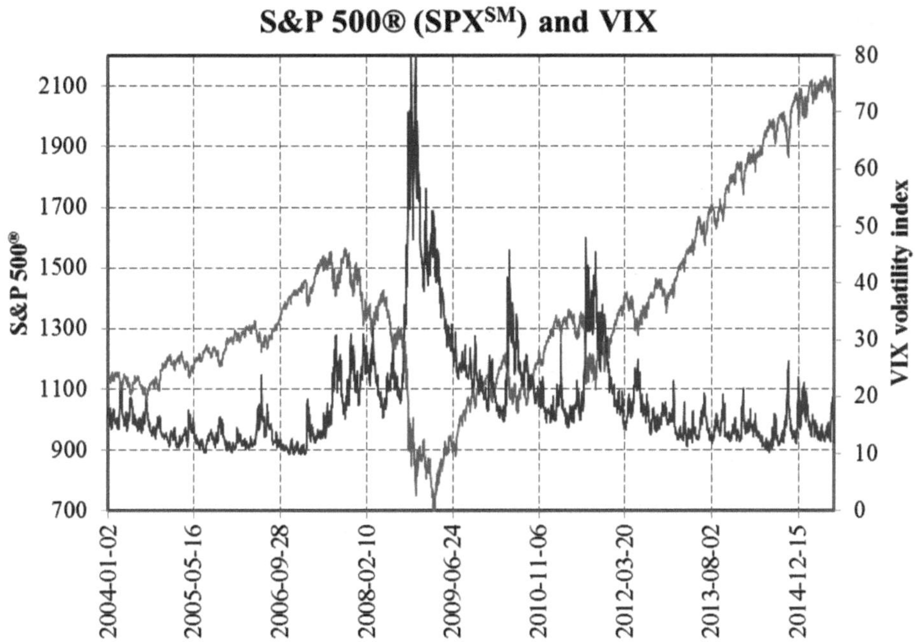

Fig. 4.4 The historical VIX volatility S&P 500 (Source: CBOE)

for the US economy. The CBOE Volatility Index® (VIX® Index) is a key measure of market expectations of near-term volatility conveyed by S&P 500 stock index option prices. Since its introduction in 1993, VIX has been considered by many to be the world's premier barometer of investor sentiment and market volatility. As we can see in Fig. 4.4, the volatility increases when the market falls. The VIX Index is therefore sometimes referred as a "market fear index".

When we see the behaviour of the volatility, we realize that a stochastic volatility model is more reasonable for option pricing. It can explain the basic shapes of the smile patterns and allow for more realistic theories of the term structure of implied volatility. A particular case is that volatility can be described with a GARCH model (see Sect. 4.4.4). In GARCH models, the variance is written as a function of past returns, but with exponentially smoothing and a certain time-decay factor. One more important feature of GARCH is that the constant term in the recursive equation allows GARCH to capture the notion that the volatility is mean reverting and allows the model to be used for forecasting volatility.

Autocorrelation Structures

We take a slight detour to introduce the definition of the *autocorrelation function*. The *correlation function* between two time series, X and Y, is given by the expression

$$Corr(X,Y) = \frac{\langle (X-\mu_X)(Y-\mu_Y) \rangle}{\sigma_X \sigma_Y}$$

where μ and σ are the mean and variance estimates of X and Y respectively, and $<\ldots>$ denotes the mean value of the expression inside the brackets. The autocorrelation function is calculated by setting $Y = X(t + \delta)$, where δ is some forward time lag of the time series X. Hence, the autocorrelation function may be expressed as

$$Corr(\delta) = \frac{\langle X(t)X(t+\delta) \rangle}{\sigma_X{}^2}.$$

The autocorrelation function is an *average* measure of the correlations that exist within a time series. The form of this volatility autocorrelation has been empirically suggested to be either exponentially decaying, or exhibiting long-range memory (power-law decay).

4.4.3 ARCH Models

An *autoregressive conditional heteroskedasticity* (ARCH(m)) model was introduced by Engle (1982) to model the volatility of UK inflation. As the name suggests, the model has the following properties:

1. Autoregression—Uses previous estimates of volatility to calculate subsequent (future) values. Hence volatility values are closely related.
2. Heteroscedasticity—The probability distributions of the volatility varies with the current value.

In order to introduce ARCH processes, let us assume that we have a time series of asset price quotes P_i for each time step i. We calculate the fractional change in the price of the asset between time step i and $i + 1$ using

$$x_i = \frac{P_i - P_{i-1}}{P_{i-1}}.$$

Furthermore, we are required to determine the long-running historical volatility (e.g. over several years) denoted by S. The volatility fluctuates about some long-running mean volatility, therefore, it seems reasonable to incorporate this quantity in the ARCH model. Formally, an ARCH(m) process may be expressed mathematically as

$$\sigma_n^2 = \gamma S + \sum_{i=1}^{m} \alpha_i x_{n-i}{}^2$$

where

$\gamma \geq 0, \qquad \alpha_i \geq 0, \qquad \gamma + \sum_{i=1}^{m} \alpha_i = 1$ and m is the number of observations of x_{n-i} used to determine σ_n. The most common ARCH(m) model is the ARCH(1) model where

$$\sigma_n^2 = \gamma S + \alpha x_{n-1}{}^2 = \gamma S + (1-\gamma) x_{n-1}{}^2.$$

4.4.4 GARCH Models

Bollerslev (1986) later proposed a more generalized form of the ARCH(m) model appropriately termed the GARCH(p, q) (General-ARCH) model. The GARCH(p, q) model may be written as

$$\sigma_n^2 = \gamma S + \sum_{i=1}^{p} \alpha_i x_{n-i}{}^2 + \sum_{j=1}^{q} \beta_j \sigma_{n-j}{}^2.$$

The p and q denote the number of past observations of x_{n-i} and σ_{n-j}, respectively, used to estimate σ_n. The simplest GARCH(p, q) model is GARCH(1,1) given by:

$$\begin{cases} \sigma_{t+1}{}^2 = \kappa\vartheta + (1-\kappa)\left((1-\lambda)y_t^2 + \lambda\sigma_t{}^2\right) \\ y_t = \dfrac{\log\left(\dfrac{S_t}{S_{t-1}}\right) - \left(r_t - \dfrac{1}{2}\sigma_t{}^2\right)\Delta t}{\sqrt{\Delta t}} \end{cases}.$$

The equation gives the evolution of the variance as the weighted average with weights κ and $1-\kappa$, of two parts, one being the constant ϑ and the other being a weighted average of y_t^2 and σ_t^2. Whatever the variance might be at time i, the variance of y_j at any date j far into the future, computed without knowing the intervening $y_{i+1}, y_{i+2}, \ldots$, will be approximately the constant ϑ. The constant ϑ is called the unconditional variance, whereas σ_t^2 is the conditional variance of y_t.

To understand the unconditional variance, it is useful to consider the variance forecasting equation. Specifically, we can calculate $E_i\left[\sigma_{i+n}^2\right]$, which is the estimate made at date i of the variance of y_{i+n}; we estimate the variance without having observed $y_{i+1}, \ldots, y_{i+n-1}$. Note that by definition $E_i\left[y_{i+1}^2\right] = \sigma_{i+1}^2$, so the above equation implies

$$E_i\left[\sigma_{i+2}^2\right] = \kappa\vartheta + (1-\kappa)\left((1-\lambda)E_i\left[y_{i+1}^2\right] + \lambda\sigma_{i+1}{}^2\right) = \kappa\vartheta + (1-\kappa)\sigma_{i+1}{}^2.$$

Likewise,

$$E_i\left[\sigma_{i+3}^2\right] = \kappa\vartheta + (1-\kappa)E_i\left[\sigma_{i+2}^2\right] = \kappa\vartheta(1+(1-\kappa)) + (1-\kappa)^2\sigma_{i+1}{}^2.$$

This generalizes to

$$E_i\left[\sigma_{i+n}^2\right] = \kappa\vartheta\left(1 + (1-\kappa) + \cdots + (1-\kappa)^{n-2}\right) + (1-\kappa)^{n-1}\sigma_{i+1}{}^2.$$

Thus, there is decay at rate κ in the importance of the current volatility $\sigma_{i+1}{}^2$ for forecasting the future volatility. Furthermore, as $n \to \infty$, the geometric series $1 + (1-\kappa) + \cdots + (1-\kappa)^{n-2}$ converges to $1/\kappa$, so, as $n \to \infty$ we obtain

$$E_i\left[\sigma_{i+n}^2\right] \to \vartheta.$$

This means that our best estimate of the conditional variance, at some date far in the future, is approximately the unconditional variance ϑ.

The most interesting feature of the volatility equation is that large returns in absolute value lead to an increase in the variance and hence are likely to be followed by more large returns. This is the "volatility clustering", which is observable in actual markets. This feature also implies that the distribution of the returns will be "fat-tailed". This means that the probability of the extreme returns is higher than under a normal distribution with the same standard deviation. It is agreed that daily and weekly returns in most markets have this 'fat-tailed' property.

The constants are determined by finding the maximum probability distribution of the observed changes in the daily closing prices.

4.4.5 EWMA

The exponentially weighted moving average model (EWMA) is a special case of the GARCH(1,1) model where $\gamma = 0$. Thus

$$\sigma_n^2 = \alpha x_{n-1}{}^2 + \beta \sigma_{n-1}{}^2 = \alpha x_{n-1}{}^2 + (1 - \alpha)\sigma_{n-1}{}^2.$$

The EWMA model differs from ARCH and GARCH models since it does *not* mean-revert. The preference between these different models is dependent upon many factors. For example, the asset class, the forecasting time frame, and the efficiency with which the weighting parameters may be calibrated to the time series. Whilst the *maximum likelihood estimator* method may be the most obvious method to select for calibration with empirical data, more efficient algorithms have also been put forward.

Since these volatility forecasting models were introduced, there have been many alternatives/modifications proposed to these models to better their use in volatility forecasting.

4.5 Parity Relations

To study parity relations we introduce the following notation:

$c(t, S, K, T, r, \sigma)$	the price on a European call option.
$p(t, S, K, T, r, \sigma)$	the price on a European put option.
S	the value of the underlying stock.
$B = 1$	the value of one cash unit.
$C_K(S) = (S-K)^+$	the call option value.
$P_K(S) = (K-S)^+$	the put option value.

Let T be the time to maturity and consider different T-contracts X: $\Phi(S) = X$. If

$$\Phi = \alpha \cdot S + \beta \cdot B + \sum_{i=1}^{n} \gamma_i C_{K_i}$$

the price is given by:

$$\Pi_t[\Phi] = \alpha \cdot \Pi_t[S] + \beta \cdot \Pi_t[B] + \sum_{i=1}^{n} \gamma_i \cdot \Pi_t[C_{K_i}] =$$
$$= \alpha \cdot S_t + \beta \cdot e^{-r(T-t)} + \sum_{i=1}^{n} \gamma_i \cdot c(t, S_t, K_i, T, r, \sigma).$$

For each Φ we can construct a replicating portfolio with constant shares over time. The result is only interesting if there exist a class of such contract functions Φ, given by linear combinations of the base functions. This is the case in reality.

The put call parity is obtained from the fact that

$$\max\{K - S_T, 0\} = K - S_T + \max\{S_T - K, 0\}.$$

If we denote by $C(t, S(t))$ the price at time t for a European call option with strike price K and exercise time T written on the stock, and by $P(t, S(t))$ the corresponding put option. Then

$$\begin{aligned} P(t, S(t)) &= e^{-r(T-t)} E^Q[\max\{K - S_T, 0\} | \mathcal{F}_t] \\ &= e^{-r(T-t)} E^Q[K - S_T + \max\{S_T - K, 0\} | \mathcal{F}_t] \\ &= e^{-r(T-t)} K - e^{rt} E^Q\left[\frac{S_T}{e^{rT}} | \mathcal{F}_t\right] + e^{-r(T-t)} E^Q[\max\{S_T - K, 0\} | \mathcal{F}_t] \\ &= e^{-r(T-t)} K - e^{rt} \frac{S_t}{e^{rt}} + C(t, S(t)) \end{aligned}$$

where we used the martingale property that the discounted stock price is martingale. This gives the put call parity

$$P(t, S(t)) = Ke^{-r(T-t)} + C(t, S(t)) - S_t.$$

For American options, the put call parity is given by

$$S - K \leq C_A - P_A \leq S - Ke^{-rT}.$$

We leave the proof to the reader.

4.6 A Practical Guide to Pricing

In the previous chapters we have described a number of techniques to price derivatives. In this section we will study in detail two similar problems and see how to price these contracts. With these examples in mind the reader should get the needed understanding to be able to solve various problems by her own.

We will value two simple and similar contracts where the pay-out function at maturity T is given by

$$\Phi(X) = S^2(T)$$

and

$$\Phi(X) = \begin{cases} S^2(T) & \text{if } S(T) > K \\ 0 & \text{else} \end{cases}$$

where K is a given strike price and S the price of an underlying stock.

4.6.1 Method 1, Without Using the Solution to *S*

We start with the first problem and define a new variable Z_t as

$$Z_t = S_t^2 \equiv S^2(t).$$

If we use Itô on Z_t we find

$$dZ_t = \frac{\partial Z_t}{\partial S} dS + \frac{1}{2}\frac{\partial^2 Z_t}{\partial S^2}(dS)^2 = 2S_t dS + \frac{1}{2} \cdot 2(dS)^2.$$

We know that the stock prices in Black–Scholes world follow a stochastic process given as a geometrical Brownian motion (GBM)

$$\begin{cases} dS = rSdt + \sigma SdW \\ S(0) = s \end{cases}$$

driven by the risk-free interest rate r. If we substitute this expression into dZ we get

$$dZ_t = 2S_t(rS_tdt + \sigma S_tdW_t) + \frac{1}{2}\cdot 2(\sigma S_t)^2dt = (2r+\sigma^2)S_t^2dt + 2\sigma S_t^2dW_t$$
$$= (2r+\sigma^2)Z_tdt + 2\sigma Z_tdW_t$$

where σ is the volatility and W_t a Wiener process. We have also used the property of the Wiener process that $(dW)^2 = dt$. If we integrate and take expectation value we find

$$E^Q\left[\int_0^t dZ_u\right] = (2r+\sigma^2)E^Q\left[\int_0^t Z_udu\right] + 2\sigma E^Q\left[\int_0^t Z_udW_u\right].$$

The last integral will vanish due to the property of increments of the Wiener process. We then get

$$E^Q[Z_t] - E^Q[Z_0] = (2r+\sigma^2)\int_0^t E^Q[Z_u]du.$$

We also know that $E^Q[Z_0]$ is a constant given by $S_0{}^2$. If we let $E^Q[Z_t] = m_t$ and take the derivative with respect to t, we find

$$\begin{cases} \frac{dm(t)}{dt} = (2r+\sigma^2)m(t) \\ m(0) = S^2(0) \end{cases}.$$

This, simple ordinary differential equation can be solved. The solution is

$$m(t) = S^2(0)e^{(2r+\sigma^2)t} = E^Q[Z_t] = E^Q\left[S^2(t)\right].$$

The price of the contract is then given as the discounted value

$$\Pi(0,S) = e^{-rT}E^Q\left[S^2(T)\right] = e^{-rT}S_0^2e^{(2r+\sigma^2)T} = S_0^2e^{(r+\sigma^2)T}.$$

4.6.2 Method 2, by Using the Solution to *S*

If we use the solution to the Brownian motion of the stock price

$$S_t = S_0 \cdot \exp\left\{\left(r - \frac{1}{2}\sigma^2\right)t + \sigma W_t\right\}$$

we have

$$S_t^2 = S_0^2 \cdot \exp\{(2r - \sigma^2)t + 2\sigma W_t\}.$$

By taking the expectation value we get

$$E^Q\left[S_t^2\right] = E^Q\left[S_0^2\right] \cdot E^Q\left[e^{\left(2r-\sigma^2\right)t}\right] \cdot E^Q\left[e^{2\sigma W_t}\right] = S_0^2 e^{\left(2r-\sigma^2\right)t} \cdot E^Q\left[e^{2\sigma W_t}\right].$$

Now we can calculate the expectation value by integration. We know that the Wiener process is normal distributed $N[0, 1]$, so that $e^{2\sigma W}$ is normal distributed $N[0, t]$ giving

$$E^Q\left[e^{2\sigma W_t}\right] = \int_{-\infty}^{\infty} e^{2\sigma\sqrt{t}z}\frac{1}{\sqrt{2\pi}}e^{-z^2/2}dz = \frac{e^{2\sigma^2 t}}{\sqrt{2\pi}}\int_{-\infty}^{\infty} e^{-\left(z-2\sigma\sqrt{t}\right)^2/2}dz = e^{2\sigma^2 t}$$

I.e.,

$$E^Q\left[S_t^2\right] = S_0^2 e^{\left(2r-\sigma^2\right)t} e^{2\sigma^2 t} = S_0^2 e^{\left(2r+\sigma^2\right)t}$$

giving

$$\Pi(0, S) = e^{-rT} E^Q\left[S^2(T)\right] = e^{-rT} S_0^2 e^{\left(2r+\sigma^2\right)T} = S_0^2 e^{\left(r+\sigma^2\right)T}.$$

The Expectation of *W* via an ODE

Another method to calculate

$$E^Q\left[e^{2\sigma W_t}\right]$$

is, by the use of Itô. We then introduce a new variable Z_t given by $Z_t = e^{2\sigma w}$. We then get

$$dZ_t = \frac{\partial Z_t}{\partial W_t} dW_t + \frac{1}{2} \frac{\partial^2 Z_t}{\partial W_t^2} (dW_t)^2 = 2\sigma Z_t dW_t + \frac{1}{2} \cdot 4\sigma^2 Z_t dt = 2\sigma^2 Z_t dt + 2\sigma Z_t dW_t.$$

Integration gives

$$\int_0^t dZ_u = 2\sigma^2 \int_0^t Z_u du + 2\sigma \int_0^t Z_u dW_u.$$

Taking expectation value, the last integral vanish and

$$E^Q[Z_t] - E^Q[Z_0] = 2\sigma^2 \int_0^t E^Q[Z_u] du.$$

If we let $E^Q[Z_t] = m_t$ and take the derivative with respect to t, we find

$$\begin{cases} \dfrac{dm(t)}{dt} = 2\sigma^2 m(t) \\ m(0) = 1 \end{cases}.$$

This, simple ordinary differential equation can be solving, giving

$$m(t) = e^{2\sigma^2 t}$$

and we get the same result as above.

$$E^Q\left[S_t^2\right] = S_0^2 e^{\left(2r - \sigma^2\right)t} e^{2\sigma^2 t} = S_0^2 e^{\left(2r + \sigma^2\right)t}$$

i.e.,

$$\Pi(0, S) = e^{-rT} E^Q\left[S^2(T)\right] = e^{-rT} S_0^2 e^{\left(2r + \sigma^2\right)T} = S_0^2 e^{\left(r + \sigma^2\right)T}.$$

4.6.3 Introducing the Strike

If we introduce the strike, we have to integrate the probability distribution. Since we have done the same several times we start with the solution to the Brownian motion

$$S_t = S_0 \exp\left\{\left(r - \frac{1}{2}\sigma^2\right)t + \sigma W_t\right\} = S_0 \cdot e^y.$$

First, we define the following variables

$$\widetilde{r} = r - \frac{1}{2}\sigma^2, \qquad \sigma W_t = \sigma\sqrt{t}z.$$

Then we have

$$S_t = S_0 \cdot \exp\{\widetilde{r}t + \sigma\sqrt{t}z\} = S_0 \cdot e^y$$

and

$$S_T{}^2 = S_0{}^2 \cdot \exp\left\{2\widetilde{r}T + 2\sigma\sqrt{T}z\right\} = S_0{}^2 \cdot e^{2y}.$$

By the above definitions we have

$$y = \widetilde{r}T + \sigma\sqrt{T}z \quad \Rightarrow \quad z = \frac{y - \widetilde{r}T}{\sigma\sqrt{T}}.$$

The strike gives the integration boundary

$$S_T > K \quad \Rightarrow \quad S_0 e^y > K \quad \Rightarrow \quad y_0 = \ln\left\{\frac{K}{S_0}\right\}$$

$$z_0 = \frac{y_0 - \widetilde{r}T}{\sigma \cdot \sqrt{T}} = \frac{\ln\{K/S_0\} - \widetilde{r}T}{\sigma \cdot \sqrt{T}}.$$

The probability distribution of S is a $N[(r - \sigma^2/2)t, \sigma^2 t]$-distribution, where the probability density function $g(S)$ is given by

$$g(S)=\frac{1}{\sigma S\sqrt{2\pi T}}\exp\left\{-\frac{(\ln\{S/S_0\}-\tilde{r}T)^2}{2\sigma^2 T}\right\}=\frac{1}{\sigma S\sqrt{2\pi T}}\exp\left\{-\frac{(y-\tilde{r}T)^2}{2\sigma^2 T}\right\}$$
$$=\frac{1}{\sigma S\sqrt{2\pi T}}\exp\left\{-\frac{z^2}{2T}\right\}.$$

We can now start to integrate to get the price of the call option

$$\begin{aligned}
\Pi &= e^{-r\cdot T}\int_{-\infty}^{\infty}\Phi(S)\cdot g(S)dS=\left\{dS=\frac{\partial S}{\partial y}dy=Sdy\right\}\\
&= e^{-r\cdot T}\int_{y_0}^{\infty}S\cdot\Phi(y)\cdot g(y)dy=\left\{dy=\frac{\partial y}{\partial z}dz=\sigma\sqrt{T}dz\right\}
\end{aligned}$$

$$\begin{aligned}
&= e^{-rT}\int_{z_0}^{\infty}\Phi(z)\varphi(z)dz = S_0{}^2 e^{-rT}\int_{z_0}^{\infty}e^{2\tilde{r}T+2\sqrt{T}z}\varphi(z)dz\\
&= \frac{S_0{}^2e^{-rT}}{\sqrt{2\pi}}\int_{z_0}^{\infty}e^{2\tilde{r}T+2\sigma\sqrt{T}z-z^2/2}dz\\
&= \frac{S_0{}^2e^{-rT}e^{(2r-\sigma^2)T}}{\sqrt{2\pi}}\int_{z_0}^{\infty}e^{2\sigma\sqrt{T}z-z^2/2}dz\\
&= \frac{S_0{}^2e^{(r-\sigma^2)T}}{\sqrt{2\pi}}e^{2\sigma^2T}\int_{z_0}^{\infty}e^{-(z-2\sigma\sqrt{T})^2/2}dz\\
&= S_0{}^2e^{(r+\sigma^2)T}N\left[-z_0+2\sigma\sqrt{T}\right]\\
&= S_0{}^2e^{(r+\sigma^2)T}N[d]
\end{aligned}$$

where

$$d=\frac{\ln\{S_0/K\}+(r-\sigma^2/2)T}{\sigma\sqrt{T}}+2\sigma\sqrt{T}=\frac{\ln\{S_0/K\}+(r+3\sigma^2/2)T}{\sigma\sqrt{T}}.$$

4.6.4 The General Problem, a Summary

As we have seen we can use two methods to calculate the price of a derivative. We can use the probabilistic method with the following schema:

1. Use Itô formula on the contract function to find the dynamics.
2. Integrate this expression and take the expectation value. The stochastic part will then vanish.
3. Introduce a new variable for the expectation value (m above).
4. Take derivative of this variable with respect to time. This gives an ordinary differential equation.
5. Solve the ODE above.
6. Discount the ODE solution with the risk free interest rate r, to a present value.
7. Done!

The second method is to use the analytical approach and integrate. Then we are using the following schema

1. Express the contract function in term of the solution to the GBM for the stock price.
2. Introduce the simplified variables and find the integration limits depending on the conditions (strike).
3. Write down the integrals and change the integrating variables as above.
4. Rewrite the integral as:
$$I = \frac{f(r,T,\sigma\ldots)}{\sqrt{2\pi}} \int_{z_0}^{\infty} \exp\left\{ -\frac{(z-m)^2}{2} \right\} dz$$
5. The integral is now equal to:
$I = f(r, T, \sigma \ldots) \cdot N[-z_0 + m]$
6. Discount the integral solution with the risk free interest rate r, to a present value.
7. Done!

The relation between the two methods is given by the Feynman–Kač representation

$$F(t,x) = e^{-r(T-t)} E^Q_{t,x}[\phi(X(T))].$$

4.6.5 A General Approach for Pricing European Call Options

The payoff of a European call options is given by

$$C_T = \max(S_T - K, 0) = (S_T - K) I_{\{S_T > K\}}$$

where S_T is the stock price at maturity, K the strike price and $I_{\{S_T > K\}}$ a indicator function equal to 1 if $S_T > K$ and 0 else. We then have the arbitrage free price as

$$\begin{aligned}
C_t &= e^{-r(T-t)} E^Q[\max(S_T - K, 0)] \\
&= e^{-r(T-t)} E^Q\left[(S_T - K, 0) I_{\{S_T > K\}}\right] \\
&= e^{-r(T-t)} \left(E^Q\left[S_T I_{\{S_T > K\}} | \mathcal{F}_t\right] - K \cdot E^Q\left[I_{\{S_T > K\}}\right]\right) \\
&= e^{-r(T-t)} \left(E^Q\left[S_T I_{\{S_T > K\}} | \mathcal{F}_t\right] - K \cdot Q(S_T > K)\right)
\end{aligned}$$

where $Q(S_T > K)$ is the probability that the option is in-the-money at maturity. If we change the measure (numeraire) we can rewrite the first term as

$$e^{-r(T-t)} E^Q\left[S_T I_{\{S_T > K\}} | \mathcal{F}_t\right] = S_t E^{\widetilde{Q}}\left[I_{\{S_T > K\}} | \mathcal{F}_t\right] = S_t \widetilde{Q}(S_T > K).$$

Therefore

$$C_t = S_t \cdot \widetilde{Q}(S_T > K) - e^{-r(T-t)} K \cdot Q(S_T > K).$$

Remark Both of the probabilities in this formula are that the option expire in-the-money. The difference is that we are calculating the probabilities under different probability measures.

Since we know the expression for the stock price:

$$S_T = S_t \exp\left\{\left(r - \frac{1}{2}\sigma^2\right)(T - t) + \sigma(W_T - W_t)\right\} = S_t \exp\{\widetilde{r}\tau + \sigma\sqrt{\tau}z\}.$$

The normal processes are given by:

$$Q: \quad N[\widetilde{r}\tau, 1]$$
$$\widetilde{Q}: \quad N[\widetilde{r}\tau, \sigma^2\tau]$$

since we have to use

$$S_T > K \quad \Rightarrow \quad S_t \exp\{\widetilde{r}\tau + \sigma\sqrt{\tau}z\} > K \quad \Rightarrow \quad z_0 = \frac{\ln\{K/S_t\} - \widetilde{r}\tau}{\sigma\sqrt{\tau}}$$

to calculate the expectation on S_t above. Here z_0 is the value of z where we hit the strike K, so

$$Q(S_T > K) = N[-z_0]$$

and

$$\widetilde{Q}(S_T > K) = N\left[-z_0 + \sigma\sqrt{\tau}\right].$$

Finally, we get the Black–Scholes formula

$$C_t = S_t \cdot N[d_1] - e^{-r(T-t)}K \cdot N[d_2],$$

where

$$d_1 = \frac{\ln\{S/K\} + (r + \sigma^2/2)(T-t)}{\sigma \cdot \sqrt{T-t}}, \quad d_2 = d_1 - \sigma \cdot \sqrt{T-t}.$$

4.7 Currency Options and the Garman–Kohlhagen Model

In 1983 Garman and Kohlhagen extended the Black–Scholes model to cope with the presence of two interest rates, one for each currency. These also called foreign exchange option or FX options.

Suppose that r_d is the risk-free interest rate to expiry of the domestic currency and $r_f$is the foreign currency risk-free interest rate where the domestic currency is the currency in which we obtain the value of the option. The formula also requires that FX rates—both strike and current spot—be quoted in terms of "units of domestic currency per unit of foreign currency".

We consider the model *geometric Brownian motion*:

$$dS_t = (r_d - r_f)S_t + \sigma S_t dW_t$$

for the underlying exchange rate quoted in FOR-DOM (foreign-domestic), which means that one unit of the foreign currency costs FOR-DOM units of the domestic currency. In the case of EUR-USD with a spot of 1.2000, this means that the price of one EUR is 1.2000 USD. The notion of *foreign* and *domestic* does not refer to the location of the trading entity, but only to this quotation convention. We denote the (continuous) foreign interest rate by r_f and the (continuous) domestic interest rate by r_d . In an equity scenario, r_f would represent a continuous dividend rate. The volatility is denoted by σ, and W_t is a standard Brownian motion.

Applying Itô's rule to ln S_t yields the following solution for the process S_t

$$S_t = S_t \exp\left\{\left(r_d - r_f - \frac{1}{2}^{\sigma 2}\right)t + \sigma \cdot W_t\right\}.$$

which shows that S_t is log-normally distributed, more precisely, ln S_t is normal with mean ln S_0 + $(r_d - r_f - ½\sigma^2)t$ and variance $\sigma^2 t$.

The payoff for a vanilla option (European put or call) is given by

$$\Phi = [\varepsilon(S_T - K)]^+,$$

where the contractual parameters are the strike K, the expiration time T and the type ε, a binary variable which takes the value +1 in the case of a call and −1 in the case of a put.

In the Black–Scholes model the value of the payoff F at time t it the spot x is denoted by $V(t, x)$ and can be computed either as the solution to the Black Scholes partial differential equation

$$\begin{aligned} &\frac{\partial V}{\partial t} + (r_d - r_f)x\frac{\partial V}{\partial x} + \frac{1}{2}\sigma^2 x^2 \frac{\partial^2 V}{\partial x^2} - r_d V = 0 \\ &V(T, x) = F. \end{aligned}$$

Or equivalently by the Feynman–Kač theorem as the discounted expected value of the payoff function

$$V(x, K, t, T, \sigma, r_d, r_f, \varepsilon) = e^{-r_d \cdot (T-t)} E[F|\mathcal{F}_t].$$

Then the domestic currency value of a call option into the foreign currency is

$$V_0 = \varepsilon \cdot e^{-r_d \cdot T_d}\{f \cdot N(\varepsilon \cdot d_1) - K \cdot N(\varepsilon \cdot d_2)\},$$

where

$$d_1 = \frac{\ln\{S_0/K\} + (r_d - r_f) \cdot T_d + \sigma^2 \cdot T_e/2}{\sigma \cdot \sqrt{T_e}}, \quad d_2 = d_1 - \sigma \cdot \sqrt{T_e}$$

and

f	the forward price of the underlying $= E[S_T \mid S_t = x] = x \exp\{(r_d - r_f)T_d\}$
x	spot FX rate denoted in domestic units per unit of foreign currency, i.e., the price of the underlying.
K	strike using the same quotation as the spot rate
T_e	time from today until expiry of the option
T_d	time from spot until delivery of the option
r_d	domestic interest rate corresponding with period T_d
r_f	foreign interest rate corresponding with period T_d
σ	volatility corresponding with strike K and period T_e
ε	1 for a call, −1 for a put
$N(.)$	cumulative normal distribution.

Hence V_0 is the value of the option expressed in domestic currency on a notional of one unit of foreign currency.

The *forward price f* is the strike, which makes the time zero value of the *forward contract*

$$F = S_T - f$$

equal to zero. The situation $r_d > r_f$ is called *contango*, and the situation $r_d < r_f$ is called *backwardation.*

The Black–Scholes delta also called spot delta of the option is equal to

$$\Delta_{BS} = \frac{\partial V}{\partial x} = \varepsilon \cdot e^{-r_f \cdot T_d} N(\varepsilon \cdot d_1).$$

The dual delta is defined by

$$\Delta_{BS}^{dual} = -\varepsilon \cdot e^{-r_d \cdot T_d} N(\varepsilon \cdot d_2).$$

In all currency markets, except the EuroDollar market, the premium in the foreign currency is included in the delta. This "premium-included" delta has to be calculated as follows

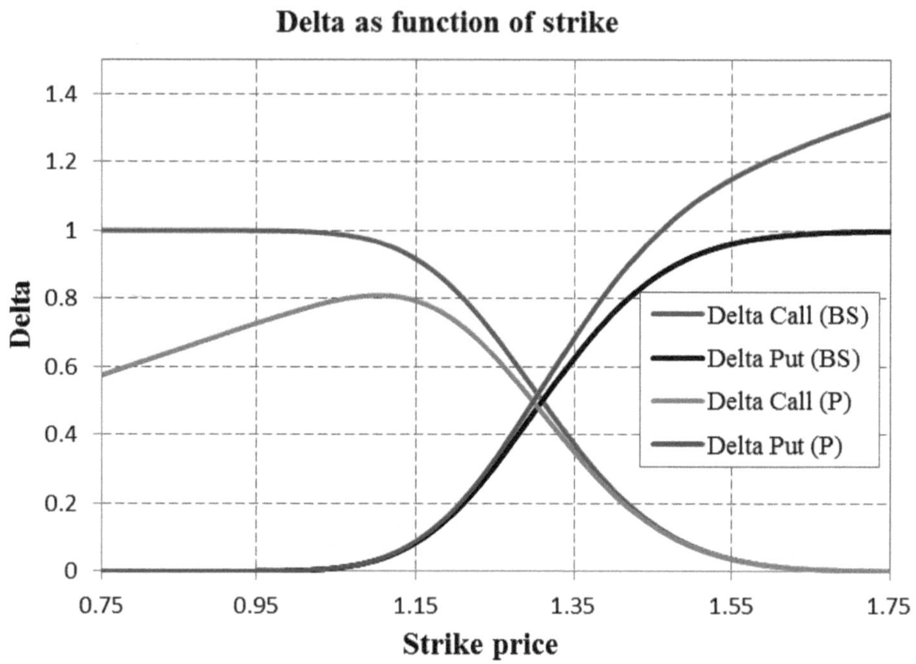

Fig. 4.5 Black–Scholes and premium-included delta as function of strike

$$\Delta_p = \Delta_{BS} - \frac{V}{x} = \varepsilon \cdot \frac{K}{x} e^{-r_d \cdot T_d} N(\varepsilon \cdot d_1)$$

The logic of this premium-included delta can be illustrated with an example. Consider a bank that sells a call on the foreign currency. This option can be delta hedged with an amount of delta of the foreign currency. However, the bank will only have to buy an amount equal to the premium-included delta when it receives the premium in foreign currency.

It can be observed from the above formula that the premium-included delta for a call is not strictly decreasing in strike like the Black–Scholes call delta. Therefore, a premium-included call delta can correspond to two possible strike prices (see the Fig. 4.5).

For emerging markets (EM) and for maturities of more than 2 years, it is usual for forward deltas to be quoted. These are defined as follows

$$\Delta^F{}_{BS} = e^{r_f \cdot T_d} \Delta_{BS} \quad \text{and}$$
$$\Delta^F{}_P = e^{r_f \cdot T_d} \Delta_P$$

The ATM strike refers to the strike of a zero delta straddle, that is, the strike for which the call delta is equal to the put delta. This strike can be calculated analytically.

4.7.1 Symmetry Relations

For FX options, the put call parity is given by

$$V(x,K,t,T,\sigma,r_d,r_f,+1) - V(x,K,t,T,\sigma,r_d,r_f,-1) \\ = x \cdot e^{-r_f(T-t)} - Ke^{-r_d(T-t)}.$$

We also have a put call delta parity given by

$$\frac{\partial V(x,K,t,T,\sigma,r_d,r_f,+1)}{\partial x} - \frac{\partial V(x,K,t,T,\sigma,r_d,r_f,-1)}{\partial x} = e^{-r_f(T-t)}.$$

In particular, we learn that the absolute value of a put delta and a call delta do not exactly add up to one, but only to a positive number $e^{-r_f(T-t)}$. They add up to one approximately if either the time to expiration $T-t$ is short or if the foreign interest rate r_f is close to zero.

Whereas the choice $K=f$ produces identical values for call and put, we seek the *delta-symmetric strike* K^* which produces absolutely identical deltas (spot, forward or driftless). This condition implies $d_1 = 0$ and thus

$$K^* = f \cdot e^{\frac{\sigma^2}{2}T}$$

in which case the absolute delta is $e^{-r_f(T-t)}/2$. In particular, we learn, that always $K^* > f$, i.e., there can't be a put and a call with identical values *and* deltas. Note that the strike K^* is usually chosen as the middle strike when trading a straddle or a butterfly. Similarly, the dual-delta-symmetric strike

$$K^* = f \cdot e^{\frac{\sigma^2}{2}T}$$

can be derived from the condition $d_2 = 0$.

If we wish to measure the value of the underlying in a different unit we can use an obviously effect the option pricing formula:

$$aV(x,K,t,T,\sigma,r_d,r_f,\varepsilon) = V(ax,aK,t,T,\sigma,r_d,r_f,\varepsilon); \quad a > 0$$

Differentiating both sides with respect to a and then setting $a = 1$ yields

$$V = x \cdot \frac{\partial V}{\partial x} + K \cdot \frac{\partial V}{\partial K}.$$

This *space-homogeneity* is the reason behind the simplicity of the delta formulas, whose tedious computation can be saved this way.

We can perform a similar computation for the time-affected parameters and obtain the obvious equation

$$V(x, K, t, T, \sigma, r_d, r_f, \varepsilon) = V\left(x, K, \frac{t}{a}, \frac{T}{a}, \sqrt{a}\sigma, ar_d, ar_f, \varepsilon\right); \quad a > 0.$$

Differentiating both sides with respect to a and then setting $a = 1$ yields

$$0 = (T - t) \cdot \frac{\partial V}{\partial t} + \frac{1}{2}\sigma \cdot \frac{\partial V}{\partial \sigma} + r_d \cdot \frac{\partial V}{\partial r_d} + r_f \cdot \frac{\partial V}{\partial r_f}.$$

By the *put call symmetry* we understand the relationship

$$V(x, K, t, T, \sigma, r_d, r_f, +1) = \frac{K}{f} V\left(x, \frac{f^2}{K}, t, T, \sigma, r_d, r_f, -1\right).$$

The strike of the put and the strike of the call result in a geometric mean equal to the forward f. The forward can be interpreted as a *geometric mirror* reflecting a call into a certain number of puts. Note that for ATM options ($K = f$) the put call symmetry coincides with the special case of the put call parity where the call and the put have the same value.

Direct computation shows that the *rates symmetry*

$$\frac{\partial V}{\partial r_d} + \frac{\partial V}{\partial r_f} = -(T - t) \cdot V$$

holds for vanilla options. This relationship, in fact, holds for all European options and a wide class of path-dependent options.

One can also directly verify the relationship the *foreign-domestic symmetry*

$$\frac{1}{x} \cdot V(x, K, t, T, \sigma, r_d, r_f, \varepsilon) = K \cdot V\left(\frac{1}{x}, \frac{1}{K}, t, T, \sigma, r_d, r_f, -\varepsilon\right).$$

This equality can be viewed as one of the faces of put call symmetry. The reason is that the value of an option can be computed both in a domestic as well as in a foreign scenario. We consider the example of St modelling the exchange rate of EUR/USD. In New York, the call option $(S_T - K)^+$ costs

$V(x, K, t, T, \sigma, r_{usd}, r_{eur}, 1)$ USD and hence $V(x, K, t, T, \sigma, r_{usd}, r_{eur}, 1)/x$ EUR. This EUR-call option can also be viewed as a USD-put option with payoff $K(1/K-1/S_T)^+$. This option costs $K\cdot V(1/x, 1/K, t, T, \sigma, r_{eur}, r_{usd}, -1)$ EUR in Frankfurt, because S_t and $1/S_t$ have the same volatility. Of course, the New York value and the Frankfurt value must agree, which leads to the equation above. This symmetry is just one possible result based on *change of numeraire*.

4.7.2 Volatility and Quotation

The quotation of FX options is a constantly confusing issue, so let us clarify this here. The exchange rate means how much of the *domestic* currency is needed to buy one unit of *foreign* currency. For example, if we take EUR/USD as an exchange rate, then the default quotation is EUR–USD, where USD is the domestic currency and EUR the foreign currency. The term *domestic* is in no way related to the location of the trader or any country. It merely means the *numeraire* currency. The terms *domestic*, *numeraire* or *base currency* are synonyms as are *foreign* and *underlying*.

EUR/USD can also be quoted in either EUR–USD, which then means how many USD are needed to buy one EUR, or in USD–EUR, which then means how many EUR are needed to buy one USD. There are certain market standard quotations listed in Table 4.2.

We call one million a *buck*, one billion a *yard*. This is because a billion is called "milliarde" in French, German and other languages. For the British pound one million is also often called a *quid*.

Certain currency pairs have names. For instance, GBP/USD is called *cable*, because the exchange rate information used to be sent through a cable in the Atlantic ocean between America and England. EUR/JPY is called the *cross*, because it is the cross rate of the more liquidly traded USD/JPY and EUR/USD.

Certain currencies also have names: for example, the New Zealand dollar (NZD) is called a *kiwi*, the Australian dollar (AUD) is called *Aussie*, and the Scandinavian currencies (DKR, NOK and SEK) are called *Scandies*.

Exchange rates are generally quoted up to five relevant figures, e.g. in EUR–USD we could observe a quote of 1.2375. The last digit '5' is called the *pip*, the middle digit '3' is called the *big figure*, as exchange rates are often displayed in trading floors and the big figure, which is displayed in bigger size, is the most relevant information. The digits left to the big figure are known anyway, the pips right of the big figure are often negligible. To make it clear, a rise of USD–JPY 108.25 by 20 pips will be 108.45 and a rise by 2 big figures will be 110.25.

Table 4.2 Quotes of exchange rates

Currency pair	Default quotation	Sample quote
GBP/USD	GPB-USD	1.8000
GBP/CHF	GBP-CHF	2.2500
EUR/USD	EUR-USD	1.2000
EUR/GBP	EUR-GBP	0.6900
EUR/JPY	EUR-JPY	135.00
EUR/CHF	EUR-CHF	1.5500
USD/JPY	USD-JPY	108.00
USD/CHF	USD-CHF	1.2800

4.7.3 Delta and Premium Convention

The spot delta of a European option without premium is well known. It will be called *raw spot delta* δ_{raw} now. It can be quoted in either of the two currencies involved. The relationship is

$$\delta_{raw}^{reverce} = -\delta_{raw} \cdot \frac{S}{K}.$$

The delta is used to buy or sell spot in the corresponding amount in order to hedge the option up to first order.

For consistency, the premium needs to be incorporated into the delta hedge, since a premium in foreign currency will already hedge part of the option's delta risk. To make this clear, let us consider EUR–USD. In the standard arbitrage theory, $V(x)$ denotes the value or premium in USD of an option with 1 EUR notional, if the spot is at x, and the raw delta V_x denotes the number of EUR to buy for the delta hedge. Therefore, xV_x is the number of USD to sell. If now the premium is paid in EUR rather than in USD, then we already have V_x EUR, and the number of EUR to buy has to be reduced by this amount, i.e. if EUR is the premium currency, we need to buy $V_x - V/x$ EUR for the delta hedge or equivalently sell $xV_x - V$ USD.

The entire FX quotation story becomes generally a mess, because we need to first sort out which currency is domestic, which is foreign, what the notional currency of the option is, and what is the premium currency. Unfortunately, this is not symmetrical, since the counterpart might have another notion of domestic currency for a given currency pair. Hence in the professional interbank market there is one notion of delta per currency pair. Normally it is the left hand side delta of the *Fenics* screen (http://www.gfigroup.com/gfifenics.aspx) if the option is traded in left hand side premium, which is normally the standard and right-hand side delta if it is traded with right-hand side premium, e.g. EUR/USD lhs, USD/JPY lhs, EUR/JPY lhs, AUD/USD rhs, etc. Since OTM options are traded most of time the difference is not huge and hence does not create a huge spot risk.

Additionally, the standard delta per currency pair [left hand side delta in *Fenics* for most cases] is used to quote options in volatility. This has to be specified by currency.

This standard interbank notion must be adapted to the real delta-risk of the bank for an automated trading system. For currencies where the risk–free currency of the bank is the base currency of the currency it is clear that the delta is the raw delta of the option and for risky premium this premium must be included. In the opposite case the risky premium and the market value must be taken into account for the base currency premium, so that these offset each other. And for premium in underlying currency of the contract the market value needs to be taken into account. In that way the delta hedge is invariant with respect to the risky currency notion of the bank—for example, the delta is the same for a USD-based bank and a EUR-based bank.

Examples 4.11

Consider two Examples in the Tables below to compare the various versions of deltas that are used in practice.

1y EUR call USD put strike $K = 0.9090$ for a EUR-based bank

Delta ccy	Prem ccy	Fenics	Formula	Delta
%EUR	EUR	lhs	$\delta_{now} - P$	44.72
%EUR	USR	rhs	$\delta_{now}-$	49.15
%USD	EUR	rhs [flip F4]	$-(\delta_{now} - P)S/K$	−44.72
%USD	USD	lhs [flip F4]	$-(\delta_{now})S/K$	−49.15

Market data: spot $S = 0.9090$, volatility $\sigma = 12$ %, EUR rate $r_f = 3.96$ %, USD rate $r_d = 3.57$ %. The raw delta is 49.15 % EUR and the value is 4.427 % EUR.

1y EUR call USD put strike $K = 0.7000$ for a EUR-based bank

Delta ccy	Prem ccy	Fenics	Formula	Delta
%EUR	EUR	lhs	$\delta_{now} - P$	72.94
%EUR	USD	rhs	$\delta_{now}-$	94.82
%USD	EUR	rhs [flip F4]	$-(\delta_{now} - P)S/K$	−94.72
%USD	USD	lhs [flip F4]	$-(\delta_{now})S/K$	−123.13

Market data: spot $S = 0.9090$, volatility $\sigma = 12$ %, EUR rate $r_f = 3.96$ %, USD rate $r_d = 3.57$ %. The raw delta is 94.82 % EUR and the value is 21.88 % EUR.

Volatility

The only unobserved input on the market is the volatility. We can also invert the relation and calculate which so-called implied volatility that should be used

Table 4.3 Example of volatility quotation

	Vols	25 Delta strangles	25 Delta Risk Revs
1W	8.08		
1M	8.18	0.19	0.59
3M	8.26	0.21	0.52
6M	8.38		
1Y	8.48	0.22	0.41
2Y	8.67		

to result in a certain price. If all Black–Scholes assumptions would hold, the implied volatility would be the same for all European vanilla options on a specific underlying FX rate. In reality we will find different implied volatilities for different strikes and maturities. In fact, all assumptions of the standard Black–Scholes model that do not hold express them in the so-called implied volatility surface. Thus, the Black–Scholes model effectively acts as a quotation convention.

In the Table 4.3 we give an example of how in the FX market implied volatilities are quoted.

We see above "Vols" the volatilities to be used for ATM options of various maturities. Furthermore, we encounter in this quotation strangles (STR) and risk reversals (RR). A strangle is a long position in an out-of-the-money (OTM) call and an OTM put. A strangle is a bet on a large move of the underlying either upwards or downwards. Note that where the ATM indicates the level of the smile, the STR can be regarded as a measure of the curvature or convexity of the volatility smile. A risk reversal is a combination of a long OTM call and a short OTM put. A RR can be seen as a measure of skewness, namely the slope of the smile. When RRs are positive, the market favours the foreign currency.

The implied volatilities correspond to 25-delta and ATM options. Delta is the sensitivity of the option to the spot FX rate and is always between 0 % and 100 % of the notional. It can be shown that an ATM option has a delta around 50 %. A 25-delta call (put) corresponds to an option with a strike above (below) the strike of an ATM option.

A 25-delta RR quote is the difference between the volatility of a 25-delta call and a 25-delta put. A 25-delta STR is equal to the average volatility of a 25-delta call and put minus the ATM volatility. Therefore, the volatility of a 25-delta call and put can be obtained from these quotes as follows:

$$\sigma_{C,25} = \sigma_{ATM} + STR_{25} + \frac{1}{2}RR_{25}$$
$$\sigma_{P,25} = \sigma_{ATM} + STR_{25} - \frac{1}{2}RR_{25}.$$

Usually quotes also exist for 10-delta RRs and STRs, although these options are not as liquid.

To derive the value for European vanilla options for other deltas one needs to interpolate between and extrapolate outside the available quotes. But inter- and extrapolation is also required for the derivation of prices for European style-derived products, such as European digitals. European digitals pay out a fixed amount if the spot at maturity ends above (or below) the strike and otherwise nothing.

Linear interpolation cannot be used for volatilities instead the following formula can be used:

$$\sigma(t) = \sigma(t_1) + \frac{\sqrt{t} - \sqrt{t_1}}{\sqrt{t_2} - \sqrt{t_t}}(\sigma(t_2) - \sigma(t_1)); \qquad t_1 \leq t \leq t_2$$

Another interpolation is the linear total variance method for the implied volatility. If we consider local volatility to be a function of time, $\varepsilon(t)$, then the implied volatility for time T is given by

$$\sigma^2(T) = \frac{1}{T}\int_0^T \varepsilon^2(t)dt$$

Let $T_0 < T < T_1$. Then

$$\sigma^2(T) = \frac{1}{T}\left[\int_0^{T_0} \varepsilon^2(t)dt + \int_{T_0}^{T} \varepsilon^2(t)dt\right] = \frac{1}{T}\left[T_0\sigma^2(T_0) + \int_{T_0}^{T} \varepsilon^2(t)dt\right].$$

The question that remains is how we should approximate the last term inside the parentheses in the above equation. A reasonable guess is that it should be set equal to

$$\left(\frac{T-T_0}{T_1-T_0}\right)\int_{T_0}^{T_1}\varepsilon^2(t)dt.$$

Intuitively, this is the proportion of the area under the local volatility curve from T_0 to T_1 that goes up to T. Therefore, we may write

$$\begin{aligned}\sigma^2(T) &= \frac{1}{T}\left[T_0\sigma^2(T_0)+\left(\frac{T-T_0}{T_1-T_0}\right)\int_{T_0}^{T_1}\varepsilon^2(t)dt\right]\\ &= \frac{1}{T}\left[T_0\sigma^2(T_0)+\left(\frac{T-T_0}{T_1-T_0}\right)\left(\int_0^{T_1}\varepsilon^2(t)dt-\int_0^{T_0}\varepsilon^2(t)dt\right)\right]\\ &= \frac{1}{T}\left[T_0\sigma^2(T_0)+\left(\frac{T-T_0}{T_1-T_0}\right)\left(T_1\cdot\sigma^2(T_1)-T_2\cdot\sigma^2(T_0)\right)\right]\end{aligned}$$

So

$$\sigma_t=\frac{1}{\sqrt{t}}\left[T_1\sigma_1^2+\left(\frac{t-T_1}{T_2-T_1}\right)\left(T_2\sigma_2^2-T_1\sigma_1^2\right)\right]^{1/2}.$$

4.7.4 Volatility in Terms of Delta

In the FX market implied volatilities are quoted in terms of delta. There are various definitions of delta. Hence, for the correct interpretation of the implied volatility quotes it is important to know what definition is used.

The mapping $\sigma\to\Delta=\varepsilon\exp\{-r_f(T-t)\}N(\varepsilon d_1)$ is not one-to-one. The two solutions are given by

$$\sigma_\pm=\frac{1}{\sqrt{T-t}}\left\{\varepsilon\cdot N^{-1}\left(\varepsilon\cdot\Delta\cdot e^{r_f(T-t)}\right)\pm\sqrt{\left[N^{-1}\left(\varepsilon\cdot\Delta\cdot e^{r_f(T-t)}\right)\right]^2-\sigma\sqrt{T-t}\cdot(d_1-d_2)}\right\}.$$

Thus using only the delta to retrieve the volatility of an option is not advisable.

The determination of the volatility and the delta for a given strike is an iterative process involving the determination of the delta for the option using ATM volatilities in a first step and then using the determined volatility to redetermine the delta and to continuously iterate the delta and volatility until the volatility does not change more than $= 0.001$ % between iterations. More precisely, one can perform the following algorithm. Let the given strike be K.

1. Choose $\sigma_0 =$ ATM volatility from the volatility matrix.
2. Calculate $\Delta_{n+1} = \Delta(\text{Call}(K, \sigma_n))$.
3. Take $\sigma_{n+1} = \sigma(\Delta_{n+1})$ from the volatility matrix, possibly via a suitable interpolation.
4. If $|\sigma_{n+1} - \sigma_n| < \varepsilon$, then quit, otherwise continue with step 2.

4.7.5 Options on Commodities

Commodity options are also handled in a similar way to FX options. Here, instead of a foreign interest rate, we have a cost of carry. In holding a commodity we have the carry of cost, *cc*, and the Black–Scholes is given by

$$\frac{\partial F}{\partial t} + (r + cc)S\frac{\partial F}{\partial S} + \frac{1}{2}\sigma^2 S^2 \frac{\partial^2 F}{\partial S^2} - rF = 0.$$

The cost of carry or carrying charge can be considered as a cost of storing the physical commodity, such as grain or metals, over a period of time. The carrying charge includes insurance, storage and interest on the invested funds as well as other incidental costs.

In the interest rate futures markets, it refers to the differential between the yield on a cash instrument and the cost of the funds necessary to buy the instrument.

For a long position, the cost of carry is the cost of interest paid on a margin account.

For a short position, the cost of carry is the cost of paying dividends, or rather the opportunity cost; the cost of purchasing a particular security rather than an alternative.

For most investments, the cost of carry generally refers to the risk-free interest rate that could be earned by investing currency in a theoretically safe investment vehicle such as a money-market account minus any future cash flows that are expected from holding an equivalent instrument with the same risk (generally expressed in percentage terms and called the convenience yield). Storage costs (generally expressed as a percentage of the spot price) should be added to the cost of carry for physical commodities such as corn, wheat, or gold.

4.7.6 Black–Scholes and Stochastic Volatility

Suppose that the volatility follows a stochastic process. This is a very realistic model, which can be seen by studying the dynamic of historical prices. With a stochastic volatility we will now try to derive a modified Black–Scholes differential equation. We start with the following model

$$\begin{aligned} dS &= \mu \cdot Sdt + \sigma \cdot SdW_1 \\ d\sigma &= p(S,\sigma,t)dt + q(S,\sigma,t)dW_2. \end{aligned}$$

As we see, we have two Wiener processes. We define a correlation so that $dW_1dW_2 = \rho dt$, where ρ is a measure of the correlation between the two processes.

The question we want to answer is the following: with known functions p and q, is it possible to create a risk-free portfolio? To get an answer we will try to hedge an option $C(S, t)$ in some portfolio. We can't hedge against σ since there is no offer on volatility on the market. For this reason we try to hedge against another option $\hat{C}(S, t)$ on the same underlying. We start with a portfolio Π

$$\Pi = C - \Delta S - \widehat{\Delta}\widehat{C}$$

and use the Itô formula

$$\begin{aligned} dC &= \frac{\partial C}{\partial t}dt + \frac{\partial C}{\partial S}dS + \frac{1}{2}\frac{\partial^2 C}{\partial S^2}dS^2 + \frac{\partial C}{\partial \sigma}d\sigma + \frac{1}{2}\frac{\partial^2 C}{\partial \sigma^2}d\sigma^2 + \frac{\partial^2 C}{\partial S \partial \sigma}dSd\sigma \\ &= C_t dt + \mu S C_s dt + \sigma S C_s dW_1 + \frac{1}{2}\sigma^2 S^2 C_{ss} dt + pC_\sigma dt \\ &\quad + qC_\sigma dW_2 + \frac{1}{2}q^2 C_{\sigma\sigma} dt + \sigma p q C_{s\sigma} dt \\ &= \left\{ C_t + \mu S C_s + \frac{1}{2}\sigma^2 S^2 C_{ss} + pC_\sigma + \frac{1}{2}q^2 C_{\sigma\sigma} + \sigma \rho q C_{s\sigma} \right\} dt \\ &\quad + \sigma S C_s dW_1 + qC_\sigma dW_2. \end{aligned}$$

From out stochastic differential equations we have to the lowest order

$$\begin{aligned} dS^2 &= \sigma^2 S^2 dt \\ d\sigma^2 &= q^2 dt \\ dSd\sigma &= \sigma \rho q S dt. \end{aligned}$$

If we substitute these into the expression for dC and a similar for $d\hat{C}$ we will find, via

$$d\Pi = dC - \Delta dS - \widehat{\Delta} d\widehat{C}$$

to be able to eliminate the randomness from the Wiener processes we have to make the choice

$$\sigma SC_s dW_1 + qC_\sigma dW_2 - \Delta\sigma S dW_1 - \widehat{\Delta}\sigma S\widehat{C}_s dW_1 + \widehat{\Delta} q\widehat{C}_\sigma dW_2 = 0$$

i.e.,

$$\begin{aligned} &\widehat{\Delta}\widehat{C}_\sigma = C_\sigma \\ &C_S - \Delta - \widehat{\Delta}\widehat{C}_S = 0. \qquad (*) \end{aligned}$$

With use of the arbitrage condition on Π and to equalize this investment with an investment in the risk-free interest rate r, we found

$$\begin{aligned} d\Pi &= r\Pi dt = r\left(C - \Delta S - \widehat{\Delta}\widehat{C}\right)dt = dC - \Delta dS - \widehat{\Delta} d\widehat{C} \\ &= dC - \Delta rSdt - \widehat{\Delta} d\widehat{C} \\ &\Rightarrow r\left(C - \widehat{\Delta}\widehat{C}\right)dt = dC - \widehat{\Delta} d\widehat{C} \\ &\Rightarrow r\left(C - \widehat{\Delta}\widehat{C}\right) \\ &= C_t + \mu SC_s + \frac{1}{2}\sigma^2 S^2 C_{ss} + pC_\sigma + \frac{1}{2}q^2 C_{\sigma\sigma} + \sigma\rho q C_{s\sigma} \\ &\qquad - \widehat{\Delta}\left(\widehat{C}_t + \mu S\widehat{C}_s + \frac{1}{2}\sigma^2 S^2 \widehat{C}_{ss} + p\widehat{C}_\sigma + \frac{1}{2}q^2\widehat{C}_{\sigma\sigma} + \sigma\rho q\widehat{C}_{s\sigma}\right). \end{aligned}$$

Rearranging and using the expressions (*) we can eliminate the terms C_σ and C_S, then

$$\begin{aligned} &C_t + \frac{1}{2}\sigma^2 S^2 C_{ss} + \frac{1}{2}q^2 C_{\sigma\sigma} + \sigma\rho q C_{s\sigma} - rC \\ &\quad = \widehat{\Delta}\left(\widehat{C}_t + \frac{1}{2}\sigma^2 S^2\widehat{C}_{ss} + \frac{1}{2}q^2\widehat{C}_{\sigma\sigma} + \sigma\rho q\widehat{C}_{s\sigma} - r\widehat{C}\right). \end{aligned}$$

This is a risk-neutral partial differential equation with derivatives on C and $\hat{C}$. If we define a differential operator D such as

$$Df = \frac{1}{f_\sigma}\left(f_t + \frac{\sigma^2 S^2 f_{SS}}{2} + \rho\sigma q S f_{S\sigma} + \frac{q^2 f_{\sigma\sigma}}{2} - rf\right)$$

and use $\widehat{\Delta}\widehat{C}_\sigma = C_\sigma$ we can express this by $DC = \hat{DC}$. This means, that we only have derivatives on each of the option on each side of the equation. This means, since the options can have different strike prices and different times to maturity, the equation is independent of the contracts. We can therefore put this equal to a function in the independent variables S, σ and t. Finally, for some arbitrary function $\lambda(S, \sigma, t)$ we have

$$DC = -(p - \lambda q).$$

Here $p{-}\lambda q$ is called the risk-neutral drift of the volatility and the function λ is called the market price of volatility.

4.7.7 The Black–Scholes Formulas

The Black–Scholes model was first developed for European options on non-dividend paying stocks. The model has subsequently been extended to cope with American options and other underlings. The basic assumptions remain the same but the valuation methodology gets more complicated.

The Black–Scholes model is widely used also for the pricing of option elements in interest rate OTC instruments. Clearly, some of the basic assumptions are highly unrealistic, and have to be modified. These modifications will be described for the instruments where the Black–Scholes model can be used.

The basic assumptions in the Black–Scholes world are:

- The underlying is a log-normally distributed stochastic variable
- The volatility of the underlying is constant
- Interest rates are constant
- There are no transaction costs in any capital markets
- Borrowing and lending can be done at constant interest rate
- There is continuous trading in all instruments.

The most important unobservable parameter in the Black–Scholes model (and in other option models) is the volatility. If it is possible to make a good estimation of the volatility, the model can be used for almost all types of options. The problem is to relate the volatility given for one type of instrument or maturity to other instruments and maturities.

When pricing bond options, the volatility for the underlying bond must be given and when pricing caps, the volatility for the forward is needed.

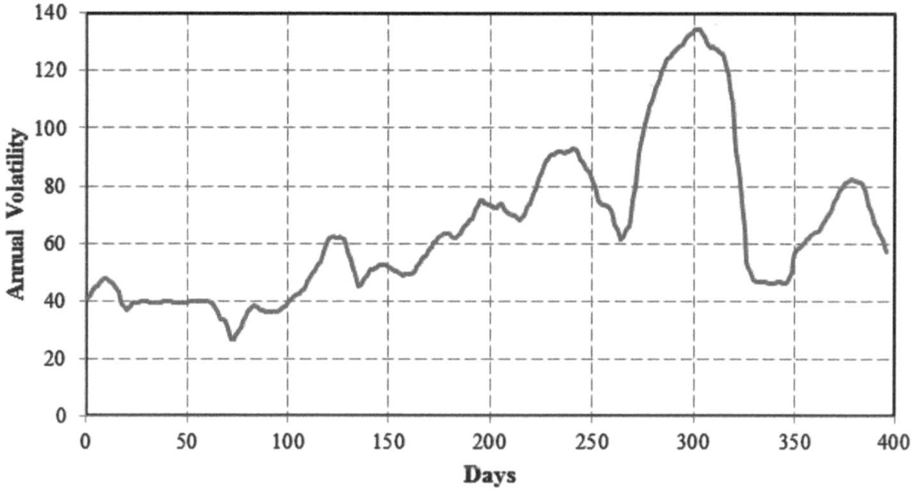

Fig. 4.6 The 3-month volatility of the Ericsson stock for a period of 400 days

To see the difficulty to estimate the volatility we plot in the Fig. 4.6 the 3-month volatility for the Ericsson stock for a time period 400 days. The data are taken from Nasdaq 2015-10-28.

The graph shows the 3-month volatility as function of time. This volatility should be an estimate of the volatility for an option with 3 months to maturity. As we can see, the value can be any between 50 % and 95 %. It can increase and decrease very fast.

A general formulation of the Black–Scholes formula for European options can then be written as

$$P_{call} = Se^{-qT}N(d_1) - Ke^{-rT}N(d_2)$$
$$P_{put} = Ke^{-rT}N(-d_2) - Se^{-qT}N(-d_1)$$

where

$$d_1 = \frac{\ln\left(\frac{S}{K}\right) + \left(r - q + \frac{\sigma^2}{2}\right)T}{\sigma \cdot \sqrt{T}}, \qquad d_2 = d_1 - \sigma \cdot \sqrt{T}.$$

P_{call} = The value of a call option
P_{put} = The value of a put option
S = The price of the underlying security
K = The strike price

r = The risk-free interest rate (typically a Treasury bond with the same maturity)
q = The dividend yield [%],
T = The time to maturity
σ = The volatility

$$N(x) = \frac{1}{\sqrt{2 \cdot \pi}} \int_{-\infty}^{x} e^{-y^2/2} dy$$
$$N'(x) = \frac{1}{\sqrt{2 \cdot \pi}} e^{-y^2/2}.$$

Mostly, $q = 0$. A simple approximation (6-digit accuracy) of the normal distribution is given by

$$N(x) = 1 - N'(x)\left(a_1 y + a_2 y^2 + a_3 y^3 + a_4 y^4 + a_5 y^5\right); \quad \text{för} \, x \geq 0$$
$$N(x) = 1 - N(-x); \quad \text{för} \, x < 0$$

where

$$y = \frac{1}{1 + \gamma x}$$
$$\gamma = 0.2316419$$
$$a_1 = 0.319381530$$
$$a_2 = -0.356563782$$
$$a_3 = 1.781477937$$
$$a_4 = -1.821255978$$
$$a_5 = 1.330274429.$$

Generally, the Black–Scholes formula for a European call options can be written as

$$P_{call} = \{S - PV(D)\} \cdot N(x_1) - PV(K) \cdot N(x_2),$$

where

$$x_1 = \frac{\ln\left(\frac{S - PV(D)}{PV(K)}\right)}{\sigma \cdot \sqrt{T}} + \frac{1}{2}\sigma\sqrt{T}, \qquad x_2 = \frac{\ln\left(\frac{S - PV(D)}{PV(K)}\right)}{\sigma \cdot \sqrt{T}} - \frac{1}{2}\sigma\sqrt{T}$$

where PV is the present value and D the dividends of the stock prior to maturity. If we think in terms of risk neutrality, then we can write the value of a call option as

$$\begin{aligned} P_{call} &= PV(P_{call}) = PV(\{E[S_T|S_T > K] - K\} \cdot P(S_T > K)) \\ &= e^{-rT}(\{E[S_T|S_T > K] - K\} \cdot P(S_T > K)). \end{aligned}$$

Or, in other words, the value of a call option is its present value, which is the expected stock price at maturity conditioned that the stock price is above the strike price, minus the strike and times the probability that the stock price at maturity is above the strike.

Another way to express the Black–Scholes formula for a call option is as

$$P_{call} = e^{-rT}\left[Se^{(r-q)T}N(d_1) - KN(d_2)\right].$$

In terms of the Black–Scholes formula, $N(d_2)$ can be interpreted as the probability that the call option will be in the money at maturity (Prob$(S_T > K)$). We also observe that $S \cdot e^{(r-q)T}$ is the expected future price of the underlying, which is the same as for a future in the underlying. This is the no arbitrage condition so the term $S \cdot e^{(r-q)T} \cdot N(d_1)$ is the value of the expected terminal stock price conditional upon the call option being in the money at expiration times the probability that the call will be in the money at expiration. The term $KN(d_2)$ is the value of the cost of exercising the option at expiration, times the probability that the call will be in the money at expiration. Finally, we have the factor e^{-rT}, which discounts the values to a present value. So the Black–Scholes formula for call option has a fairly simple interpretation. The call price is simply the discounted expected value of the cash flows at expiration.

If $q = 0$ we can write d_1 as

$$d_1 = \frac{\ln\left(\frac{S}{K}\right) + \left(r + \frac{\sigma^2}{2}\right)T}{\sigma \cdot \sqrt{T}} = \frac{\ln(S/Ke^{-rT})}{\sigma \cdot \sqrt{T}} + \frac{1}{2}\sigma\sqrt{T}.$$

Here S/Ke^{-rT} is a measure of the *moneyness* of the option, that is, the distance between the exercise price and the stock price and $\sigma\sqrt{T}$ the time adjusted volatility, that is, the volatility of the return on the underlying asset between now and maturity.

In Fig. 4.7 we show the relationship between the call value and the spot price where the initial stock price and the strike is 100, the discount rate equal 2 %, the volatility 40 % and time to maturity 6 months. It is important to notice the time value in relation to the intrinsic value. For the call option the time value is always greater than zero, but for a put option the time value can be greater than zero or less than zero.

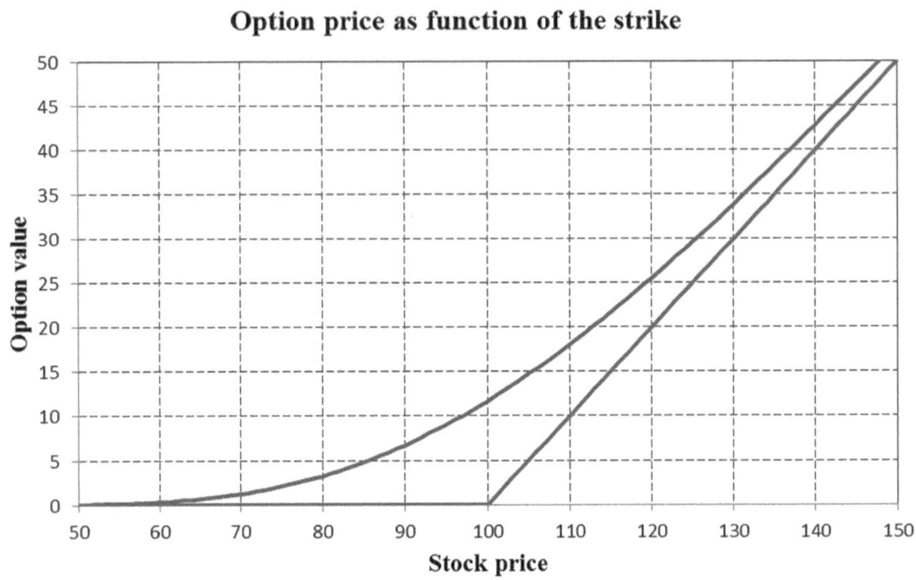

Fig. 4.7 The call option price as function of the underlying stock price

In Fig. 4.8 we show the same relationship as in Fig. 4.7 for three different maturities, 0.50, 0.25 and 0.10 years. As we see, for shorter maturities we get closer to the intrinsic value with is represented as the "hockey-stick".

In Fig. 4.9 we show the same relationship as in Fig. 4.7 for three different maturities, 0.50, 0.25 and 0.10 years. As we see, for shorter maturities we get closer to the intrinsic value with is represented as the "hockey-stick".

4.7.8 Digital Options

For digital options the Black–Scholes formulas can be simplified

$$
\begin{aligned}
P_{call} &= e^{-rT} K N(d) \\
P_{put} &= e^{-rT} K N(-d)
\end{aligned}
$$

where

$$
d = \frac{\ln\left(\frac{S}{K}\right) + \left(r + \frac{\sigma^2}{2}\right)T}{\sigma \cdot \sqrt{T}}.
$$

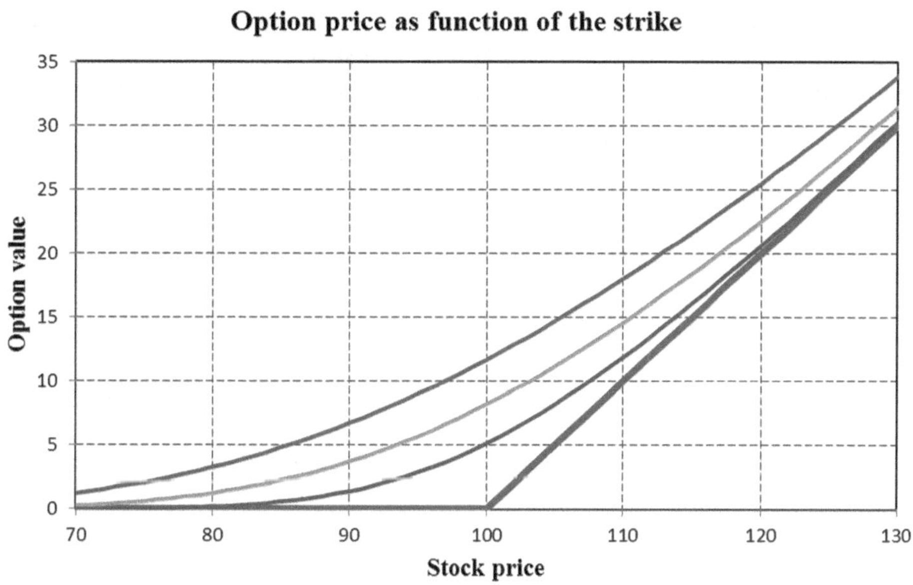

Fig. 4.8 The call option price as function of the underlying stock price for time to maturity 0.50, 0.25 and 0.10 year

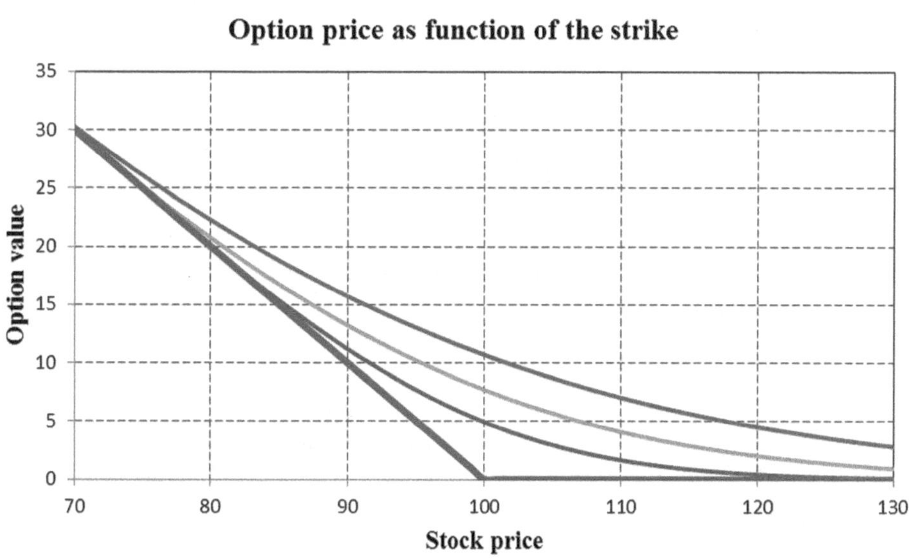

Fig. 4.9 The put option price as function of the underlying stock price for time to maturity 0.50, 0.25 and 0.10 year

Digital options, (sometimes called binary options) as these above is called cash-or-nothing since they pay a given amount, the strike, if the underlying price reach a certain level. Another digital option is the asset-or-nothing where the price is given by

$$P_{call} = SN(d)$$

and

$$P_{put} = SN(-d).$$

As we see, an asset-or-nothing in combination of a cash-or-nothing are the two terms of the Black–Scholes formula.

There also exist American types of digitals. Reiner and Rubinstein derived formula for such options in 1991.

4.7.9 Black-76 and Options on Forwards and Futures

For options on forwards and futures the Black–Scholes formula is reduced to the Black-76 formula

$$\begin{aligned} P_{call} &= e^{-r(T-t)}(F \cdot N(d_1) - K \cdot N(d_2)) \\ P_{put} &= e^{-r(T-t)}(K \cdot N(-d_2) - F \cdot N(-d_1)) \end{aligned}$$

where

$$d_2 = \frac{\ln(F/K) - (\sigma^2/2)T}{\sigma\sqrt{T}} = d_1 - \sigma\sqrt{T}.c$$

This is derived directly from Black–Scholes model with

$$F = e^{r(T-t)} \cdot S$$

As we can see here, since the only terms including the rate is the discounting, an American put option on a Forward/Future is never optimal to exercise. The Black formula is also used fir interest rates derivatives. Therefore, we see Black's formula as an extension to Black–Scholes.

4.7.10 The Hedge Parameters

The hedge parameters, or the Greeks, measure the sensitivities of the option prices with respect to the dependent variables. These describes the change in the option value if any of the variables S, T, r or σ is changing when all the others remain the same. The hedge parameters are defined by the partial derivatives

$$\Delta = \frac{\partial P}{\partial S}, \quad \Gamma = \frac{\partial^2 P}{\partial S^2}, \quad \Theta = \frac{\partial P}{\partial T}, \quad \upsilon = \frac{\partial P}{\partial \sigma} \text{ and } \rho = \frac{\partial P}{\partial r}.$$

To hedge a holding of the underlying we use the value of Δ (delta hedge), to calculate the optimal number of options (or vice versa). The Black–Scholes Δ is given by

$$\begin{array}{rl} \Delta = N(d_1) & \text{for a call option} \\ \Delta = [N(d_1) - 1] & \text{for a put option} \end{array}.$$

As we can see, Δ for call options is in the interval [0, 1] and for put options is negative, [−1, 0]. If $\Delta = \pm 1/2$ the options are ATM.

4.7.11 Some Graphs

In Fig. 4.10a we show the values of call- and put options as function of the underlying stock price. In Fig. 4.11b we show how the values of the underlying have to grow to keep the value if the call options until maturity. We will now also show how the Greeks vary in the same situations. We use as before, the values $K = S = 100$, $T = 0.5$ years, $r = 2$ % and $\sigma = 40$ %.

Delta

Delta, measures the sensitivity to changes in the option price with respect to the underlying stock price. This can also be represented by the derivative of the option price with respect to the underlying stock price. As we have seen, delta is given by $\Delta = N(d_1)$ for a call- and $\Delta = [N(d_1) - 1]$ for a put option. In Figs. 4.12 and 4.13 we show the relationship between delta for a call and a put option respectively. The initial stock price and strike is 100, the discount rate is 2 %, the volatility 40 %. The time to maturity are shown as 6 months,

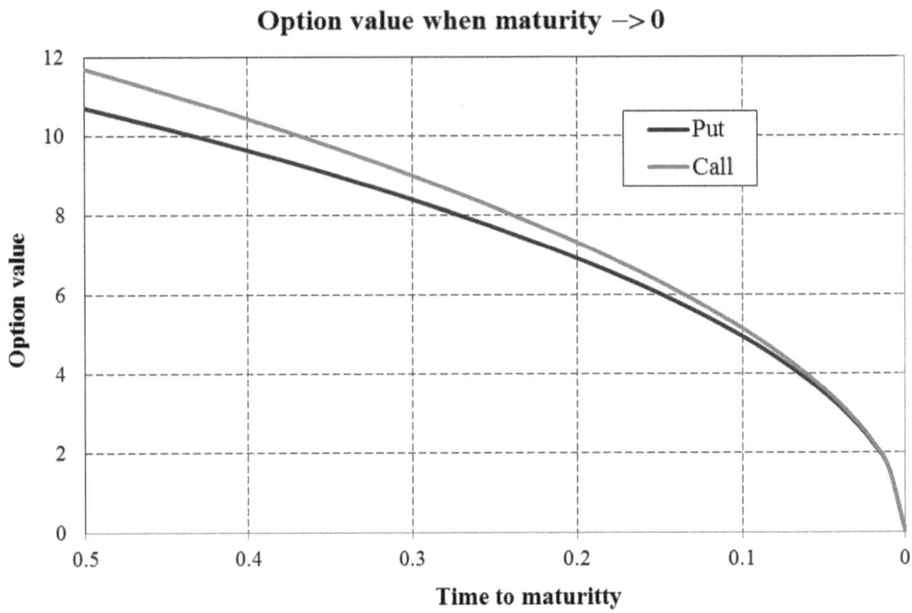

Fig. 4.10 Here we see how the time value goes to zero for a call and a put option when time goes to maturity

1 month and 1day. For the call option delta is in the interval [0, 1] and for the put [−1, 0].

Gamma

Gamma, measures the sensitivity to changes in the option price with respect to delta. This can also be represented by the second order derivative of the option price with respect to the underlying stock price. As we have seen, gamma is given by

$$\Gamma = \frac{e^{-d_1^2/2}}{S \cdot \sigma \cdot \sqrt{2 \cdot \pi \cdot (T - t)}}$$

Sometime, a Greek speed is used to measure the third order sensitivity to price. The speed is the third derivative of the value function with respect to the underlying price

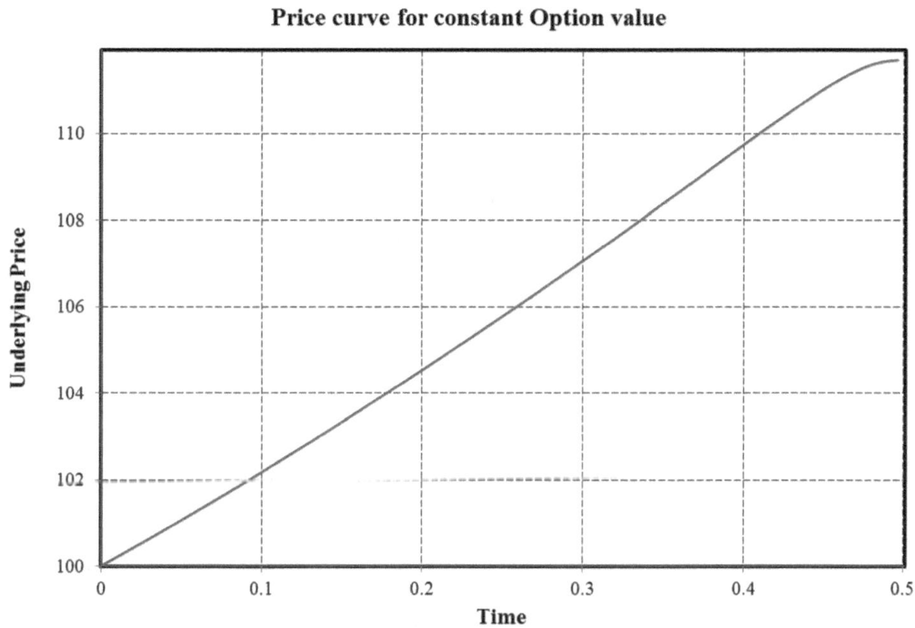

Fig. 4.11 Here we see how the value of the underlying stock must change in time to keep the value of the call option constant when time goes to maturity

$$\gamma = \frac{\partial^3 P}{\partial S^3}.$$

The change in price, $\Delta\Pi$ in a Δ-neutral portfolio is given by

$$\Delta\Pi = \Theta \cdot \Delta t + \frac{1}{2}\Gamma(\Delta S)^2.$$

In Fig. 4.14 we see gamma for three different times to maturities, 6, 3 and 1 months.

Theta

Theta Θ, measure the sensitivity to the passage of time, and is given by the derivative of the option value with respect to the amount of time to expiry, and is given by

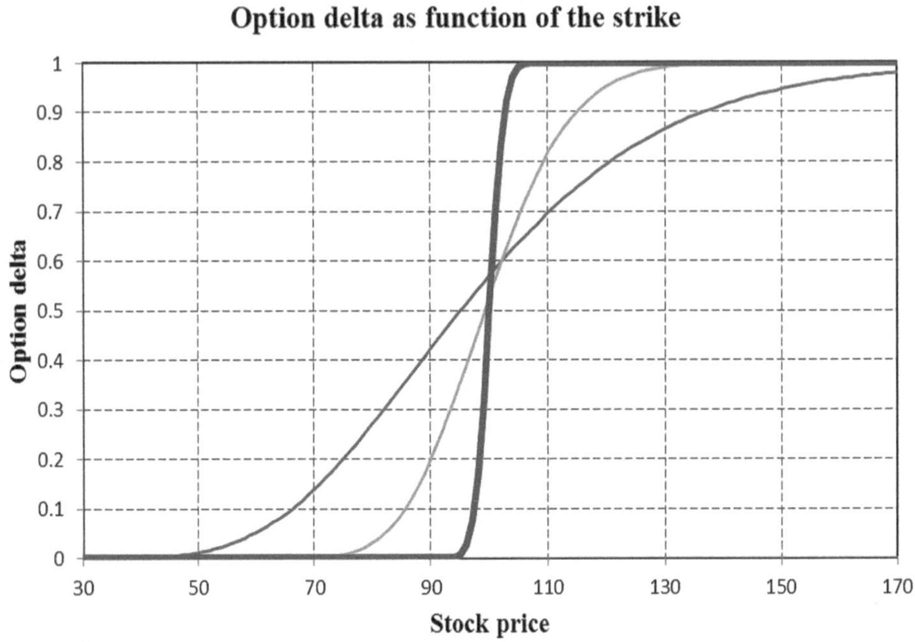

Fig. 4.12 Delta for a call option price as function of the underlying stock price for time to maturity 6 months, 1 month and 1 day. The fat line represents the option with maturity in one day. We observe that delta converge to a Heaviside step function near maturity

$$\Theta_{\text{call}} = -\frac{S \cdot e^{-d_1^2/2} \cdot \sigma \cdot e^{-q(T-t)}}{2 \cdot \sqrt{2 \cdot \pi \cdot (T-t)}} + q \cdot S \cdot N(d_1) \cdot e^{-q(T-t)} - r \cdot X \cdot N(d_2) \cdot e^{-r(T-t)}$$

$$\Theta_{put} = -\frac{S \cdot e^{-d_1^2/2} \cdot \sigma \cdot e^{-q(T-t)}}{2 \cdot \sqrt{2 \cdot \pi \cdot (T-t)}} - q \cdot S \cdot N(-d_1) \cdot e^{-q(T-t)} + r \cdot X \cdot N(-d_2) \cdot e^{-r(T-t)}.$$

In Figs. 4.15 and 4.16 we show theta for a call- and a put option respectively as function of the underlying stock price for three different maturities, 6, 3 and 1 months. We use $S = K = 100$, $r = 0.02$ % and volatility 40 %.

Vega

Vega ν, which is not a Greek letter (ν, *nu* is used instead), measure the sensitivity to volatility. The vega is the derivative of the option value with respect to the volatility of the underlying. The term *kappa*, κ, is sometimes

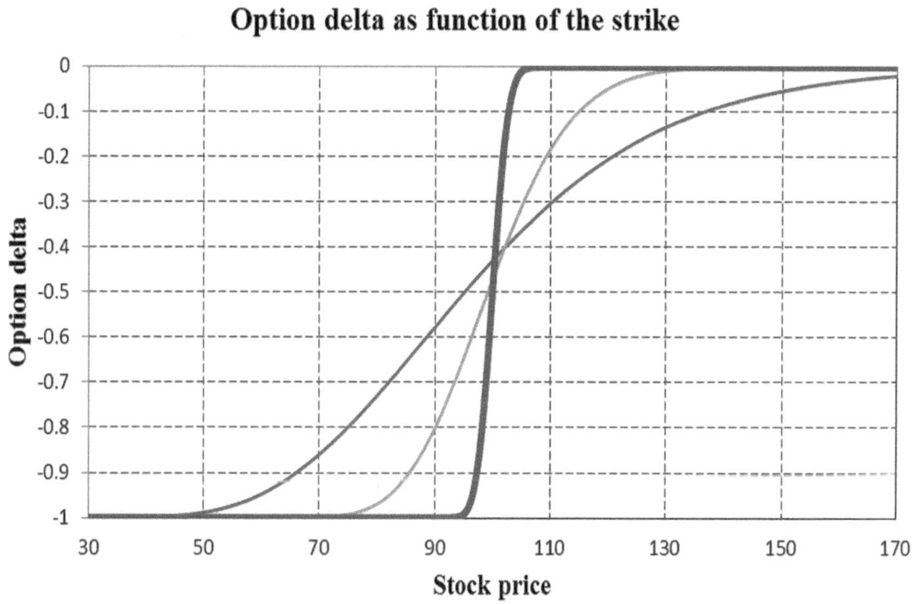

Fig. 4.13 Delta for a put option price as function of the underlying stock price for time to maturity 6 months, 1 month and 1 day. The fat line represents the option with maturity in 1 day

used instead of *vega*, and some trading firms have also used the term *tau*, τ is given by:

$$\nu = S\sqrt{\frac{T-t}{2\cdot\pi}}\cdot e^{-d_1^2/2}\cdot e^{-q\cdot(T-t)}$$

In Fig. 4.17 we show vega as function of the underlying stock price for three different maturities, 6, 3 and 1 months. We use $S = K = 100$, $r = 0.02$ % and volatility 40 %.

Rho

Rho ρ, measure the sensitivity to the applicable interest rate. The ρ is the derivative of the option value with respect to the risk free rate is given by:

$$\rho_{call} = (T-t)\cdot X\cdot e^{-r\cdot(T-t)}N(d_2)$$
$$\rho_{put} = -(T-t)\cdot X\cdot e^{-r\cdot(T-t)}N(-d_2)\text{.}$$

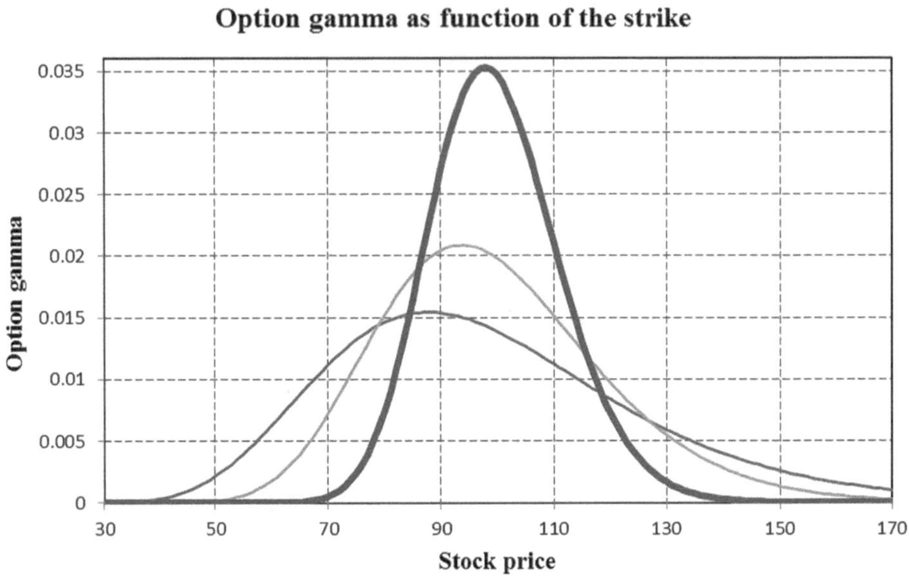

Fig. 4.14 Gamma as function of the underlying stock price for time to maturity 6, 3 and 1 months. The fat line represents the option with maturity in 1 month. We observe that Gamma tends to concentrate near maturity

In Figs. 4.18 and 4.19 we show rho for a call and a put option respectively as function of the underlying stock price for three different maturities, 6, 3 and 1 months. We use $S = K = 100$, $r = 0.02$ % and volatility 40 %.

Other Greeks

There are also some less commonly used Greeks. *Lambda* λ is the percentage change in option value per change in the underlying price, or the logarithmic derivative

$$\lambda = \frac{1}{P}\frac{\partial P}{\partial S}.$$

The *vega gamma* or *volga* measures second order sensitivity to implied volatility. This is the second derivative of the option value with respect to the volatility of the underlying

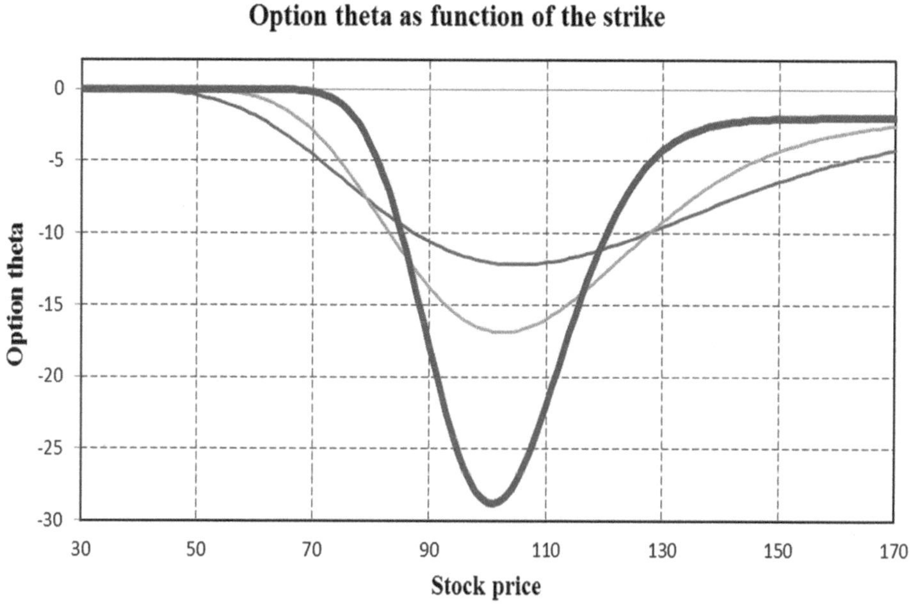

Fig. 4.15 Theta for a call option as function of the underlying stock price for time to maturity 6, 3 and 1 months. The fat line represents the option with maturity in 1 month

$$\vartheta = \frac{\partial^2 P}{\partial \sigma^2} = S \cdot N'(d_1)\frac{d_1 d_2}{\sigma} = \nu \frac{d_1 d_2}{\sigma}.$$

Vanna measures the cross-sensitivity of the option value with respect to change in the underlying price and the volatility

$$\frac{\partial^2 P}{\partial S \partial \sigma} = -N'(d_1)\frac{d_2}{\sigma} = \frac{\nu}{S}\left[1 - \frac{d_1}{\sigma\sqrt{T-t}}\right].$$

Vanna can also be interpreted as the sensitivity of delta to a unit change in volatility. The *delta decay*, or *charm*, measures the time decay of delta

$$\frac{\partial \Delta}{\partial T} = \frac{\partial^2 V}{\partial S \partial T}.$$

This can be important when hedging a position over a weekend. For a call option the charm is given by

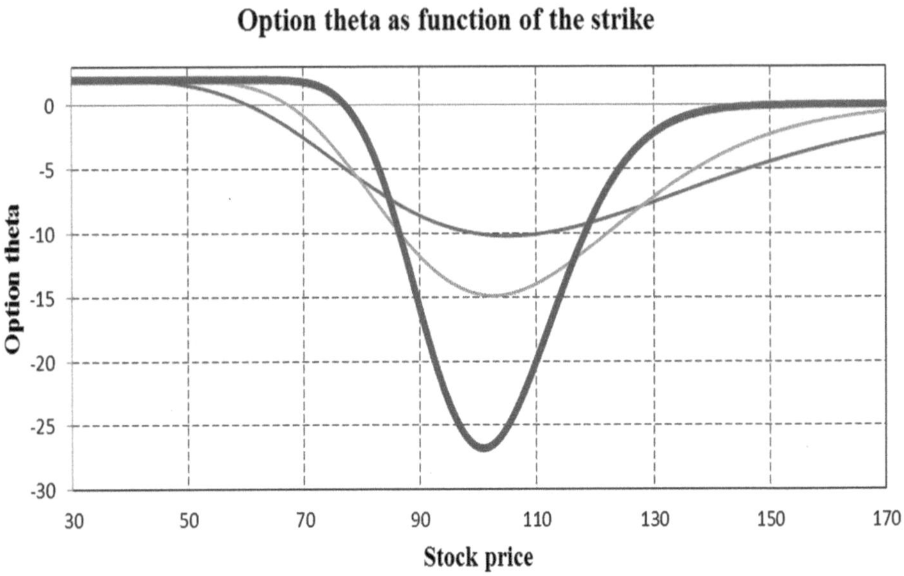

Fig. 4.16 Theta for a put option as function of the underlying stock price for time to maturity 6, 3 and 1 months. The fat line represents the option with maturity in 1 month

$$\frac{\partial \Delta_C}{\partial T} = N'(d_1)\frac{2 \cdot r \cdot t - d_2 \cdot \sigma \cdot \sqrt{T-t}}{2 \cdot (T-t) \cdot \sigma \cdot \sqrt{T-t}}.$$

For a put option, the charm is given by

$$\frac{\partial \Delta_P}{\partial T} = -N'(d_1)\frac{2 \cdot r \cdot t - d_2 \cdot \sigma \cdot \sqrt{T-t}}{2 \cdot (T-t) \cdot \sigma \cdot \sqrt{T-t}}.$$

The *colour* measures the sensitivity of the *charm*, or *delta decay* to the underlying asset price. It is the third derivative of the option value, twice to underlying asset price and once to time

$$\frac{\partial^3 V}{\partial S^2 \partial T} = \frac{N'(d_1)}{2S \cdot (T-t) \cdot \sigma\sqrt{T-t}}\left[1 - \frac{2r \cdot (T-t) - d_2 \cdot \sigma\sqrt{T-t}}{2(T-t) \cdot \sigma\sqrt{T-t}}\right]$$

For further information about Greeks, see Espen Garder Haug (1997).

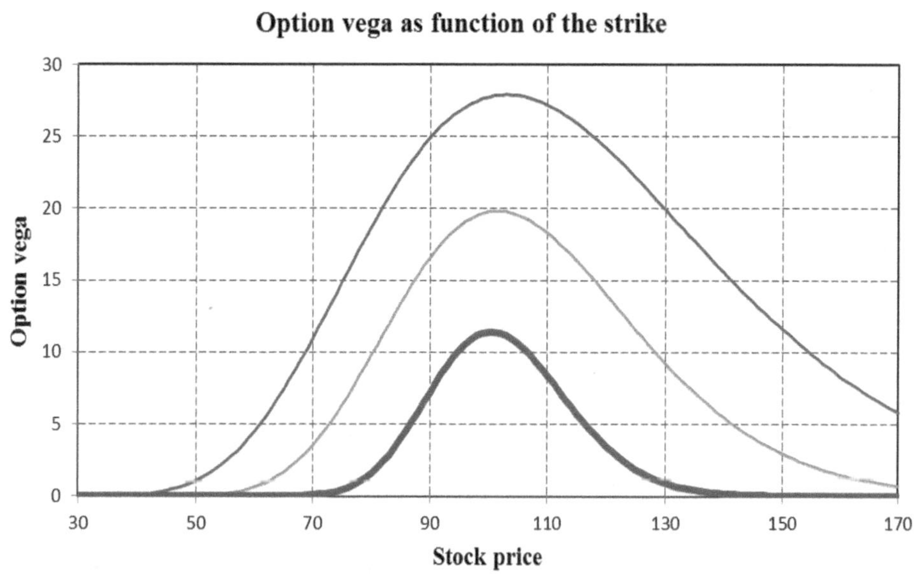

Fig. 4.17 Vega as function of the underlying stock price for time to maturity 6, 3 and 1 months. The fat line represents the option with maturity in 1 month

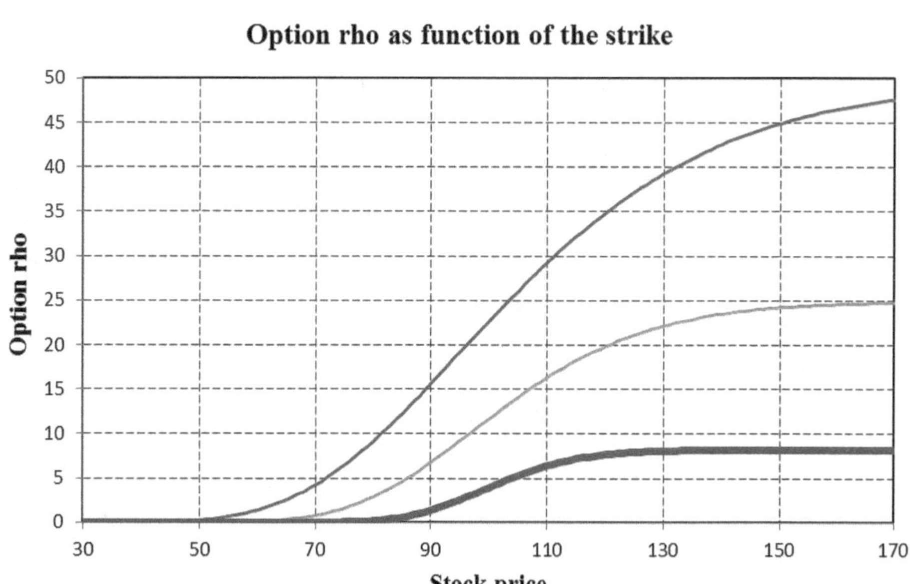

Fig. 4.18 Rho for a call option as function of the underlying stock price for time to maturity 6, 3 and 1 months. The fat line represents the option with maturity in 1 month

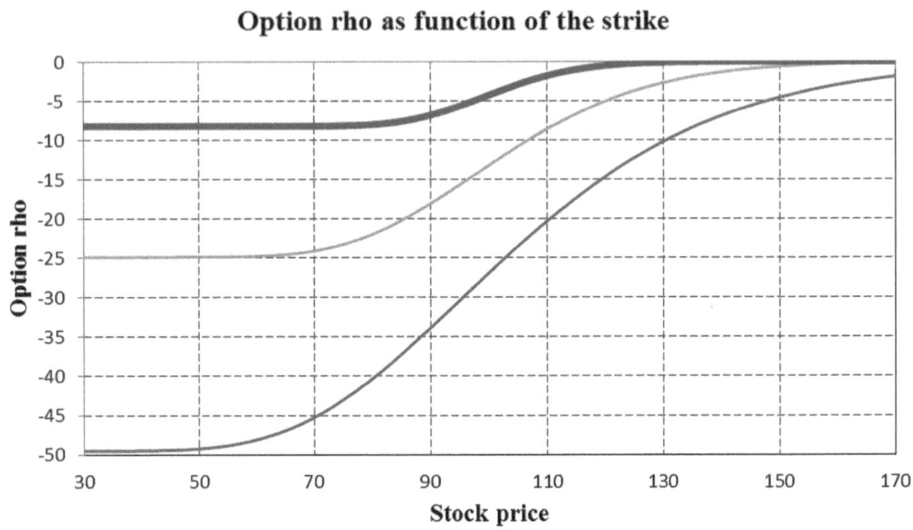

Fig. 4.19 Rho for a put option as function of the underlying stock price for time to maturity 6, 3 and 1 months. The fat line represents the option with maturity in 1 month

4.7.12 American Versus European Options

As we have seen, we have been able to solve the Black–Scholes PDE for European options. One may ask why we can't find a solution for American contracts as well. The answer is hidden in the boundary condition, which is not well defined at maturity. Since we are allowed to exercise the American option any time during the option lifetime, we have a floating boundary condition. For the same reason we cannot find a put call parity for an American option. There exist a number of approximations, but any general closed form solution does not exist.

The possibility of early exercise gives the American option a higher price than a European. This is called the *early exercise premium*. The difference in price can be seen in Fig. 4.20 where we compare an American put option with a European, where $S = 100$, $K = 110$, $T = 2$ years, $r = 0.02$ % and $\sigma = 40$ %.

Notice that the cost for the American put is more than its European counterpart when in-the-money, and the two curves tend to fall on top of one another when out-of-the-money. This is because the early exercise premium tends to zero the more out-of-the-money it gets since the option is unlikely to be exercised early. Therefore, a deep out-of-the-money American and put option have almost the same value.

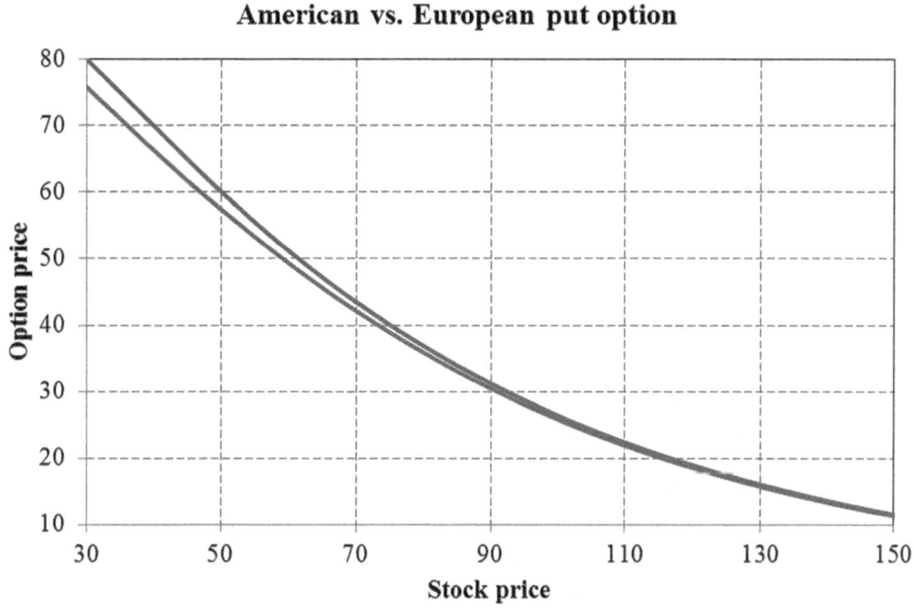

Fig. 4.20 The American option price vs. the European. Here $S = 100$, $K = 110$, $T = 2$ years, $r = 0.02$ % and $\sigma = 40$ %

So,the freedom to exercise an American option whenever the holder wishes introduces a boundary problem to solving the Black–Scholes equation as done before for vanillas.

There are several factors to consider on an early exercise decision. First, the early exercise premium is lost if exercised early. Secondly, early exercise leads to a loss or gain in time value of the asset depending on if a put or call. Therefore, one immediately sees that early exercise on an America call option tends to be less favourable, whereas for an American put it is more likely (assuming $r > 0$ here). However, a holder of an American put options on dividend paying assets generally prefers not be exercised early since the dividend payments are lost if the asset is sold early. For a discrete dividend-paying asset, a decision on early exercise is also influenced by the size of the dividend payments. One may exercise an American call option early if the dividend payments are higher than initially forecasted, if not exercised the ex-div value *may* fall below the expectation price due the loss in asset value through the payouts. All these factors have an influence on early exercise of an American option, and the decision on when to exercise seems almost a subjective one.

The contract holder will ideally, of course, only exercise the option prior to the expiry date if the present *payoff* at time t exceeds the discounted

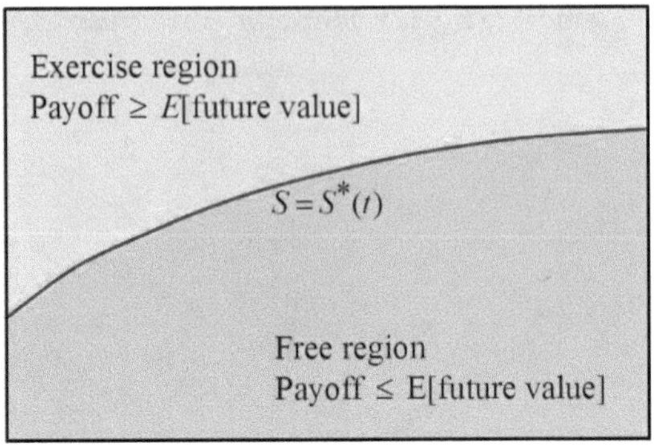

Fig. 4.21 The exercise and free area of an American option

expectation of the possible *future* values of the option from time t to T. So only if what the option-holder gets out of exercising early exceeds the markets' view of the expected future return in keeping the option alive will early exercising result. Otherwise, he/she will continue to hold on to the option. At every time t there will be a region of values of S whereby it is best to exercise the option, the *exercise region* and a complimentary region whereby it is best to keep the option, the *free region*. There will also be a particular value $S^*(t)$ which defines the *optimal exercise boundary* separating the two regions. We have already stated what factors can determine this boundary. (Fig. 4.21).

With a simple arbitrage analysis we can examine this problem more closely. Consider a non-dividend paying American call option worth $C(S, t)$ at time t on an underlying stock $S(t)$ and strike value K. We buy this together with a bond guaranteed to pay K at time the same maturity time T of the option. Let's further consider the case that when the option is exercised early and the owner is forced to keep the underlying up to maturity. Prior to maturity if $S(t) > K$ we may be tempted to exercises. However, by exercising at time t the value of our portfolio is $S(t) - K + Ke^{-r(T-t)}$ which is less than $S(t)$, and thus by keeping this to time T we would be left with $S(T)$. Instead, if we wait to expiry our portfolio value *may* be worth $\max[S - K, 0] + K = \max[S(T), K]$. Clearly, in this case it is best to keep the option alive up to maturity. Of course $S(t)$ may well be greater than $\max[S(T), K]$, and we are not forced to keep it up to time T, but can cash its value in. However, in this case it would be better to sell the option and cash in its value as an insurance at time t worth more than $S(t)$.

If we own an American call option and our intention is to buy the underlying stock, there is no profit in an early exercise. The reason is that the price we have to pay, the strike price, is constant. If we wait to exercise at expiration we can earn interest on the money we will use to buy the underlying.

4.7.13 American Call Options with No Dividends

For the reasons above we can price American calls with on a non-dividend paying stock because of a key attribute; that it is never beneficial to exercise the option prior to expiry. The detailed reasons behind this won't be considered here, but two primary reasons exist;

1. Firstly, holding the call option instead of exercising it and holding the stock is an insurance factor. An adverse stock movement (fall) would result in losses for the stockholder, but holding the call would enable the holder of the call to insure against any falls.
2. Secondly, there is the concept of time value of money. Paying the strike price earlier rather than later means that the holder of option loses out on the time value the money can achieve for the remainder of the option.

The attribute of non-exercise means that the American option can be priced via the standard Black–Scholes European call option formula and forcing dividends to 0.

The Perpetual American Put

There is an interesting American put option that merits study in detail. This is the perpetual American put option. This option will never expire, which why it is called perpetual. The payoff is given by $\max\{X-S, 0\}$ and we want to calculate the value of this option. Since the option never expires the value is independent of time and therefore dependent only on the underlying price. We also note that since this is an American option the option value can never go below the early exercise payoff. This is because of the non-arbitrage condition. Since the option is independent of time, it has to satisfy

$$\frac{1}{2}\sigma^2 S^2 \frac{d^2F(S)}{dS^2} + rS\frac{dF(S)}{dS} - rF(S) = 0.$$

This is an ODE, which easily can be solved. First we make a change in variables by setting $V(S) = F(\ln(S)) = F(y)$. The derivatives are then changed as

$$\frac{dF}{dS} = \frac{dV dy}{dy dS} = \frac{1 dV}{S dy} \equiv \frac{1}{S}V'$$

$$\frac{d^2F}{dS^2} = \frac{d}{dS}\left(\frac{1dV}{Sdy}\right) = -\frac{1}{S^2}V' + \frac{1}{S}\frac{d}{dy}\frac{dF}{dS} = \frac{1}{S^2}\left(V'' - V'\right).$$

We then have the following linear ODE

$$\frac{1}{2}\sigma^2 V'' + \left(r - \frac{1}{2}\sigma^2\right)V' - rV = 0$$

or

$$V'' + \left(\frac{2r}{\sigma^2} - 1\right)V' - \frac{2r}{\sigma^2}V = 0.$$

If we make the ansatz $V = C \cdot e^{\lambda y}$ we get a second order polynomial in λ

$$\lambda^2 + \left(\frac{2r}{\sigma^2} - 1\right)\lambda - \frac{2r}{\sigma^2}V = 0$$

with two solutions (1 and $-2r/\sigma$) . Therefore we can write the general solution in terms of S as

$$F(S) = aS + bS^{-2r/\sigma^2}$$

where a and b are arbitrary constants. Here we have used that

$$V(y) \sim e^{\lambda \cdot y} = e^{\lambda \cdot \ln S} = e^{(\ln s)^{\lambda}} = S^{\lambda}.$$

The first part of this solution is the asset itself, and we know that the asset itself satisfies the Black–Scholes equation. If we can find a and b, we have found the solution for the perpetual American put. The coefficient a must be zero as $S \to \infty$ the value of the option must become zero. If S is to low, we will immediately exercise the option, receiving X–S. (Common sense tell us not to exercise if $S > X$.) Suppose that we decide to exercise if $S = S^*$. How do we choose S^*? We then have that $F(S^*) = X - S^*$. If it is less, we have an arbitrage opportunity, if it is more we will not exercise. Therefore

$$F(S^*) = b(S^*)^{-2r/\sigma^2} = X - S^*.$$

But since both b and S^* are unknown, we need one more equation. If we eliminate b we have

$$F(S) = (X - S^*)\left(\frac{S}{S^*}\right)^{-2r/\sigma^2}.$$

We are now going to choose S^* to maximize the option value at any time before exercise. In other words, what choice of S^* makes F as large as possible? We find this by differentiating with respect to S^* and setting the resulting expression equal to zero

$$\frac{\partial}{\partial S^*}(X - S^*)\left(\frac{S}{S^*}\right)^{-2r/\sigma^2} = \frac{1}{S^*}\left(\frac{S}{S^*}\right)^{-2r/\sigma^2}\left(-S^* + \frac{2r}{\sigma^2}(X - S^*)\right) = 0.$$

We find that when

$$S^* = \frac{X}{1 + \frac{\sigma^2}{2r}}$$

it is optimal to exercise.

A Further Theoretical View

We will now end the discussion with a mathematical view of American options. We suppose we have a market free of arbitrage, with a risk-free security B with the deterministic interest rate r. We also have an n-dimensional equity-price vector S. The equity prices are supposed to be given by a given filtration F and all equities are prices by a martingale measure Q. Now, let the function

$$\Phi : R_+ \times R^n \to R$$

and a fix time T be given.

Definition 4.12 An *American contract on* Φ *with maturity* T is a contract with the following properties: the holder of a contract has the possibility at any time in the interval $[0, T]$ exercise the contract. If the holder exercise at time t, he receives $\Phi(t, S(t))$ cash units. If the holder not exercise under the interval $[0, T)$ the holder receives $\Phi(t, S(T))$ at time T. The decision by the holder to exercise at time t or not is only based upon the information known at the time t, i.e., on F_t. We denote such a contract with the symbol $\Phi_{A,T}$ (or sometimes only Φ_A).

The most common American contract is the American call and put options with $n = 1$ given by the well-known expressions

$$\begin{cases} \Phi(t,x) = \max[x - K, 0] \\ \Phi(t,x) = \max[K - x, 0] \end{cases}.$$

The American contract has a decision problem, namely whether or not we shall exercise. This is related to stopping times in such a way that the holder can choose a stochastic stopping time τ to exercise and have $\Phi(\tau, S(\tau))$ in cash.

To hold an American contract is equivalent to holding a family of consumption plans

$$\Phi_A = \{A^\tau; 0 \le \tau \le T\}$$

where τ is a stopping time and A^τ is defined by

$$A^\tau(t) = \begin{cases} 0, & 0 \le t < \tau \\ \Phi(\tau, S(\tau)) & \tau \le t \le T \end{cases}.$$

Our question is how to price the contract Φ_A. If we suppose we are at time $t = 0$ and have chosen the exercise strategyτ. The price is then given by

$$\pi_0[A^\tau] = E^Q[e^{-r\tau} \cdot \Phi(\tau, S(\tau))].$$

But, on an effective market, no one will pay more than

$$\sup_{0 \le \tau \le T} E^Q[e^{-r\tau} \cdot \Phi(\tau, S(\tau))]$$

for the contract. We can then define the following price process.

Definition 4.13 The price process for the American contract Φ_A is given by

$$\pi_t[\Phi_A] = \sup_{t \le \tau \le T} E^Q\left[e^{-r(\tau - t)} \cdot \Phi(\tau, S(\tau)) | \mathcal{F}_t\right] \quad 0 \le t \le T.$$

This is an abstract price and gives us no information at all of the value of the contract.

In a wider perspective the American contracts is a special case of the class of optimal stopping-time problems. The problem is to calculate

$$\max_{0 \leq \tau \leq T} E^Q[X(\tau)]$$

where τ vary over the class of stopping times. In continuous time this is a difficult problem.

There is at least one simple problem that can be solved: Consider an American call option with no dividends. Suppose we are at time $t = 0$, then

$$X(t) = e^{-rt} \cdot \max[S(t) - K, 0] = \max[e^{-rt} \cdot S(t) - e^{-rt} \cdot K, 0]$$

where we want to solve the problem

$$\max_{0 \leq \tau \leq T} E^Q[X(\tau)]$$

where Q is a martingale measure, i.e., Q makes the following process a martingale

$$Z(t) = e^{-rt} S(t).$$

The second term $(e^{-rt}K)$ is deterministic so we can write

$$X(t) = f(T(t))$$

where $y(t) = e^{-rt}S(t) - e^{-rt}K$ is a sub-martingale and $f(y) = \max[y, 0]$. Therefore we have proven that the price of an American call option and a European call option is the same.

4.8 Analytical Pricing Formulas for American Options

During the last years there have emerged a number of analytical models for American options. These methods are all approximations in some since. We will here present some of the most common analytical models for American contracts.

4.8.1 The Roll–Geske–Whaley Model

This is a model for a call option with a single dividend. Such American call option can be considered to be a series of call options which expire at the ex-dividend dates, and this case becomes a compound option or (an option on an option) with a closed-form solution as follows:

$$C_D = (S_0 - D_1 e^{-rt_1})N(b_1) + (S_0 - D_1 e^{-rt_1})\, M\left(a_1, -b_1; -\sqrt{t_1/T}\right)$$
$$- Xe^{-rT} M\left(a_2, -b_2; -\sqrt{t_1/T}\right) - (X - D_1)\, e^{-rt_1} N(b_2),$$

With the variables defined as

$$a_1 = \frac{\log[(S_0 - D_1 e^{-rt_1})/X] + (r + \sigma^2/2)\ T}{\sigma\sqrt{T}},$$
$$a_2 = a_1 - \sigma\sqrt{T},$$

$$b_1 = \frac{\log[(S_0 - D_1 e^{-rt_1})/S^*] + (r + \sigma^2/2)\ t_1}{\sigma\sqrt{t_1}},$$
$$b_2 = b_1 - \sigma\sqrt{t_1},$$

where $M(a, b, \rho)$ is the bivariate cumulative normal distribution function and S^* is the critical stock price for which the following equation is satisfied

$$c(S^*) = S^* + D_1 - X,$$

where $c(S^*)$ is the price given by Black–Scholes and $T - t_1$ time to maturity. The critical stock price can be solved iteratively via the Bisectional method.

4.8.2 The Barone–Adesi–Whaley Model

Barone–Adesi–Whaley model gave a quadratic approximation to price American options, and the pricing of the option is essentially a European option with adjusted for an early exercise premium. If $S_t < S^*$:

$$C_t = c_t + A_2\left(\frac{S_t}{S^*}\right)^{q_2}$$

and else

$$C_t = S_t - X.$$

The European option is valued using the Black–Scholes formula. Defining the variables as

$$A_2 = \frac{S^*\left[1 - e^{\sigma(T-t)}N(d_1)\right]}{q_2}$$

where

$$q_2 = \frac{1 - n + \sqrt{(n-1)^2 - 4k}}{2}$$

and

$$n = \frac{2(r-\delta)}{\sigma^2} \quad \text{and} \quad k = \frac{2r}{\sigma^2\left(1 - e^{-r(T-t)}\right)}$$

The critical value of S^* is defined as

$$S^* - X = c_t\left(S^*, X, T-t\right) + \left\{1 - e^{-\sigma(T-t)}N(d_1)\right\} \cdot \frac{S^*}{q_2}$$

and can be solved using the Newton–Raphson method and specifying appropriate seed values. For corresponding put values, we have a set of formulas to determine the value of an American put. If $S_t > S^{**}$

$$P_t = p_t + A_1\left(\frac{S_t}{S^{**}}\right)^{q_1}$$

and else

$$C_t = S_t - X$$

where the variables are defined as

$$A_1 = \frac{S^{**}\left[1 - e^{\sigma(T-t)}N(-d_1)\right]}{q1}$$

where

$$q_1 = \frac{1 - n1\sqrt{(n-1)^2 + 4k}}{2}$$

n and k being the same as for a call. This approximation is suitable and fast for practical pricing of American options and gives a very close value when compared with closed form Black-76 (see Haug 1997).

4.8.3 The Bjerksund, Stensland Model

The Bjerksund–Stensland approximation assumes that the exercise is initiated to a corresponding "flat" boundary, making use of a trigger price. This approximation is computational inexpensive and the method is fast, with evidence indicating that the approximation may be more accurate in pricing long dated options than the Barone–Adesi–Whaley model

$$\begin{aligned} C = \alpha S - \alpha\phi(S,T,\beta,I,I) + \phi(S,T,1,I,I) \\ - \phi(S,T,1,X,I) - X\phi(S,T,0,I,I) + X\phi(S,T,0,X,I) \end{aligned}$$

where

$$\begin{aligned} \alpha &= (I - X)I^{\beta} \\ \beta &= \left(\frac{1}{2} - \frac{r-D}{\sigma^2}\right) + \sqrt{\left(\tfrac{r-D}{\sigma^2} - \tfrac{1}{2}\right)^2 + 2\left(\tfrac{r}{\sigma^2}\right)}. \end{aligned}$$

The function ϕ is given as

$$\phi(S,T,\gamma,H,I) = e^{\lambda}S^{\gamma}\left[N(d) - \left(\frac{I}{S}\right)^{\kappa} N\left(d - \frac{2\ln(I/S)}{\sigma\sqrt{T}}\right)\right]$$

where

$$\lambda = \left[\gamma(r-D) - r + \frac{1}{2}\gamma(\gamma-1)\sigma^2\right]T$$

$$d = \frac{\ln(S/H) + (r - D + (\gamma - 0.5)\sigma^2)\ T}{\sigma\sqrt{T}}.$$

$$\kappa = \frac{2(r-D)}{\sigma^2} + (2\gamma - 1)$$

The trigger price I is given as the following

$$
\begin{aligned}
I &= B_0 + (B_\infty - B_0) \cdot \left(1 - e^f\right) \\
f &= -\left\{T(r-D) + 2\sigma\sqrt{T}\right\} \cdot \left(\frac{B_0}{B_\infty - B_0}\right) \\
B_\infty &= \frac{\beta}{\beta - 1} \cdot X \\
B_0 &= X \cdot \max\left\{1, \frac{r}{D}\right\}
\end{aligned}
.
$$

To price an American put, we consider the Bjerksund–Stensland approximation for the call option and apply put call parity in the form of

$$P(S,X,T,r,r-D,\sigma) = C(S,X,T,D,D-r,\sigma)$$

4.8.4 The Geske–Johnson Model

Geske & Johnson (1984) give an accurate approximation for an American put option by considering it as a series of Bermudan options, with the value of the American option given when the number of exercise dates for the Bermudan option tends to infinity or an infinite series of multivariate normal terms.

4.8.5 Trinomial Trees by Boyle

The trinomial tree is similar to the binomial method in that it employs a lattice-type method for pricing options. The exceptions are that the trinomial method arises at an accurate value faster than its binomial counterpart due to the use of a 3-proned path compared to the 2-proned path seen with binomial trees. The probabilities of the price going up at the next time period are given as

$$
\begin{aligned}
p_u &= \left(\frac{e^{(r-D)\cdot(T-t)/2} - e^{-\sigma\sqrt{(T-t)/2}}}{e^{\sigma\sqrt{(T-t)/2}} - e^{-\sigma\sqrt{(T-t)/2}}}\right)^2 \\
p_d &= \left(\frac{e^{\sigma\sqrt{(T-t)/2}} - e^{(r-D)\cdot(T-t)/2}}{e^{\sigma\sqrt{(T-t)/2}} - e^{-\sigma\sqrt{(T-t)/2}}}\right)^2 \\
p_m &= 1 - p_d - p_u.
\end{aligned}
$$

The respective American call and put can now be priced via backwards inductioncall

$$C_{i,j} = \max\left\{ \begin{array}{l} S \cdot u^{\max(0,j-1)} \cdot d^{\max(0,i-j)} - X, \\ e^{-r(T-t)}\left[p_u \cdot C_{i+1,j+2} + p_m \cdot C_{i+1,j+1} + p_d \cdot C_{i+1,j}\right] \end{array} \right\}$$

Put

$$P_{i,j} = \max\left\{ \begin{array}{l} X - S \cdot u^{\max(0,j-1)} \cdot d^{\max(0,i-j)}, \\ e^{-r(T-t)}\left[p_u \cdot P_{i+1,j+2} + p_m \cdot P_{i+1,j+1} + p_d \cdot P_{i+1,j}\right] \end{array} \right\}.$$

4.9 Poisson Processes and Jump Diffusion

A Poisson process is a pure jump process: a process that changes instantaneously from one value to another at random times. The following is a simulation of a standard Poisson process (where the jump sizes are restricted to 1) (Fig. 4.22).

The model for such a process extends from the discrete time Poisson distribution. This states that the number of Poissonly distributed events (N) in a time interval $(0, T]$ is distributed according to

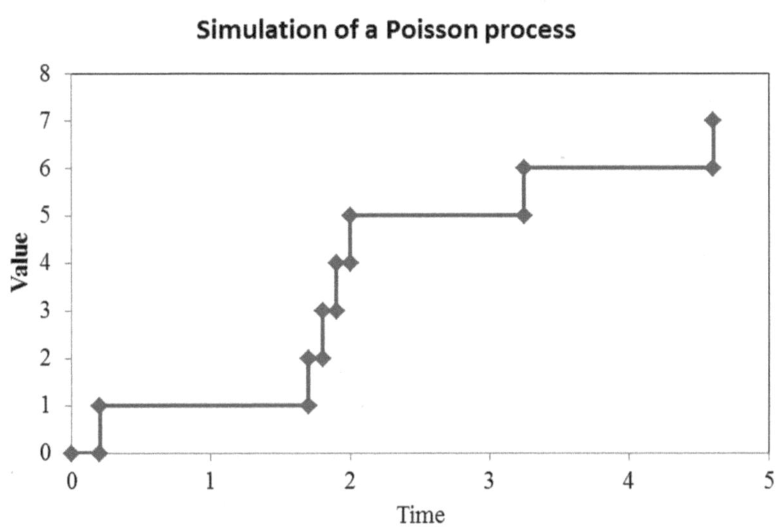

Fig. 4.22 A simulation of a standard Poisson process

$$P(N = x) = \frac{(\lambda T)^x e^{-\lambda T}}{x!}; \quad x = 1, 2, \ \ldots, \ \infty.$$

Here, the number of events corresponds to the number of jumps and λ is the intensity of the Poisson process; a measure of the 'frequency' of jumps, often scaled to units of per unit time. The Poisson process is a discrete probability distribution and has been successfully used to model the arrival times of certain events, or the occurrences of certain events, over a pre-defined period. The difference from most discrete distributions is that the number of occurrences can in theory (and with non-zero probability) be infinite.

Of particular relevance to finance (default modelling) is the waiting time between the arrivals of each event/jump. This is given by the exponential distribution and we will often be interested in the first arrival time τ. Although the Poisson distribution is a discrete one, the inter-arrival and first arrival times are continuous exponentially distributed random variables. The PDF is

$$P(\tau) = \lambda e^{-\lambda T}.$$

The probability of the first jump occurring in the time interval (0,*s*] is then

$$P(\tau < s) = \int_0^s \lambda e^{-\lambda T} dt = 1 - e^{-\lambda s}.$$

It is important to note that the Poisson process is consistent with the Poisson distribution for,

$$P(N = 0) = \frac{(\lambda T)^0 e^{-\lambda s}}{0!} = e^{-\lambda s} = P(\tau > s).$$

An important property of the Poisson process is the Markov property. Stated briefly, this is the "loss in memory" property where the distribution of the Poisson process in the future is independent of the past. For e.g. at time 0 the probability of not observing a jump over a time horizon T is simply $\exp(-\lambda T)$ from the derivation above. Now assume that we return to the process after a time s and the process is still at 0 (i.e. no jump has yet occurred). The probability of not observing a jump for a further time T (i.e. no jump until time I) given that no jump has occurred until time s is

$$P(\tau > T + s | \tau > s) = \frac{P(\tau > T + s)}{P(\tau > s)} = e^{-\lambda s}.$$

Using the expressions derived earlier this probability is just the same as the time 0 probability of not observing a jump over a time horizon T. This illustrates the Markov property—the fact that the process has not jumped until time s (whatever s might be) does not dictate the probability of future jumps.

How is a Poisson process mathematically characterized? Quite simply, the value of a standard Poisson process after a time T has elapsed is simply:

$$N(T) = N(0) + \sum_{s<T} [N(s) - N(s^-)].$$

The expression looks more daunting than it is. $N(0)$ is simply the initial condition (set to zero in a standard Poisson process). The latter term is the mathematical expression for "the number of jumps in the time interval $(0, T]$". Since the process jumps finitely in infinitesimal time, the time s corresponds to an infinitesimal time step before time s, and where a jump is observed $[N(s) - N(s-)]$ is 1; otherwise it is 0. In its more useful form, the process can also be expressed as $dN(t)$ which models the change in the Poisson process over a time step dt. Using the Markov property the value of $dN(t)$ at any time t does not depend on the history of the Poisson process. Furthermore,

$$P(\tau < dt) = 1 - e^{-\lambda dt} = 1 - \lambda dt + \frac{1}{2}(\lambda dt)^2 - \cdots \sim \lambda dt$$

because dt is very small. Thus, $dN(t)$ can be thought of as a random variable that increases by 1 over a time step dt with probability λdt and is zero with probability $1 - \lambda dt$.

4.9.1 Jump Diffusion

If X is a stochastic diffusion process that can jump as well then it is called jump diffusion

$$dX = A(t, X)dt + B(t, X)dW + C(t, X)dN.$$

The first two terms are the usual drift and white noise that have been used extensively to model stock prices in finance. The last term introduces the possibility of a jump occurring. dN constitutes a standard Poisson process; over a time interval dt a jump of size 1 can be observed with probability λdt. The scaling by $C(x,t)$ allows the jump size to vary.

Such models are becoming increasingly important in modelling stocks as they result in distributions with 'fatter tails' than the standard Ito processes. They are also being used to model energy and power prices where the jump behavior is very often observed. The key for mathematical finance is to now derive the SDE for a function $F(X)$. The key is to consider the process X as the sum of 2 processes:

$$\begin{aligned} dX^c &= A(t,X)dt + B(t,X)dW \\ dY &= C(t,X)dN. \end{aligned}$$

Then, to consider the Taylor series expansion of $F(x)$ by first considering the contribution from the continuous process and then the jump process,

$$dF = \frac{dF}{dX}dX^c + \frac{1}{2}\frac{d^2F}{dX^2}(dX^c)^2 + [F(X + C(t,X)) - F(X)]dN.$$

The last term arises from the jump component. $[x + C(x,t)]$ denotes the value of the process x just after a jump. The majority of the times the last term is zero because $dN = 0$. Only in those cases when a jump occurs the last term is non-zero and the jump in x is also observed in the function F.

5

Black–Scholes – Diffusion Models

To better understand the Black–Scholes world and to be able to handle more complex instruments, we will now continue with diffusion processes and some theorems. We will in this chapter explain the concept of changing measure and relative pricing. By changing measure we can value any securities relative to a given security. In most cases we value relative the money-market account.

5.1 Martingale Representation

Theorem 5.1 *Let W be a Wiener process on* $[0, 1]$ *and M a martingale such as*

(i) *M is $\mathcal{F}_t^W$-adapted*
(ii) $E[M^2(t)] < \infty \ \forall \ t \in [0, T]$.

Then, there exist a $\mathcal{F}_t^W$-adapted process g such as

$$E\left[\int_0^T g(s)dW(s)\right] < \infty \tag{5.1}$$

$$M(T) = M(0) + \int_0^T g(s)dW(s) \tag{5.2}$$

We can also write equation (5.2) as a differential

J.R.M. Röman, *Analytical Finance: Volume I*,
DOI 10.1007/978-3-319-34027-2_5

$$dM(t) = g(t)dW(t).$$

The martingale representation theorem (Theorem 5.1) is an abstract existence result. It guarantees the existence of the process $g(t)$ but it does not tell us what the process looks likes. $g(t)$ will in the following section be a very important function used in the method of changing measure, i.e., in the relative pricing theory.

5.2 Girsanov Transformation

Consider a stochastic binomial process S with $p = 0.75$ and $q = 0.25$. Here p is the probability that the underlying stock price increase and q the probability it will decrease. We ask ourselves if such a process is a fair game. Obviously is this not the case and for that reason is the process S in the continuous limit not a martingale. However, in the continuous limit we can model the stochastic process as

$$\begin{aligned} dS &= \mu dt + \sigma dX \\ \mu &= E[S] = (+1) \cdot p + (-1) \cdot q = 0.5 \\ \sigma^2 &= E[S^2] - (E[S])^2 = (+1)^2 \cdot p + (-1)^2 \cdot q - (p - q)^2 \\ &= 1 - (p - q)^2 = (p + q)^2 - (p - q)^2 = 4pq = 0.75 \end{aligned}$$

This is possible since the fluctuation of the original process is normal distributed and only the first two moments are of importance. This means that with known μ and σ we can imitate S with a fair game via a normal distributed process dX. But, for this a transformation is needed. Therefore, let dW be another normal process such as

$$dX = dW + \gamma dt.$$

Obviously, the process dW is not a fair game with respect to the process dX, but with a shift γ the process will be a fair game in another reference system. For each unique outcome in dX there exists a unique outcome in dW and vice versa. We can then write

$$dS = \mu dt + \sigma(dW + \gamma dt).$$

When we change a Weiner process like this, we say that we are making a Girsanov transformation. If we now choose $\gamma = -(\mu/\sigma)$ we get $dS = \sigma dW$.

This means that we have removed the drift by a transformation so that dS is a martingale under a new probability measure. Now, dS is a fair game with respect to the process following the deterministic evolution $(\mu/\sigma)t$. The same argument can be used for the log-normal process whereby dS/S becomes martingale with respect to dW:

$$dS = \sigma S dW.$$

Now, dS is a fair game with respect to the process following the deterministic evolution $e^{(\mu/\sigma)t}$.

In purely financial terms, we will see that if $S(t)$ represents the value on a security at the time t, dS is not a fair game. The reason is that μ/σ is the expected payoff with respect to the risk we take. But, if we discount with the risk-free interest rate we will get a fair game. We therefore have to choose

$$\gamma = -\frac{\mu - r}{\sigma}.$$

This is called the *market price of (volatility) risk* or the *sharp ratio* and is interpreted as the minimum extra payoff needed to take the extra risk in terms of the volatility.

Lemma 5.2 *Let g be a $\underline{\mathcal{F}}$-adapted process with*

$$P\left(\int_0^T g^2(t)dt < \infty\right) = 1$$

then

$$\begin{cases} dL(t) = g(t)L(t)dW(t) \\ L(0) = 1 \end{cases}$$

have a unique solution $L > 0$

$$L(t) = \exp\left\{\int_0^t g(s)dW(s) - \frac{1}{2}\int_0^t g^2(s)ds\right\}$$

The proof is left to the reader. Use the Itô formula.

If we now define a market via $Z = (S/B)$, where B is the money-market account with a unique martingale measure Q. This means that we define a market where we study any security S relative B. We write

$$Q(A) = \int_{\Omega} Z dP(A).$$

If we apply the Itô formula on Z we get

$$\begin{aligned} dZ(t) &= \frac{\partial Z}{\partial S} dS + \frac{\partial Z}{\partial B} dB \\ &= \frac{1}{B(t)} \{\mu(t)S(t)dt + \sigma(t)S(t)dW(t)\} - \frac{S(t)}{B^2(t)} r(t)B(t)dt \\ &= \{\mu(t) - r(t)\}Z(t)dt + \sigma(t)Z(t)dW(t). \end{aligned}$$

With a Girsanov transformation $dW(t) = g(t)dt + dV(t)$ we get

$$\begin{aligned} dZ(t) &= \{\mu(t) - r(t)\}Z(t)dt + \sigma(t)Z(t)\{g(t)dt + dV(t)\} \\ &= \{\mu(t) - r(t) + \sigma(t)g(t)\}Z(t)dt + \sigma(t)Z(t)dV(t). \end{aligned}$$

Here we define the likelihood function $L(t)$

$$\begin{cases} dL(t) = g(t)L(t)dW(t) \\ L(0) = 1 \end{cases}$$

with the solution

$$L(t) = \exp\left\{ \int_0^t g(s)dW(s) - \frac{1}{2}\int_0^t g^2(s)ds \right\}.$$

Note that L is a Radon–Nikodym derivative. We call the function $g(t)$ the Girsanov kernel. $g(t)$ is given by

$$g(t) = \frac{r(t) - \mu(t)}{\sigma(t)}.$$

We call the quotient, as above

$$\frac{\mu(t) - r(t)}{\sigma(t)}$$

the market price of risk or risk premium per unit risk or the sharp ratio. The numerator $\mu(t) - r(t)$ is called the risk premium of the stock and denotes the excess rate return over the risk-free rate of return on the market. $\mu(t)$ is then the expected return of the stock. We can now give the Girsanov theorem.

Theorem 5.3 (***The Girsanov theorem***) *Suppose that we are given a probability space* $(\Omega, \mathcal{F}, P)$, *where* P *is the market probability measure. Let* X *be a* $(\mathcal{F}, P)$-*Wiener process (a Brownian motion) and let* $\mathcal{F}(t)$ *be the filtration generated by this Wiener process. Also let* L *and* g *be as above* ($g(t)$ *is adapted to* $\mathcal{F}(t)$). *Furthermore, suppose that* $E[L(T)] = 1$ *and define* Q *via* $dQ = L(T)dP$ *on* $\mathcal{F}(t)$. *Then:*

$$W(t) = X(t) - \int_0^t g(s)ds$$

is a $(\mathcal{F}, Q)$-*Wiener process (Brownian motion under* Q).

Interpretation X is a Q-Wiener process with the drift g.

One important point about Girsanov's theorem is its converse that every equivalent measure is given by a drift change. This implies that in the Black–Scholes world there is only one equivalent risk-neutral measure. If this were not the case there would be multiple arbitrage-free prices.

Theorem 5.4 (***The reverse of Girsanov theorem***) *Given* $(\Omega, \mathcal{F}, P)$, X *and suppose* $Q << P$ *on* $\mathcal{F}_T^X$, *then there exist a unique* $\{\mathcal{F}_t^X\}$-*adapted process* g *such as:*

$$dQ(t) = L(t)dP(t)$$

where L *is given by*

$$\begin{cases} dL(t) = g(t)L(t)dW(t) \\ L(0) = 1 \end{cases}$$

5.2.1 The Market Price of Risk

In classical economic theory, no rational person would invest in a risky asset unless they expect to beat the return from holding a risk-free asset. Typically, risk is measured by standard deviation of returns, or volatility. The market price of risk for a stock is measured by the ratio of expected return in excess of the risk-free interest rate divided by the standard deviation of return. Interestingly, this quantity is not affected by leverage. If you borrow at the risk-free interest rate to invest in a risky asset both the expected return and the risk increase, such that the market price of risk is unchanged. This ratio, when suitably annualized, is also the Sharpe ratio.

If a stock has a certain value for its market price of risk then an obvious question to ask is what is the market price of risk for an option on that stock? In the famous Black–Scholes world in which volatility is deterministic and you can hedge continuously and costless, then the market price of risk for the option is the same as that for the underlying equity. This is related to the concept of a complete market in which options are redundant because they can be replicated by stock and cash.

In derivatives theory we often try to model quantities as stochastic, that is, random. Randomness leads to risk, and risk makes us ask how to value risk, that is, how much return should we expect for taking risk. By far the most important determinant of the role of this market price of risk is the answer to the question "is the quantity you are modelling traded directly in the market?"

If the quantity is traded directly, the obvious example being a stock, then the market price of risk does not appear in the Black–Scholes option-pricing model. This is because you can hedge away the risk in an option position by dynamically buying and selling the underlying asset. This is the basis of risk-neutral valuation. Hedging eliminates exposure to the direct that the asset is going and also to its market price of risk. You will see this if you look at the Black–Scholes equation. There, the only parameter taken from the stock random walk is its volatility, there is no appearance of either its growth rate or its price of risk.

On the other hand, if the modelled quantity is not directly traded then there will be an explicit reference in the option-pricing model to the market price of risk. This is because you cannot hedge away associated risk. And because you cannot hedge the risk you must know how much extra return is needed to compensate for taking this unhedgeable risk. Indeed, the market price of risk will typically appear in classical option-pricing models any time you cannot hedge perfectly. So expect it to appear in the following situations:

- When you have a stochastic model for a quantity that is not traded. examples: stochastic volatility; interest rates (this is a subtle one, the spot rate is not traded); risk of default.
- When you cannot hedge. Examples: jump models; default models; transaction costs.

When you model stochastically a quantity that is not traded then the equation governing the pricing of derivatives is usually of diffusion form, with the market price of risk appearing in the "drift" term with respect to the non-traded quantity. To make this clear, here is a general example.

Suppose that the price of an option depends on the value of a quantity of a substance called sepofan. Sepofan is not traded but either the option's payoff depends on the value of sepofan, or the value of sepofan plays a role in the dynamics of the underlying asset. We model the value of sepofan as

$$dS = \mu dt + \sigma dW.$$

The market price of sepofan risk is λ. In the classical option-pricing models we will end up with an equation for an option with the following term

$$\ldots + (\mu - \lambda\sigma)\frac{\partial V}{\partial s} + \ldots = 0,$$

where the ... represent all the other terms that one usually gets in a Black–Scholes-type of equation. Observe the expected change in the value of sepofan, μ, has been adjusted to allow for the market price of sepofan risk. We call this the risk-adjusted or risk-neutral drift. Conveniently, because the governing equation is still of diffusive type we can continue to use Monte Carlo simulation methods for pricing. Just remember to simulate the risk-neutral random walk

$$dS = (\mu - \sigma\lambda)dt + \sigma dV$$

and not the real one.

You can imagine estimating the real drift and volatility for any observable financial quantity simply by looking at a time series of the value of that quantity. But how can you estimate its market price of risk? Market price can you estimate its market price of risk? Market price of risk is only observable through option prices. This is the point at which practice and elegant theory start to part company. Market price of risk sounds like a way of calmly assessing required extra value to allow for risk. Unfortunately, there is nothing calm about the way that markets react to risk. For example, it is quite simple to relate the slope of the yield curve to the market price of interest rate risk. But evidence from this suggests that market price of risk is itself random, and should perhaps also be modelled stochastically.

Note that when you calibrate a model to market prices of options you are often effectively calibrating the market price of risk. But that will typically be just a snapshot at one point in time. If the market price of risk is random, reflecting peoples shifting attitudes from fear to greed and back again, then you are assuming fixed something, which is very mobile, and calibration will not work.

There are some models in which the market price of risk does not appear because they typically involve using some form of utility theory approach to find a person's own price for an instrument rather than the market's.

5.2.2 Black–Scholes and the Z-economy

Given a probability space $(\Omega, \mathcal{F}, P, W, \underline{\mathcal{F}})$ and chose a fix time T^* and let $\mathcal{F} = \{\mathcal{F}_t; 0 \leq t \leq T^*\}$ to be the natural filtration, $\mathcal{F}_t = \sigma\{W_s; s \leq t\}$. In the Black–Scholes world, let's study a bond B and a stock S

$$\begin{cases} dB(t) = r \cdot B(t)dt \\ B(0) = 1 \end{cases} \Rightarrow B(t) = e^{rt}$$

$$\begin{cases} dS(t) = \alpha \cdot S(t)dt + \sigma \cdot S(t)dW(t) \\ S(0) = s \end{cases}$$

where r, α, and σ are constants, $\sigma > 0$.

Lemma 5.5 *X is martingale if and only if $dX(t) = g(t)dW(t)$.*

Now, we want to change probability measure so that Black–Scholes becomes martingale. We therefore introduce the Z-economy via $Z(t) = (Z^0(t), Z^1(t))$ so that

$$Z(t) = \frac{1}{B(t)}(B(t), S(t)) = \left(1, \frac{S(t)}{B(t)}\right)$$

is martingale. We want $S(t)/B(t) \equiv e^{-rt}S(t)$ to be martingale. Therefore, we need to find the dynamics under Q for Z^1. Itô gives

$$\begin{aligned} dZ^1(t) &= \frac{\partial Z^1(t)}{\partial t}dt + \frac{\partial Z^1(t)}{\partial S}dS = -r \cdot e^{-rt}S(t)dt + e^{-rt}dS(t) \\ &= (\alpha - r)Z^1(t)dt + \sigma Z^1(t)dW(t) \end{aligned}$$

First, we have to find a Girsanov transformation so that $dZ^1(t)$ is martingale.

Let $dQ = L_T dP$ on $\mathcal{F}_{T^*}$, L_T is called a likelihood process. We have

$$\begin{cases} dL(t) = g(t)L(t)dW(t) \\ L(0) = 1 \end{cases}$$

The Girsanov theorem gives

$$dW(t) = g(t)dt + dv(t)$$

where $v(t)$ is a Q-Wiener process where the dynamics of Q for $Z^1(t)$ is given by

$$\begin{aligned} dZ^1(t) &= (\alpha - r)Z^1(t)dt + \sigma Z^1(t)(g(t)dt + dv(t)) \\ &= (\alpha - r + \sigma g(t))Z^1(t)dt + \sigma Z^1(t)dv(t). \end{aligned}$$

This dynamic is martingale if $(\alpha - r + \sigma g(t)) = 0$, i.e. if $g(t) = (r - \alpha)/\sigma$. The function $g(t)$ is called the Girsanov kernel. To sum up, under the martingale measure Q we have a Z-economy with the dynamic given by

$$\begin{cases} dZ^0(t) = 0 \\ dZ^1(t) = \sigma Z^1(t)dv(t) \end{cases}$$

$$Z^1(t) = e^{-rt}S(t) \quad \Rightarrow \quad dS(t) = r \cdot S(t)dt + \sigma S(t)dv(t)$$

We can easily prove this as

$$S(t) = e^{rt}Z^1(t)$$

$$\begin{aligned} dS(t) &= \frac{\partial S(t)}{\partial t}dt + \frac{\partial S(t)}{\partial Z^1(t)}dZ^1(t) = re^{rt}Z^1(t)dt + e^{rt}Z^1(t)\sigma dv(t) \\ &= rS(t)dt + \sigma S(t)dv(t) \end{aligned}$$

which, as before, is independent of α.

If we define a likelihood process L via

$$dL_t = \frac{r - \alpha}{\sigma}L_t dW_t$$

we get

$$\begin{aligned} L(t) &= \exp\left\{ \int_0^t \frac{r - \alpha}{\sigma} dW(s) - \frac{1}{2}\int_0^t \left(\frac{r - \alpha}{\sigma}\right)^2 ds \right\} \\ &= \exp\left\{ \frac{r - \alpha}{\sigma} W(t) - \frac{1}{2}\left(\frac{r - \alpha}{\sigma}\right)^2 t \right\} \end{aligned}$$

and

$$dW_t = \frac{r-\alpha}{\sigma}dt + dv_t.$$

When we change measure, and as above, use $B(t)$ as numeraire, we value all other instruments in our economy in terms of the numeraire, $B(t)$. In a general economy we can use any security S_i as the numeraire and value all other instruments with respect S_i. This means that we discount all other securities with S_i. S_i itself has the value 1 at all times and is then used as a *risk-free* asset (rate, rate of return). The numeraire asset defines the unit in which other prices are measured. We can then find the processes so that all prices will become martingaled with respect to S_i.

Definition 5.6 A probability measure Q is said to be a *martingale measure* if

(i) $Q \sim P$,
(ii) Under Q the Z^1-dynamic is given by: $dZ_t^1 = \sigma Z_t^1 dv_t$

where v_t is a $(Q, \underline{\mathcal{F}})$-wiener process.

Definition 5.7 A *portfolio strategy* is a stochastic process $h = (h^0, h^1)$ such as h is
$\{\mathcal{F}_t\}$-adapted and integrable.

Definition 5.8 The *S-value process* is given by: $V^s(t) = h^0(t)B(t) + h^1(t)S(t)$.

Definition 5.9 The *Z-value process* is given by: $V^z(t) = h^0(t) + h^1(t)Z^1(t)$.

Definition 5.10 S is *self-financed* if: $dV^s(t) = h^0(t)dB(t) + h^1(t)dS(t)$.
This can also be written as

$$V^S(t) = h^0(0) \cdot V^S(0) + \int_0^T h^1(u)dS(u).$$

Definition 5.11 Z is *self-financed* if: $dV^z(t) = h^1(t)dZ^1(t)$.
This can also be written as

$$V^Z(t) = V^Z(0) + \int_0^t h^1(u)dZ(u).$$

Definition 5.12 A *contingent T-claim* is a stochastic variable X, which is $\mathcal{F}_t$-measurable and integrable.

Definition 5.13 S is said to be *S-reachable* if $dV^s(t, h) = X$.

Definition 5.14 Z is said to be *Z- reachable* if $dV^z(t, h) = X$.

Theorem 5.15 *Black–Scholes is free of arbitrage.*

Proof

$$X = V^z(T, h) = \int_0^T h^1(t)dZ^1(t) = \int_0^T h^1(t)\cdot\sigma\cdot Z^1(t)dv(t)$$

$$E^Q[X] = E^Q\left[\int_0^T h^1(t)\cdot\sigma\cdot Z^1(t)dv(t)\right] = 0 \quad \text{since } v(t) \text{ is } Q\text{-martingale.}$$

Since P and Q are equivalent measures ($P \sim Q$) this also holds in the S economy. To have an arbitrage possibility we must have $E^Q[X] > 0$ but we see from the above that $E^Q[X] = 0$.

Theorem 5.16 *Black–Scholes is complete, i.e. all contingent claims are reachable.*

Proof: We study the Z-economy. For a given X show that there is a portfolio h such as

$$V^z(t) = h^0(t) + h^1(t)Z^1(t) = X$$
$$dV^z(t) = h^1(t)dZ^1(t) = h^1(t)\sigma Z^1(t)dv(t)$$

on the probability measure Q where $v(t)$ is a Q-Wiener process. We know that if such a portfolio exists, the value process above is Q-martingale. If we use the martingale representation theorem

$$M(t) = M(0) + \int_0^t g(s)dv(s) \Rightarrow dM(t) = g(t)dv(t)$$

and define a portfolio strategy as

$$\begin{cases} h^0(t) = M(t) - h^1(t)Z^1(t) \\ h^1(t) = \dfrac{g(t)}{\sigma\cdot Z^1(t)} \end{cases}$$

Then we get that

$$M(t) = h^0(t) + h^1(t)Z^1(t) = V^Z(t)$$

and

$$dV^Z(t) = h^1(t) \cdot dZ^1(t) = h^1(t) \cdot \sigma \cdot Z^1(t) \cdot dv(t) = g(t) \cdot dv(t) = dM(t).$$

Therefore, the portfolio is self-financed and X is Z-reachable which gives us that the model is complete.

Theorem 5.17 *In the Black–Scholes model, the martingale measure is given by Q where $Q \sim P$ and*

1. *Derivative prices are given by:*

$$\Pi(t) = e^{-r\cdot(T-t)} E^Q_{r,t}[\Pi(T)|\mathcal{F}_t]$$

2. *The dynamics of Q is given by:*

$$d\Pi_t = r\Pi_t dt + \sigma_\Pi \Pi_t dv_t$$

where v_t is Q-martingale.

3. $\dfrac{\Pi(t)}{B(t)}$ *is martingale.*

The relation between the classical PDE solving theory in finance and the probabilistic financial theory is given by the Feynman–Kač representation given by condition (1).

Example 5.18

Consider a model for two countries. We then have a domestic and a foreign market. The domestic and foreign interest rates, r_d and r_f, are assumed to be given real numbers. Consequently, the domestic and foreign savings accounts satisfy

$$B_t^d = e^{r_d t} \quad B_t^f = e^{r_f t}$$

where B^d and B^f are denominated in units of domestic and foreign currency, respectively. The exchange rate process X, which is used to convert foreign payoffs into domestic currency, is modelled by the following stochastic differential equation under the objective measure P

(continued)

Example 5.18 (continued)

$$dX = \mu_X X dt + \sigma_X X dW$$

where μ_x and σ_x are assumed to be constants and W is a P-Wiener process. A domestic martingale measure, Q^d, is a measure which is equivalent to the objective measure P and which makes all a priori given price process, expressed in units of the domestic currency and discounted using the domestic risk-free rate, martingales. We assume that if we buy the foreign currency this is immediately invested in a foreign bank account. All markets are assumed to be frictionless. We will first determine the Q^d-dynamics of the process X. Under Q^d the process Z defined by

$$Z_t = \frac{X_t B_t^f}{B_t^d}$$

should then be a martingale. Itô's formula gives the following dynamics for Z under P

$$\begin{aligned} dZ_t &= \frac{\partial Z_t}{B_t^f} dB_t^f + \frac{\partial Z_t}{B_t^d} dB_t^d + \frac{\partial Z_t}{X_t} dX_t = \frac{X_t}{B_t^d} dB_t^f - \frac{X_t B_t^f}{(B_t^d)^2} dB_t^d + \frac{B_t^f}{B_t^d} dX_t \\ &= (r_f - r_d + \mu_X) Z dt + \sigma_X Z dW \end{aligned}$$

Now, make a Girsanov transformation

$$dW = g(t)dt + dW^d$$

and let

$$dQ^d = L_t dP \quad \text{on } \mathcal{F}_t$$

where

$$\begin{cases} dL = gLdW \\ L(0) = 1. \end{cases}$$

We then get

$$\begin{aligned} dZ_t &= (r_f - r_d + \mu_X) Z dt + \sigma_X Z \big(g(t)dt + dW^d\big) \\ &= (r_f - r_d + \mu_X + \sigma_X g(t)) Z dt + \sigma_X Z dW^d. \end{aligned}$$

So we make the following choice for the Girsanov kernel $g(t)$, which makes the process martingale

$$g = \frac{r_d - \mu_X - r_f}{\sigma_X}.$$

By using the Girsanov theorem we can then see that Z is a martingale under the new measure Q^d. Again using the Girsanov theorem we find that the Q^d-dynamics of X are

(continued)

Example 5.18 (continued)

$$dX = \mu_X X dt + \sigma_X X dW = \mu_X X dt + \sigma_X X\left(\frac{r_d - \mu_X - r_f}{\sigma_X}dt + dW^d\right)$$
$$= (r_d - r_f)X dt + \sigma_X X dW^d$$

where W^d is a Q^d-Wiener process.

If we now take the viewpoint of a foreign-based investor, that is an investor who consistently denominates his/her profits and losses in units of foreign currency. A foreign martingale measure, Q^f, is a measure which is equivalent to the objective measure P and which makes all a priori given price process, expressed in units of foreign currency and discounted using the foreign risk-free rate, martingales. We now find the Girsanov transformation between Q^d and Q^f.

First we need the exchange rate process Y, which is used to convert domestic payoffs into foreign currency. This process is given by $Y = 1/X$ Using Ito's formula we obtain the following dynamics under Q^d

$$dY = \frac{\partial Y}{\partial X}dX + \frac{1}{2}\frac{\partial^2 Y}{\partial X^2}(dX)^2 = -\frac{1}{X^2}dX + \frac{1}{2}\frac{2}{X^3}(dX)^2$$
$$= \left(r_f - r_d + \sigma_X^2\right)Y dt - \sigma_X Y dW^d.$$

Under Q^f the process ζ defined by

$$\varsigma_t = \frac{YB_t^d}{B_t^f}$$

should be a martingale. Itô's formula gives the following dynamics for ζ under Q^d

$$d\varsigma_t = \frac{\partial \varsigma_t}{\partial B_t^d}dB_t^d + \frac{\partial \varsigma_t}{\partial B_t^f}dB_t^f + \frac{\partial \varsigma_t}{\partial Y_t}dY_t = \frac{Y_t}{B_t^f}dB_t^d - \frac{Y_t B_t^d}{\left(B_t^f\right)^2}dB_t^f + \frac{B_t^d}{B_t^f}dY_t$$
$$= \sigma_X^2 \varsigma_t dt - \sigma_X \varsigma_t dW^d.$$

We now make a new Girsanov transformation

$$dW^d = h(t)dt + dW^f$$

and let

$$dQ^f = L_t \cdot dQ^d \quad \text{on } \mathcal{F}_t$$

where

$$\begin{cases} dL = h_t \cdot L_t \cdot dW^d \\ L(0) = 1. \end{cases}$$

(continued)

Example 5.18 (continued)

We get

$$\begin{aligned} d\varsigma_t &= \sigma_X^2 \cdot \varsigma_t \cdot dt - \sigma_X \cdot \varsigma_t \cdot \left(h(t) \cdot dt + dW^f\right) \\ &= \varsigma_t \cdot \left(\sigma_X^2 - \sigma_X \cdot h(t)\right)dt - \sigma_X \cdot \varsigma_t \cdot dW^f \end{aligned}$$

and

$$h = \sigma_X.$$

By using the Girsanov theorem we can then see that ζ is a martingale under the new measure Q^f.

The domestic (foreign) market is said to be risk-neutral if the domestic (foreign) martingale measure is equal to the objective measure P. In order for the two martingale measures to be equal (which they have to be if they are both to be equal to P) the likelihood process L must be identically equal to one (recall that $dQ^f = L_t dQ^d$ on $\mathcal{F}_t$). Since we have that

$$L_t = \exp\left\{\int_0^t h(s)dW^d(s) - \frac{1}{2}\int_0^t h^2(s)ds\right\} = \exp\left\{\sigma_X W_t^d - \frac{1}{2}\sigma_X^2 t\right\}$$

we see that $L = 1$ requires $\sigma_x = 0$ and we get that the two measures are equal if the exchange rate is deterministic. In order for Q^d to be equal to P we find, using the same technique as above, that we must have $\mu_x = r_d - r_f$

$$\begin{aligned} L_t &= \exp\left\{\int_0^t \frac{r_d - r_f - \mu_X}{\sigma_X} dW(s) - \frac{1}{2}\int_0^t \left(\frac{r_d - r_f - \mu_X}{\sigma_X}\right)^2 ds\right\} \\ &= \exp\left\{\frac{r_d - r_f - \mu_X}{\sigma_X} W_t - \frac{1}{2}\left(\frac{r_d - r_f - \mu_X}{\sigma_X}\right)^2 t\right\} \end{aligned}$$

and $dX_t = (r_d - r_f)X_t dt$.

5.2.3 Siegel's Exchange Rate Paradox

Let us again study a market with two currencies, a domestic rate, r_d and a foreign rate, r_f. The exchange rate is given by $X(t)$. The process for the exchange rate is given by

$$dX(t) = X(t)\left(r_d(t) - r_f(t)\right)dt + \sigma(t)\rho(t)X(t)dW^d(t).$$

In the formula above the mean rate of change for the exchange rate $X(t)$ is $r_d(t) - r_f(t)$ under the domestic risk-neutral measure. We also introduced a correlation factor $\rho(t)$.

From the foreign perspective, the exchange rate is $1/X(t)$, and one should expect the mean rate of change of $1/X(t)$ to be $r_f(t) - r_d(t)$. This turns out not to be as straightforward as one might expect because of the convexity of the function $f(x) = 1/x$.

Example 5.19

Let the exchange rate from EUR to USD be 0.9. Then 1 EUR = 1.1111 USD. If the dollar price of euro falls by 5 %, 1 EUR becomes 0.95 × 1.1111 = 1.0556 dollars. This is an exchange rate of 1/1.0556 = 0.9474 EUR for each USD. The change from 0.9 to 0.9474 EUR for each USD is a change of 5.26 % (= 1/0.95 − 1) increase of EUR to USD, not 5 %. To understand why, let's study the inverse, $f(x) = \frac{1}{x}$ so that $f'(x) = -\frac{1}{x^2}$ and $f''(x) = \frac{2}{x^3}$. We then obtain by Itô

$$
\begin{aligned}
d\left(\frac{1}{X}\right) &= df(X) = f'(X)dX + \frac{1}{2}f''(X)(dX)^2 \\
&= \frac{1}{X}\left[(r_f - r_d)dt - \sigma dW^d\right] + \frac{1}{X}\sigma^2\left(dW^d\right)^2 \\
&= \frac{1}{X(t)}\left[(r_f - r_d + \sigma^2)dt - \sigma dW^d\right]
\end{aligned}
$$

The mean rate of change under the domestic risk-neutral measure is $r_f - r_d + \sigma^2$ not $r_f - r_d$. If we also include the correlation above, we also observe that the correlation $\rho(t) = -1$. However, the asymmetry introduced by the convexity of $f(x) = 1/x$ is resolved if we switch to the foreign risk-neutral measure, which is the appropriate one for derivative security-pricing in the foreign currency. First, recall the relationship

$$
dW^f(t) = -\sigma(t)dt + dW^d(t) \Rightarrow dW^d(t) = \sigma(t)dt + dW^f(t).
$$

In terms of $W^f(t)$, we may write

$$
d\left(\frac{1}{X}\right) = \frac{1}{X}\left[(r_f - r_d)dt - \sigma dW^f\right].
$$

Under the foreign risk-neutral measure, the mean rate of change for $1/X$ is $r_f - r_d$, as expect. Under the actual probability measure P, however, the asymmetry remains. By studying

$$
dX(t) = \mu(t)X(t)dt + \sigma(t)X(t)dW(t)
$$

we get

(continued)

Example 5.19 (continued)

$$d\left(\frac{1}{X(t)}\right) = \left(\frac{1}{X(t)}\right)(-\mu(t) + \sigma^2(t))dt - \frac{1}{X(t)}\sigma(t)dW(t)$$

and we observe that both X and $1/X$ have the same volatility. But, their mean rates are not the negative of each other.

5.2.4 Maximum Likelihood Estimation

In this section we give a brief introduction to maximum likelihood (ML) estimation for Itô processes. This is outside the main scope of this book, but since ML theory is such an important topic and we have already developed most of the necessary machinery, we include it here. We need the concept of a statistical model.

Definition 5.20 A dynamic statistical model over a finite time interval $[0, T]$ consists of the following objects:

- A measurable space $(\Omega, \mathcal{F})$.
- A flow of information on the space, formalized by a filtration $\underline{\mathcal{F}} = \{\mathcal{F}_t\}_{t>0}$
- An indexed family of probability measures $\{P_\alpha; \alpha \in A\}$, defined on the space $(\Omega, \mathcal{F})$, where A is some index set and where all measures are assumed to be absolutely continuous on $\mathcal{F}_t$ with respect to some base measure $P_{\alpha o}$.

In most concrete applications (see examples below) the parameter α will be a, real number or a finite dimensional vector—that is, A will be the real line or some finite dimensional Euclidian space. The filtration will typically be generated by some observation process X.

The interpretation of all this is that the probability distribution is governed by some measure P_α, but we do not know which. However, we do have access to a flow of information over time, and this is formalized by the filtration above, so at time t we have the information contained in $\mathcal{F}_t$. Our problem is to try to estimate α_t of α, based upon the information contained in $\mathcal{F}_t$, that is, based on the observations over the time interval $[0, T]$. The last requirement is formalized by requiring that the estimation process should be adapted on $\underline{\mathcal{F}}$, i.e., that $\alpha_t \in \mathcal{F}_t$.

One of the most common techniques used in this context is that of finding, for each t. the ML estimate of α. Formally the procedure works as follows.

- Compute, for each a the corresponding likelihood process $L(\alpha)$ defined by

$$L_t(\alpha) = \frac{dP_\alpha}{dP_{\alpha 0}} \quad \text{on} \quad \mathcal{F}_t$$

- For each fixed t, find the value of α which maximizes the likelihood ratio $L_t(\alpha)$.
- The optimal α is denoted by $\widehat{\alpha}_t$ and is called the ML estimate of α based the information gathered over $[0, t]$.

As the simplest possible example let us consider the problem of estimating the constant but unknown drift of a scalar Wiener process. In elementary terms we could naively formulate the model by saying that we can observe a process X with dynamics given by

$$\begin{cases} dX_t = \alpha dt + dW_t \\ X_0 = 0. \end{cases}$$

Here W is assumed to be Wiener under some given measure P and the drift α is some unknown real number. Since this example is so simple, we do in fact have an obvious candidate for the estimator process, namely

$$\widehat{\alpha}_t = \frac{X_t}{t}.$$

In a naive formulation like this, we have a single underlying Wiener process W, under a single given probability measure P, and we see that for different choices of α we have different X-processes. In order to apply the ML techniques we must reformulate our problem, so that we instead have a single X-process and a family of measures. This is done as follows:

- Fix a process X which is Wiener under some probability measure P_0. In other words: under P_0, the process X has the dynamics

$$dX_t = 0 \cdot dt + dW_t^0$$

where W^0 is P^0-Wiener.

- We assume that the information flow is the one generated by observations of X, so we define the filtration by setting $\mathcal{F}_t = \mathcal{F}_t^X$. For every real number α, we then define a Girsanov transformation to a new measure P_α by defining the likelihood process $L(\alpha)$ through

$$\begin{cases} dL_t = \alpha L_t(\alpha) dX_t \\ L_0(\alpha) = 1 \end{cases}$$

- From Girsanov's theorem it now follows immediately that we can write $dW_t^0 = \alpha dt + dW_t^\alpha$ where W_t^α is a P^α-Wiener process. Thus X will have the P^α-dynamics

$$dX_t = \alpha dt + dW_t^\alpha.$$

We now have a statistical model along the general lines above, and we notice that, as opposed to the case in the naive formulation, we have a single process X, but the driving Wiener processes are different for different values of α.

To obtain the ML estimation process for α, we need to compute the likelihood process explicitly, i.e. we have to solve L_t. This is easily done and

$$L_t(\alpha) = \exp\left\{\alpha X_t - \frac{1}{2}\alpha^2 t\right\}$$

We may of course maximize $\ln[L_t(\alpha)]$ instead of maximize $L_t(\alpha)$ so our problem is to maximize (over α) the expression

$$\alpha X_t - \frac{1}{2}\alpha^2 t.$$

This trivial quadratic optimization problem can be solved by setting the α derivative equal to zero, and we obtain the optimal a as

$$\widehat{\alpha}_t = \frac{X_t}{t}$$

Thus we see that in this example the ML estimator actually coincides with our naive guess above. The point of using the ML technique is of course that in a more complicated situation we may have no naive candidate, whereas the ML technique in principle is always applicable.

Example 5.21 Let W be a standard Wiener process on $(Q, \mathcal{F}, P_0)$ where the filtration is the one generated by W. Fix a time interval $[0, T]$. Under the measure P_0, the process X has the dynamics

$$dX_t = \sigma\sqrt{X_t}dW_t$$

where σ is a known constant.

Define, for each real number α, a Girsanov transformation such that the measure P_0 is transformed into a measure P_α, such that X under P_α solves the equation

$$dX = \alpha X dt + \sigma\sqrt{X}dW^\alpha,$$

where W^α is a P_α-Wiener process. Our task is to give a precise description of this measure transformation, by specifying the dynamics of the corresponding likelihood process L^α, where

$$L_t^\alpha = \frac{dP_\alpha}{dP_0}, \quad \text{on } \mathcal{F}_t.$$

A Girsanov transformation with the Girsanov kernel

$$g_t = \frac{\alpha}{\sigma}\sqrt{X_t}$$

will change the dynamics of X in the desired way. The dynamics of the likelihood process L^α are given by

$$\begin{cases} dL_t^\alpha = \dfrac{\alpha}{\sigma}\sqrt{X_t}L_t^\alpha dW_t \\ L_0^\alpha = 1. \end{cases}$$

The likelihood process is thus given by

$$L(t) = \exp\left\{\int_0^t \frac{\alpha}{\sigma}\sqrt{X_s}dW_s - \frac{1}{2}\int_0^t \frac{\alpha^2}{\sigma^2}X_s ds\right\}.$$

Next, determine, for every $t \leq. T$, the maximum likelihood estimator $\widehat{\alpha}(t)$ for the parameter α, based on observations of X over the interval $[0, t]$, i.e. the value of α that maximizes L^α. Note that the answer shall be expressed in terms of the process X, and simplified as far as possible.

You obtain the same estimate if you maximize the logarithm of the likelihood function. The maximum likelihood estimate is thus given as the solution to the following problem

$$\max_{\alpha} \left\{ \int_0^t \frac{\alpha}{\sigma} \sqrt{X_s} dW_s - \frac{1}{2} \int_0^t \frac{\alpha^2}{\sigma^2} X_s ds \right\}.$$

Since we want the answer to be expressed in terms of X we write this as

$$\max_{\alpha} \left\{ \frac{\alpha}{\sigma^2} \int_0^t dX_s - \frac{1}{2} \int_0^t \frac{\alpha^2}{\sigma^2} X_s ds \right\} = \max_{\alpha} \left\{ \frac{\alpha}{\sigma^2} (X_t - X_0) - \frac{1}{2} \int_0^t \frac{\alpha^2}{\sigma^2} X_s ds \right\}$$

Since the objective function is concave in a the maximum will be obtained in a point where the derivative with respect to α is zero. This yields

$$\widehat{\alpha}_t = \frac{X_t - X_0}{\int_0^t X_s ds}.$$

5.2.5 Complete Markets

A slightly more mathematical, but yet still quite easily understood description of a complete market is to say a complete market is one for which there exist the same number of linearly independent securities as there are states of the world in the future.

Consider, for example, the binomial model in which there are two states of the world at the next time-step, and there are also two securities, cash and the stock. That is a complete market. Now, after two time-steps there will be three possible states of the world, assuming the binomial model recombines so that an up-down move gets you to the same place as down-up. You might think that you therefore need three securities for a complete market. This is not the case because after the first time-step you get to change the quantity of stock you are holding: this is where the dynamic part of the replication comes in.

In a complete market you can replicate derivatives with the simpler instruments. But you can also turn this on its head so that you can hedge the derivative with the underlying instruments to make a risk-free instrument. In the binomial model you can replicate an option from stock to make cash. Same idea, same equations, just move terms to be on different sides of the "equals" sign.

As well as resulting in replication of derivatives, or the ability to hedge them, complete markets also have models. In this model you specify the probability of the stock rising (and hence falling because the probabilities must add to one). It turns out that this probability does not affect the price of the option. This is a simple consequence of complete markets, since you can hedge the option with the stock, you don't care what the probabilities are. People can therefore disagree on the probability of a stock rising or falling but still agree on the value of an option, as long as they share the same view on the stock's volatility.

In probabilistic terms we say that in a complete market there is a unique martingale measure, but for an incomplete market there is no unique martingale measure. The interpretation of this is that even though options are risky instruments we don't have to specify our own degree of risk aversion in order to price them.

Enough of complete markets: where can we find incomplete markets? The answer is "everywhere". In practice all markets are incomplete because of real-world effects that violate the assumptions of the simple model.

Take volatility as an example. As long as we have a log-normal equity random walk, no transaction costs, continuous hedging, perfectly divisible assets and so on, and constant volatility, then we have a complete market. If that volatility is a known time-dependent function, then the market is still complete. It is even still complete if the volatility is a known function of stock price. But as soon as that volatility becomes random the market is no longer complete. This is because there are now more states of the world than there are linearly independent securities. In reality, we don't know what volatility will be in the future, so markets are incomplete.

We also get incomplete markets if the underlying security follows a jump diffusion process. Again we have more possible states than there are underlying securities.

Another common reason for getting incompleteness is if the underlying or one of the variables governing the behaviour of the underlying is random. Options on such acts cannot be hedged since these actions aren't traded.

We still have to price contracts, even in the incomplete markets, so what can we do? There are two main ideas here. One is to price the actuarial way, the other is to try to make all option prices consistent with each other.

The actuarial way is to look at pricing in some average sense. Even if you can't hedge the risk from each option it doesn't necessarily matter in the long run. Because in that long run you will have made many hundreds or thousands of option trades, so all that really matters is what the average price of each contract should be, even if it is risky. To some extent this relies on results from the central limit theorem. This is called the actuarial approach because it is how the insurance business works. You can't hedge the lifespan of individual policyholders but you can work out what will happen to hundreds or thousands of them on average using actuarial tables.

The other way of pricing is to make options consistent with each other. This is commonly used when we have stochastic volatility models, for example, and is also often seen in fixed-income derivatives pricing. Let's work with the stochastic volatility model to start with. Suppose we have a log-normal random walk with stochastic volatility. This means we have two sources of randomness (stock and volatility) but only one quantity with which to hedge (stock). That's like saying that there are more states of the world than underlying securities, hence incompleteness. Well, we know we can hedge the stock price risk with the stock, leaving us with only one source of risk that we can get rid of. That's like saying there is one extra degree of freedom in states of the world than there are securities.

Whenever you have risk that you can't get rid of you have to ask how that risk should be valued. The more risk, the more return you expect to make in excess of the risk-free rate. This introduces the idea of the market price of risk. Technically this case introduces the market price of volatility risk. This measures the excess expected return in relation to unhedgeable risk. Now all options on this stock with the random volatility have the same sort of unhedgeable risk—some may have more or less risk than others but they are all exposed to volatility risk. The end result is a pricing model which explicitly contains this market price of risk parameter. This ensures that the prices of all options are consistent with each other via this "universal parameter". Another interpretation is that you price options in terms of the prices of other options.

And yet it is possible to have an efficient, arbitrage-free market for which there is no unique price enforced by the market. This can happen when there is no unique way to replicate the asset cash flows. If it is not possible to replicate the asset, it becomes harder for arbitrageurs to guarantee risk-less profits by trading at a proposed valuation price.

5.2.6 A Multidimensional Diffusion Model

Given $(\Omega, \mathrm{F}, P, W, \underline{\mathcal{F}})$, where W is a k-dimensional Wiener process and $\underline{\mathcal{F}} = \{\mathcal{F}_{\mathrm{t}}|0 \leq t \leq T\}$ the natural filtration, generated by W, i.e.

$$\mathcal{F}_t = \sigma\{W_s | s \le t\}$$

The model consists of an (N + 1)-dimensional, strict positive price process S with the dynamics

$$\begin{cases} dS_t^0 = rS_t^0 dt \\ S_0^0 = 1 \end{cases}$$

$$\begin{cases} dS_t^i = \alpha_i S_t^i dt + S_t^i \sum_{j=1}^{k} \sigma_{ij} dW_t^j \\ S_0^i = s^i \end{cases},$$

where α_i and σ_{ij} are adapted processes and $i = 1, \ldots, N$. We write this as

$$dS_t^* = D(S_t^*)\alpha dt + D(S_t^*)\sigma dW_t$$

where $D(x)$ is a diagonal matrix with the components x in the diagonal and

$$\alpha = \begin{bmatrix} \alpha_1 \\ \vdots \\ \alpha_N \end{bmatrix}, \quad W = \begin{bmatrix} W_1 \\ \vdots \\ W_N \end{bmatrix}, \quad \sigma = \begin{bmatrix} \sigma_{11} & \cdots & \sigma_{1N} \\ \vdots & \ddots & \vdots \\ \sigma_{N1} & \cdots & \sigma_{NN} \end{bmatrix}.$$

To investigate the conditions of arbitrage and the completeness of this economy, we define the discounted price process as

$$Z_t = e^{-rt} S_t.$$

Using Itô's formula, we get

$$\begin{cases} dZ_t^0 = 0 \\ dZ_t^i = Z_t^i(\alpha_i - r)dt + Z_t^i \sigma_i dW_i \end{cases}$$

or on vector form

$$dZ_t^* = D(Z_t^*)(\alpha - r1_N)dt + D(Z_t^*)\sigma dW_t.$$

We start to investigate the arbitrage free conditions with a Girsanov transformation that makes the drift term to disappear. With a Radon–Nikodym derivative L_T

$$L_T = \exp\left\{\int_0^T g_t' dW_t - \frac{1}{2}\int_0^T \|g_t\|^2 dt\right\}$$

where g is a k-dimensional process and ' means transposing. If $E_P[L_T] = 1$, we can change measure via

$$dQ = L_T dP$$

where Girsanov theorem gives, under Q: $dW_t = g_t dt + dv_t$, where v_t is a k-dimensional $(Q, \underline{F})$-wiener process. We then have

$$dZ_t^* = D(Z_t^*)(\alpha - r1_N + \sigma g)dt + D(Z_t^*)\sigma dv_t.$$

We therefore must have

$$\sigma g = r1_N - \alpha.$$

The problem now, is to find the process g such as $E_P[L_T] = 1$.

Definition 5.22 A probability measure Q *is said to be a martingale* measure if

1. $Q \sim P$,
2. Under Q the Z^*-dynamic is given by

$$dZ_t^* = D(Z_t^*)\sigma dv_t$$

where v_t is a $(Q, \underline{F})$-Wiener process. The class of martingale measures denotes ***P***.

Lemma 5.23 *Let α and σ be given and suppose that*

1. *The matrix $\sigma(t, \omega)$ has RANG N for all t*
2. *There exist a $c > 0$ such as $\lambda(t, \omega) \geq c$ for all (t, ω), where λ is the smallest singular value of σ.*

Then $\mathbf{P} \neq \emptyset$. Under Q we can write

$$\begin{cases} dS_t^* = D(S_t^*) r 1_N dt + D(S_t^*) \sigma dv_t \\ dZ_t^* = D(Z_t^*) \sigma dv_t \end{cases}$$

where v_t is a $(Q, \underline{F})$-wiener process and Q a martingale measure.

We now want to show that, if $\boldsymbol{P} \neq \emptyset$, then the economy is free of arbitrage. Therefore, we choose a $Q \in \boldsymbol{P}$ and keep Q fixed. We define the class $\mathcal{H}$ of possible portfolio strategies and the set $\mathcal{K}$ of all conditioned contracts. It is impotent to notice that the definition depends on the specific choice of Q, and sometimes we therefore write $\mathcal{H}(Q)$ and $\mathcal{K}(Q)$.

Definition 5.24 Fix a $Q \in \boldsymbol{P}$. A *portfolio strategy* is a process $h = (h^0, h^1, \ldots, h^N) = (h, h^*)$, such that

$$h \text{ is } \{\mathcal{F}_t\}\text{-adapted},$$

$$\int_0^T |h_t^0| dt < \infty,$$

$$E^Q\left[\int_0^T \|h^* D(Z^*) \sigma\|^2 dt\right] < \infty$$

For a given portfolio strategy h, the value processes $V^S(h)$ and $V^Z(h)$ are given by

$$\begin{cases} V_t^S(h) = h_t S_t = h_0 S_0 + h_t^* S_t^* \\ V_t^Z(h) = h_t Z_t = h_0 + h_t^* Z_t^* \end{cases}$$

A given portfolio strategy h is said to be self-financed if

$$V_t^S(h) = V_0^S(h) + \int_0^T h_u dS_u$$

i.e., if

$$dV_t^S(h) = h_t dS_t.$$

The class of self-financed portfolio strategies is denoted by $\mathcal{H}$. A conditioned contract is a stochastic variable X such as

$$X \text{ is } \mathcal{F}_T\text{-measurable}$$
$$E^Q\left[X^2\right] < \infty.$$

The set of all such contracts are denoted by $\mathcal{K}$ and with $\mathcal{K}^+$ we denote those $X \in \mathcal{K}$ such as $P(X \geq 0) = 1$, and $P(X > 0) > 0$.

A conditioned contract X is said to be S and Z reachable if there exist a self-financed strategy h such as

$$V_T^S(h) = X \quad \text{on} \quad P$$
$$V_T^Z(h) = X \quad \text{on} \quad Q$$

respectively.

First we want to know the relation between the S and the Z economy. Therefore, we need the following lemma:

Lemma 5.25

(i) *For each portfolio strategy, we have*

$$V_t^S(h) = e^{rt} V_t^Z(h), \qquad V_t^Z(h) = e^{-rt} V_t^S(h)$$

(ii) *The contract X is S reachable if and only if $e^{-rt}X$ is Z reachable.*

(iii) *The portfolio strategy h is self-financing if and only if*

$$dV_t^z(h) = h_t^* dZ_t^*$$

(iv) *If $h \in \mathcal{H}$, then $V^Z(h)$ is a quadratic integrable Q-martingale.*

Definition 5.26 A strategy $h \in \mathcal{H}$ is said to be an *arbitrage strategy* if

$$V_0^S(h) = 0 \quad \text{and} \quad V_T^S(h) \in K^+.$$

Theorem 5.27 *Suppose $P \neq \emptyset$. Then, the model is free of arbitrage so that there will not exist any arbitrage strategies in $\mathcal{H}(Q)$ for any $Q \in P$.*

Theorem 5.28 *Suppose $P \neq \emptyset$ and that the matrix process $\boldsymbol{\sigma}(t, \omega)$ have the rang k for all t and P. Then, the model is complete so that for any $Q \in P$ all $X \in \mathcal{K}(Q)$ is S-reachable.*

From above we see that the condition of freedom of arbitrage and completeness acts on opposite directions. To have freedom of arbitrage, the matrix $\boldsymbol{\sigma}$ must have at least as many rows as columns—that is, there must be at least as many driving Wiener processes as underlying equities. This means that if we specify the stochastic base for the economy in terms of k given Wiener processes, the freedom of arbitrage generically demands at maximum k equities. This is natural since each equity can offer potentially arbitrage possibilities.

Completeness, on the other hand, demands $\boldsymbol{\sigma}$ to have at least as many columns as the number of rows—that is, there must be as many equities as driving Wiener processes. This is naturally since any new equity gives a new possibility to realize a contingent claim in terms of a self-financed portfolio strategy. This gives the following theorem:

Theorem 5.29 *Suppose*

(i) $k = N$
(ii) *The matrix process $\boldsymbol{\sigma}(t, \omega)$ is invertible for all t and P.*
(iii) *There exist a $c > 0$ such as $|\lambda(t, \omega)| \geq c$ for all t and P where $\lambda(t, \omega)$ is the smallest eigenvalue for $\boldsymbol{\sigma}(t, \omega)$.*

Then

(a) *P consists of exactly one measure Q.*
(b) *The economy is free of arbitrage.*
(c) *The economy is complete.*

We can now summarize the general model as

(i) If $P = Q$ there will not exist any definable prices. Especially, it is not possible to value a specific contract X from a generating portfolio since different portfolios which generates X can have different value processes.
(ii) If $P \neq \emptyset$ then it is possible, for all $Q \in P$ to define a price process $\boldsymbol{\pi}^Q$ on $\mathcal{K}(Q)$ via:

$$\boldsymbol{\pi}^Q[X, t] = e^{-r(T-t)} E^Q[X|\mathcal{F}_t] \qquad X \in \mathcal{K}(Q)$$

(iii) For any fixed X, $\boldsymbol{\pi}^Q[X, t]$ can take different values for different choices of $Q \in P$. In other words, there are no any unique price of X, but the price process $\boldsymbol{\pi}^Q[X, t]$ tell us the possible prices for X if we not have any arbitrage possibilities in the economy.

(iv) For a reachable X for all $Q \in P$, the $\boldsymbol{\pi}^Q[X]$ will be independent of which $Q \in P$ we choose. Furthermore

$$\pi^Q[X, t] = V_t^S(h)$$

for each h that generates X.

(v) In a complete model there exist exactly one martingale measure Q and each contract X have a unique price as above.

5.3 Securities Paying Dividends

Many securities pay some kind of dividends. Forwards or futures on indices usually pay a dividend that can be modelled as a continuous compounded dividend. Stock pays discrete types of dividends, typically once a year. When valuing derivatives we need to consider dividends, since this will affect the prices of the underlyings and this way changes the values of the derivatives as well.

5.3.1 Black–Scholes with Continuous Dividend Yield

The simplest generalization of the Black–Scholes model is to consider options when the underlying asset will pay out dividends during the lifetime. If we assume that the asset will pay a continuous dividend yield, q, then in time t, the asset receives an amount $qSdt$. Then the change of the portfolio value is given by

$$d\Pi = \frac{\partial F}{\partial t}dt + \frac{\partial F}{\partial S}dS + \frac{1}{2}\sigma^2 S^2 \frac{\partial^2 F}{\partial S^2}dt - \Delta dS - q\Delta S dt$$

The Black–Scholes PDE is then given as

$$\frac{\partial F}{\partial t} + (r - q)S\frac{\partial F}{\partial S} + \frac{1}{2}\sigma^2 S^2 \frac{\partial^2 F}{\partial S^2} - rF = 0$$

Then, the Black–Scholes formula with dividends is given as

$$P_{call} = Se^{-qT}N(d_1) - Ke^{-rT}N(d_2)$$
$$P_{put} = Ke^{-rT}N(-d_2) - Se^{-qT}N(-d_1)$$

where

$$d_1 = \frac{\ln\left(\frac{S}{K}\right) + \left(r - q + \frac{\sigma^2}{2}\right)T}{\sigma \cdot \sqrt{T}}, \qquad d_2 = d_1 - \sigma \cdot \sqrt{T}.$$

5.3.2 Securities with General Dividends

In this section we will study a market with $N + 1$ securities where the price of security i, at time t, is given by $S^i(t)$, $i = 0,\ldots, N$. Security number zero is supposed to be the risk-free one, that is, $B(t) = S^0(t)$:

$$dB(t) = r(t)B(t)dt$$

where r is the short rate. We suppose all processes are defined on a probability space $(\Omega, \mathcal{F}, P, \underline{\mathcal{F}})$ where all processes have stochastic differentials.

We also suppose that we have, another process D^i for each $i = 1,\ldots,N$. We interpret $D^i(t)$ as the total dividend generated by security number i on the interval $[0, t]$. More precisely if we hold the security i under the period $(s, t]$ we receive, during this time $D^i(t) - D^i(s)$ cash units. Note that the security 0 (B) doesn't pay any dividends. We can therefore specify the market by a $2(N + 1)$-dimensional matrix process

$$[S(t), D(t)] = \begin{bmatrix} S^0(t) & D^0(t) \\ S^1(t) & D^1(t) \\ \vdots & \vdots \\ S^N(t) & D^N(t) \end{bmatrix} = \begin{bmatrix} B(t) & 0 \\ S^1(t) & D^1(t) \\ \vdots & \vdots \\ S^N(t) & D^N(t) \end{bmatrix}.$$

Remark: If we allow jumps in prices or in dividends, we must decide how to interpret $S(t)$. Since $S(t)$ is a continuous function we have two possible choices; Either $S(t)$ is the price before or after the dividend. The choice will be of importance when we will define a self-financed portfolio. We will choose $S(t)$ to be price after the dividend since this is the most common in the literature. This problem does not exist if all trajectories are continuous.

We will now analyse self-financed portfolios by using $B(t)$ as the numeraire process and make the step from the $S(t)$ economy to the $Z(t)$ economy. Then we can define martingale measures, self-financed portfolios, contingent claims, for example.

Definition 5.30 On a given market $[S, D]$ the *gain process* is defined as:

$$G(t) = S(t) + D(t)$$

The *Z-market* $[Z, D^Z]$is defined as:

$$\begin{cases} Z(t) = \dfrac{S(t)}{B(t)} \\ D^Z(t) = \displaystyle\int_0^t \frac{1}{B(s)} dD(s) \end{cases}$$

i.e.,

$$dD^z(t) = \frac{1}{B(t)} dD(t)$$

and the *Z-gain process* G^Z is defined as

$$G^Z(t) = Z(t) + D^Z(t).$$

We said that the price $S(t)$ and the dividend $dD(t)$ at time t are discounted with the deflator $B(t)$.

Definition 5.31 A probability measure Q is said to be a martingale measure for the market $[S, D]$ if

(i) $Q \sim P$,
(ii) G^Z is a squared integrable Q-martingale.

The set of martingale measures are denoted by $\boldsymbol{P}$.

As usual, we now need to show that, if $\boldsymbol{P} \neq \emptyset$, then the economy is free of arbitrage. Therefore, we need to define the class of self-financed portfolio strategies, the class of contingent claims and the concept of free of arbitrage. We will define these on a market with a fixed martingale measure.

Definition 5.32 For a fixed $Q \in \boldsymbol{P}$, a self-financed *portfolio strategy*, is a process $h = (h^0, h^*) = (h^0, h^1, \ldots, h^N)$ (a row-vector) such as

(i) h is $\underline{\mathcal{F}}$-predictable,
(ii) The process $V^Z(h)$ defined by

$$V_t^Z(h) \underset{def}{=} h_t Z_t = \sum_{i=0}^{N} h_t^i Z_t^i$$

is squared integrable, i.e., $E^Q\left[V_t^Z(h)\right] < \infty, \quad \forall t \geq 0$. The process $V^Z(h)$ is called the *Z-value process*, while the *S-value process* $V(h)$ is defined by

$$V_t(h) \underset{def}{=} h_t S_t = \sum_{i=0}^{N} h_t^i S_t^i.$$

A portfolio strategy h is said to be *self-financed* if

$$V_t(h) = V_0(h) + \int_0^t h(s) dG(s)$$

i.e., if

$$dV_t(h) = h_t dG_t.$$

The class of all self-financed portfolios are denoted by $\mathcal{H}$.

Remark The condition (i) above is only critical if the processes contain jumps. Otherwise we only need to have left-continuous and adapted processes.

Definition 5.33 For a fixed martingale measure Q, a *contingent claim* is a stochastic process X such as

$$X \in \mathcal{F}_T$$
$$E^Q\left[X^2\right] < \infty$$

The set of all contingent claims are denoted by $\mathcal{K}(Q)$. With $\mathcal{K}^+$ we denote those $X \in \mathcal{K}$ such as

$$P(X \geq 0) = 1 \text{ and } P(X > 0) > 0.$$

A contingent claim is said to be *reachable* (in S and Z respectively) if there exist a self-financed strategy h such as

$$\begin{aligned} V_T^S(h) &= X \\ V_T^Z(h) &= X. \end{aligned}$$

The relationship between the S- and the Z-market is given by Lemma 5.34.

Lemma 5.34

(i) *For each portfolio strategy h we have:*

$$\begin{cases} V_T^S(h) = B(T) \cdot V_T^Z(h) = V_T^Z(h) \cdot \exp\left\{ \int_0^T r(s)ds \right\} \\ V_T^Z(h) = B(T)^{-1} \cdot V_T^S(h) = V_T^S(h) \cdot \exp\left\{ -\int_0^T r(s)ds \right\} \end{cases}$$

(ii) *The contract X is S-reachable if and only if the contract $B(T)^{-1}X$ is Z-reachable.*

(iii) *The portfolio strategy h is self-financed if and only if*

$$dV_t^Z(h) = h_t dG_t^Z = h_t^* dG_t^{Z*}$$

(iv) *If $h = \mathcal{H}(Q)$, then $V^Z(h)$ becomes a squared integrable Q-martingale.*

Definition 5.35 For a fixed martingale measure Q a self-financed portfolio $h \in \mathcal{H}(Q)$ is said to be an arbitrage strategy if

$$V_0^S(h) = 0, \ \text{ and } \ V_T^S(h) \in \mathcal{K}^+(Q).$$

Theorem 5.36 *Suppose $\boldsymbol{P} \neq \emptyset$. Then, the market is free of arbitrage, in the meaning that there will not exist for any $Q \in P$ any arbitrage strategies in $H(Q)$.*

We now have tools to price contingent claims in our economy. Therefore, suppose that $\boldsymbol{P} \neq \emptyset$ on the market $[S, D]$ and choose a martingale measure Q (there can be many) and consider a contingent T-claim X (this means that the delivery should be at time $t = T$). The arbitrage free price of X is then given by

$$\pi_t[X;Q] = E^Q\left[X \cdot \exp\left\{-\int_t^T r(s)ds\right\}|\mathcal{F}_t\right]$$

π^Z is then a Q-martingale and by the theorem it follows that the $[S, D]$-market adjoint with the pair of price and dividend $(\pi, 0)$ is free of arbitrage.

Theorem 5.37 *With Q and X as above, Then*

(i) *The market $[S, D]$ adjoint with $(\pi, 0)$, where π is defined as above is free of arbitrage.*
(ii) *Different choices of $Q \in P$ will generically give different price processes.*
(iii) *If a given contract is reachable all martingale measures will give the same price process.*
(iv) *Especially, for each $Q \in P$ and for each pair $[S^i, D^i]$:*

$$S^i(t) = E^Q\left[S^i(T) \cdot \exp\left\{-\int_t^T r(s)ds\right\} + \int_t^T \exp\left\{-\int_t^s r(u)du\right\} dD^i(s)\middle|\mathcal{F}_t\right]$$

Consider a given market consisting of one risky asset with price process $S(t)$ and cumulative dividend process $D(t)$ and a risk-free asset $B(t)$, with dynamics

$$\begin{cases} dB(t) = r \cdot B(t)dt \\ B(0) = 1 \end{cases}$$

where r denotes a stochastic interest rate.

Now consider a fixed contingent T-claim X. A futures contract on X with time of delivery T is a financial asset with price process Π and dividend process D with the following properties:

$$\begin{aligned} &D(t) = F(t,T,X) \\ &F(T,T,X) = X \\ &\Pi(t) = 0 \text{ for } 0 \le t \le \text{T}. \end{aligned}$$

Here $F(t, T, X)$ denotes the futures price process. Note that $F(t, T, X)$ is determined at time t. Recall that if Q is a martingale measure for this model, then the normalized gain process of any price-dividend pair $[\Pi , D]$ is a Q-martingale. Thus we have that

$$G^Z(t) = \frac{\Pi(t)}{B(t)} + \int_0^t \frac{1}{B(s)} dD(s)$$

is a Q-martingale. Now use that $\Pi(t) = 0$, and that $D(t) = F(t, T, X)$ along with the martingale representation Theorem to obtain that

$$dG^Z(t) = \frac{dF(t,T,X)}{B(t)} = h(t)dV(t)$$

for some adapted process h and the Q-Wiener process V generating the filtration. From this we see that

$$dF(t,T,X) = B(t)h(t)dV(t),$$

which means that the futures price process is a Q-martingale. Using the martingale property of the futures price process and the boundary condition $F(T, T, X) = X$, we have that the futures prices are given by

$$F(t,T,X) = E^Q[X|\mathcal{F}t]$$

Now, consider a Black–Scholes model with a constant continuous dividend yield, i.e. where B, S and D satisfy

$$\begin{cases} dB_t = rB_t dt \\ dS_t = \alpha S_t dt + \sigma S_t dW_t, \\ dD_t = \delta S_t dt \end{cases}$$

where r, α, σ and δ are assumed to be constants and W denotes a P-Wiener process. To give an explicit formula for the futures price for the case when $X = S(T)$, we recall that for this model the Q-dynamics of S are given by

$$dS(t) = (r - \delta)S(t)dt + \sigma S(t)dV(t),$$

where V denotes a Q-Wiener process. This follows from the fact that the normalized gain process

$$G^Z(t) = \frac{S(t)}{B(t)} + \int_0^t \frac{\delta S(u)}{B(u)} du$$

is a Q-martingale. Integrating this we obtain

$$S(u) = S(t) + \int_t^u (r - \delta)S(\tau)d\tau + \int_t^u \sigma S(\tau)dV(\tau).$$

Now take the conditional expectation with respect to $\mathcal{F}_t$

$$E[S_u|\mathcal{F}_t] = S_t + \int_t^u (r - \delta)E[S_\tau|\mathcal{F}_t]d\tau + 0.$$

Let $m_u = \mathrm{E}[S_u| \mathcal{F}_t]$ and take derivatives with respect to u

$$\begin{cases} \dot{m}_u = (r - \delta)m_u \\ m_t = S_t \end{cases}.$$

Solving the ODE above we get

$$m_u = S_t e^{(r-\delta)(u-t)}.$$

The futures price for this case is therefore

$$F(t, T, X) = S(t)e^{(r-\delta)(T-t)}.$$

If we suppose that the constant $\delta = 0$. The gain process G of an asset with price process Π and dividend process D is given by $G(t) = \Pi(t) + D(t)$, and a portfolio containing h of these assets is self-financing if its corresponding value process satisfies $dV(t, h) = h(t)dG(t)$. If we denote by P the price of the self-financing portfolio we have that $dP(t) = 1(d\Pi(t) + dD(t)) = dF(t, T, X)$ where we have used that $d\Pi(t) = 0$ (since $\Pi(t) = 0$ for $0 \leq t \leq T$) and that $D(t) = F(t, T, X)$. Integrating this we have that $P(t) = P(0) + F(t, T, X) - F(0, T, X)$ or, using the expression for the futures price from the previous exercise (with $\delta = 0$) $P(t) = P(0) + S(t)e^{r(T-t)} - S(0)e^{rT}$. With a slight abuse of notation we now let $P(t, x) = P_0 + s \cdot e^{r(T-t)} - S_0 e^{rT}$ denote the pricing function of the portfolio. It is now easy to compute Δ. We have

$$\Delta = \frac{\partial P}{\partial s} = e^{r(T-t)}.$$

5.3.3 Black–Scholes with Continuous Dividends

We will now see how we can use this on Black–Scholes with continuous dividends. The theory with discrete dividends is a little more complex and will not be discussed here. But in practice you can use the Black–Scholes formulas and subtract the present values (discounted by the risk-free interest rate) of the individual dividends.

As usual, we start with a market consisting of two securities, B and S with the given P-dynamics

$$\begin{cases} dB(t) = r \cdot B(t)dt \\ dS(t) = \alpha \cdot S(t)dt + \sigma \cdot S(t)dW(t) \end{cases},$$

where r, α, and σ are deterministic constants, $\sigma > 0$. We now suppose that we to S also have a dividend process:

$$dD(t) = \delta \cdot S(t)dt + \gamma \cdot S(t)dW(t),$$

where δ, and γ are deterministic constants. We begin to look for a martingale measure and investigate what we will get from a Girsanov transformation:

$$L(t) = \frac{dQ}{dP},$$

where

$$\begin{cases} dL(t) = h(t)L(t)dW(t) \\ L(0) = 1 \end{cases}.$$

Girsanov's Theorem gives us the following dynamics under Q, where $V(t)$ is a Q-Wiener process.

$$\begin{cases} dS(t) = (\alpha + \sigma \cdot h)S(t)dt + \sigma \cdot S(t)dV(t) \\ dD(t) = (\delta + \gamma \cdot h)S(t)dt + \gamma \cdot S(t)dV(t) \end{cases}.$$

The Z-dynamics can be shown to be given by

$$\begin{cases} dZ(t) = (\alpha + \sigma \cdot h - r)Z(t)dt + \sigma \cdot Z(t)dV(t) \\ dD^Z(t) = (\delta + \gamma \cdot h)Z(t)dt + \gamma \cdot Z(t)dV(t) \end{cases},$$

so that the gain process of Z, G^Z have the dynamics given by

$$dG^Z(t) = (\alpha + \sigma \cdot h - r + \delta + \gamma \cdot h)Z(t)dt + (\sigma + \gamma)Z(t)dV(t).$$

Therefore, if G^Z is a Q-martingale we have to choose the Girsanov kernel h such as

$$\alpha + \sigma \cdot h - r + \delta + \gamma \cdot h = 0$$

i.e.,

$$h = \frac{r - \alpha - \delta}{\sigma + \gamma}.$$

Therefore, we have shown the following result:

Theorem 5.38 *Given a financial market*

$$\begin{cases} dB(t) = r \cdot B(t)dt \\ dS(t) = \alpha \cdot S(t)dt + \sigma \cdot S(t)dW(t) \\ dD(t) = \delta \cdot S(t)dt + \gamma \cdot S(t)dW(t) \end{cases}$$

(i) *This market is free of arbitrage and has a unique martingale measure given by*

$$L(t) = \frac{dQ}{dP},$$

where

$$\begin{cases} dL(t) = h(t)L(t)dW(t) \\ L(0) = 1 \end{cases}$$

and

$$h = \frac{r - \alpha - \delta}{\sigma + \gamma}$$

(ii) *The arbitrage-free price of a T-contract X is given by*

$$\pi_t[X] = E^Q[X|\mathcal{F}_t]$$

(iii) *The dynamics of [S, D] under Q is given by*

$$\begin{cases} dS(t) = \left(\alpha + \sigma \cdot \dfrac{r-\alpha-\delta}{\sigma+\gamma}\right) S(t)dt + \sigma \cdot S(t)dV \\ dD(t) = \left(\delta + \gamma \cdot \dfrac{r-\alpha-\delta}{\sigma+\gamma}\right) S(t)dt + \gamma \cdot S(t)dV \end{cases}$$

(iv) *If* $\gamma = 0$ *the dynamics of [S, D] under Q is given by*

$$\begin{cases} dS(t) = (r-\delta)S(t)dt + \sigma \cdot S(t)dV \\ dD(t) = \delta S(t)dt \end{cases}$$

(v) *If* $\gamma = 0$ *and X can be written as* $X = \Phi(S(T))$, *then the arbitrage-free is given by*

$$\pi_t[X] = F(t, S(t)),$$

where F(t, S(t)) is the solution to the following partial differential equation

$$\begin{cases} F_t(t,S) + (r-\delta)SF_S(t,S) + \frac{1}{2}\sigma^2 S^2 F_{SS}(t,S) - rF(t,S) = 0 \\ F(T,S) = \Phi(S) \end{cases}.$$

With (v) above, the price can be written as

$$F(t,S) = e^{-r(T-t)} \int_{-\infty}^{\infty} \Phi\left(s \cdot \exp\left\{\left(r - \delta - \frac{1}{2}\sigma^2\right)(T-t) + \sigma y \sqrt{T-t}\right\}\right)\varphi(y)dy$$

where φ is the density function of an N(0, 1) distribution.

A European call option on S with maturity T and strike K is given by:

$$C(t,s) = s \cdot e^{-\delta(T-t)} N[d_1(t,s)] - Ke^{-r(T-t)} N[d_2(t,s)]$$

where

$$\begin{cases} d_1 = \dfrac{1}{\sigma\sqrt{T-t}}\left\{\ln\left(\dfrac{s}{K}\right) + \left(r - \delta + \frac{1}{2}\sigma^2\right)(T-t)\right\} \\ d_2 = d_1 - \sigma\sqrt{T-t} \end{cases}$$

5.4 Hedging

An important concept when dealing with finance is hedging. When you are dealing with derivatives—for example, selling a call option— the counterparty might exercise the option. If and when the counterparty exercises the option you have the obligation to deliver the underlying stock. This can be very expensive if you need to buy the underlying stock at a high price and sell it at a much lower price. One of your possibilities is to hedge the option position by buying the underlying stock before or when you enter the option contract. This is the simplest kind of hedging, called delta-hedging. The name come from the fact that you can use the option delta to calculate the number of stocks you need to buy to be hedged. Since delta might vary during the option life time you can need to rebalance your portfolio by buying or selling more stocks.

5.4.1 Delta-Hedging

We will now in detail study how to hedge using delta. Suppose that we take a short position in European call options on 100,000 stocks in ACME Inc. Suppose we have

$$\begin{aligned} S &= 365, \\ X &= 370, \\ \sigma &= 20\%, \\ r &= 2\% \text{and} \\ T - \mathrm{t} &= 0.25\,\text{year}. \end{aligned}$$

Via Black–Scholes formula, we get the option price 13.09497 so the total value is 1,309,497. As we will see, we are exposed for a risk.

First, suppose we have a naked position, e.g. we have no ownership in the underlying stock.

Study two cases:

1. At maturity, the stock price is < 370, (the option has no value) so we make a profit of 1,309,497.
2. At maturity, the stock price is 395 so we have to buy 100,000 stocks for 395 cash units each and then sell them at a price of 370. The cost will be $100{,}000(395 - 370) = 2.5$ million, so we lose 1,190,503.

Next, suppose we have a covered position, i.e. we buy the stock at 365 each.
Study two cases:

1. At maturity, the stock price is 360, so we lose 500,000 selling the stocks, but get a total profit of 1,024,738. If the stock price goes below 350 we will make a negative profit totally.
2. At maturity, the stock price is 380. We sell the stocks at 380 each and make a total profit of 2,024,738.

We will now see how we can protect ourselves with a hedge. We will therefore calculate the number of stocks we have to buy, to hedge the options. If $F(t, S)$ is the option value, N_c, the number of options and N_s the number of stocks, the total portfolio value is given by:

$$V = -N_c F(t, S) + N_s S$$

With a delta-hedge

$$\frac{\partial V}{\partial S} = 0 \quad \Rightarrow \quad N_s = N_c \cdot \frac{\partial F}{\partial S} = N_c \cdot \Delta = N_c \cdot N[d_1]$$
$$d_1 = \frac{1}{\sigma\sqrt{\Delta t}}\left\{\ln\left(\frac{S}{K}\right) + \left(r + \frac{1}{2}\sigma^2\right)\Delta t\right\} = -0.03606....$$

This gives $N[d_1] = 0.485619$, so we need $N_s = 48{,}562$ stocks to hedge the options.

5.4.2 Delta-gamma-Hedging

If we also want to be Γ-neutral (i.e. have $\Gamma = 0$) we have to user a second option in our hedge. Remember that gamma is given by

$$\Gamma = \frac{N'[d_1]}{S\sigma\sqrt{\Delta t}}$$

for both call and put options. Suppose there exist a put option p, at strike price of 355 cash units. Then our portfolio is

$$V = N_p F_p(t, S) - N_c F_c(t, S) + N_s S$$

and

$$\frac{\partial V}{\partial S}=\frac{\partial^2 V}{\partial S^2}=0 \quad \Rightarrow \quad \{\Delta_p=\Delta_c-1\} \quad \Rightarrow$$
$$\begin{cases} N_s=N_c\cdot\Delta_c-N_p\cdot\Delta_p \\ 0=N_c\cdot\Gamma_c-N_p\cdot\Gamma_p \end{cases}.$$

If we also want to remove the sensitivity in Δ, we also must have $\Gamma=0$. Since $\Gamma=0$ for stocks we have to use one more option. Given a portfolio Π with a stock S and two derivatives, F and G. We want to choose X_F and X_G so that the total portfolio becomes both Δ- and Γ-neutral:

$$\begin{cases} N_p=100{,}000\cdot\dfrac{\Gamma_c}{\Gamma_p} \\ N_s=100{,}000\cdot\Delta_c-100{,}000\Delta_p\cdot\dfrac{\Gamma_c}{\Gamma_p} \end{cases}.$$

Giving

$$\begin{cases} N_p=107{,}328 \\ N_s=-20{,}901 \end{cases}$$

Example 5.39

At the writing moment the stock price of some share is 35. Now, we want to hedge 1000 stocks of this share. On the market there exist options with strikes 30 and 37. The risk-free interest rate is estimated to 4.5 % and the time to maturity of the options is 102 days. The volatility is estimated to 37.5 %. If we use the formulas above on a call option with strike 30 and a put option with strike 37, we will find that if we buy 548 put options and go short in 847 call options we will hedge our 1000 stocks. In Fig. 5.1 we illustrate the total portfolio value when the stock price varies between 10 and 70. As we can see, the hedge is very good in a region between 28 and 42. We also observe that we earn 3150 in the hedge.

If we in detail see how the hedge works, we can plot all three instruments in the same graph. This is shown in Fig. 5.2.

If we switch the two options in the strike and use a call option with strike 37 and a put option with strike 30. The total portfolio value looks like the curve in Fig. 5.3. In this hedge we earn 1417. We have to sell 1386 call options and buy 2142 put options.

If we in detail see how this hedge works, we can plot all three instruments in the same graph. This is shown in Fig. 5.4.

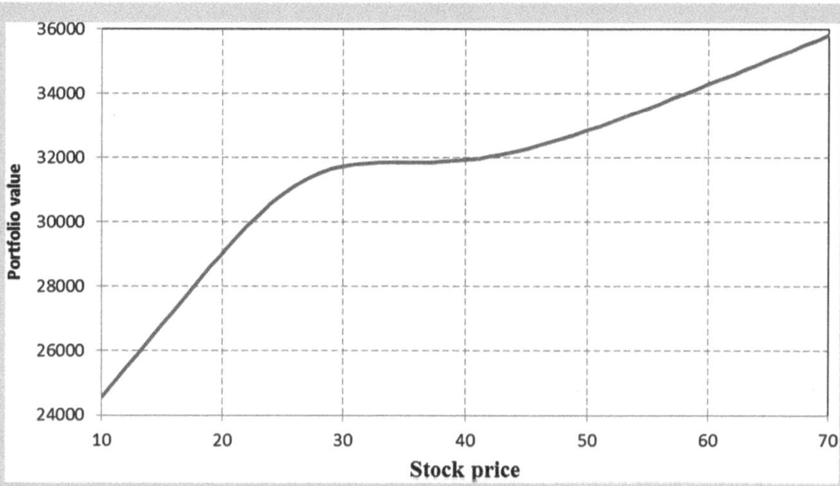

Fig. 5.1 Illustration of the delta-gamma hedge of 1000 stocks

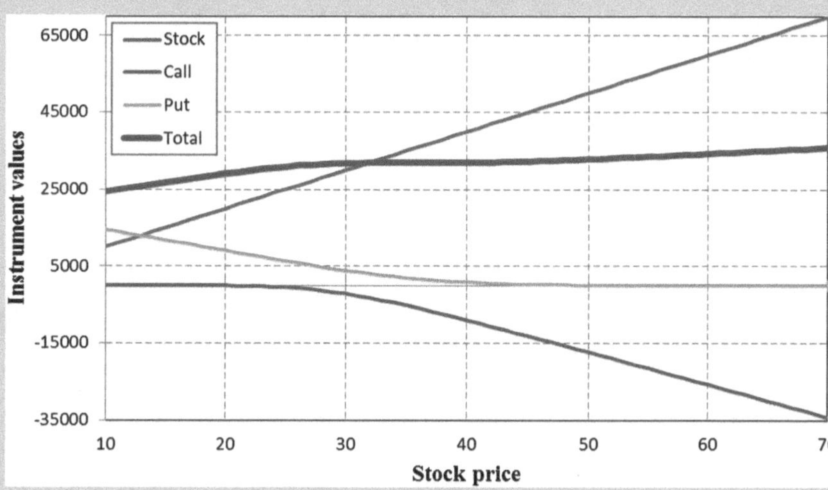

Fig. 5.2 Illustration of how the delta-gamma hedge of 1000 stocks is made up by the two options and the stock itself. The fat line represents the total portfolio shown in Fig. 5.1

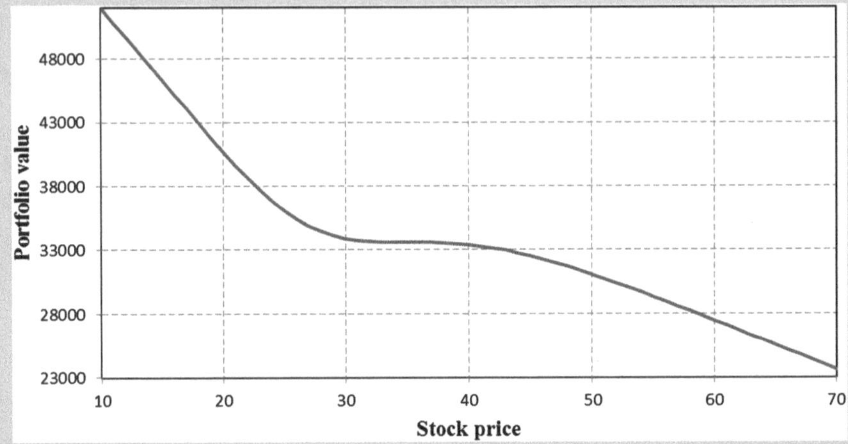

Fig. 5.3 Illustration of the delta-gamma hedge of 1000 stocks with switched option strikes

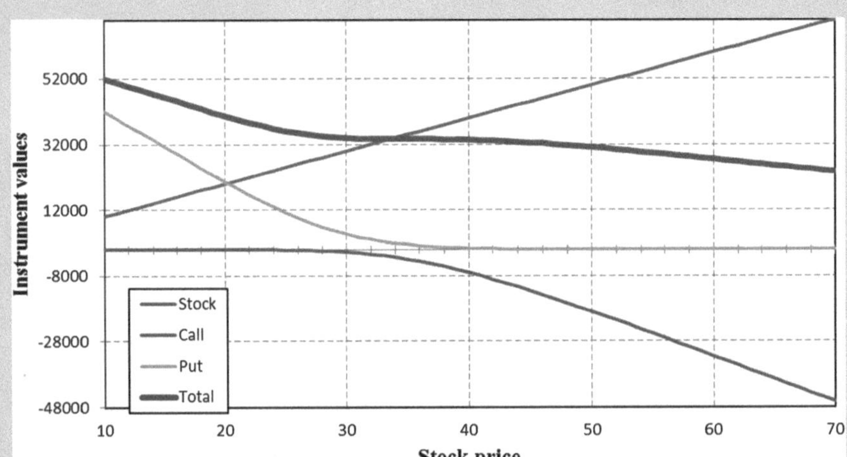

Fig. 5.4 Illustration of how the delta-gamma hedge of 1000 stocks is made up by the two options and the stock itself. The fat line represents the total portfolio shown in Fig. 5.3

6

Exotic Options

A standard option has some well-defined properties:

Type	Call or put
Style	European, Bermudan or American
Strike	A given price X
Expiry date	The time of maturity
Settlement type	Physical or cash delivery
Underlying	Stock, currency, index, etc.

For exotic options one or some of the above properties are defined differently or additional properties are added. Exotic options can be constructed in many different ways. To understand how to construct these kinds of option we give an overview of some common types:

- Cash-or-nothing options
- Knock-out and knock-in options
- Barrier options
- Lookback options
- Asian options
- Chooser options
- Options on two underlyings
- Options on options
- Currency options
- Forward options

J.R.M. Röman, *Analytical Finance: Volume I*,
DOI 10.1007/978-3-319-34027-2_6

6.1 Contract for Difference: CFD

Before we explore exotic options we describe a new instrument that has been very popular, the contract for difference (CFD).

A contract for difference is a contract between two parties, typically described as "buyer" and "seller", stipulating that the seller will pay to the buyer the difference between the current value of an asset and its value at contract time. If the difference is negative, then the buyer pays the seller. For example, when applied to equities, such a contract is an equity derivative that allows investors to speculate on share price movements without the need for ownership of the underlying shares.

Contracts for difference allow investors to take long or short positions, and unlike futures contracts they have no fixed expiry date, standardized contract or contract size.

Investors in CFDs are required to maintain a certain amount of margin as defined by the brokerage or market maker (usually ranging from 1 % to 30 %). One advantage to investors is that they don't have to pay collateral for the full notional value. Therefore, a given quantity of capital can control a larger position, amplifying the potential for profit or loss. On the other hand, a leveraged position in a volatile CFD can expose the buyer to a margin call in a downturn, which often leads to losing a substantial part of the assets.

As with many leveraged products, the maximum exposure is not limited to the initial investment since it is possible to lose more than one put in. These risks are typically mitigated through the use of stop orders and other risk-reduction strategies. For the most risk-averse, guaranteed stop loss orders are available at a certain cost.

6.1.1 History

CFDs were originally developed in the early 1990s in London. Based on equity swaps, they had the additional benefit of being traded on margin and being exempt from stamp duty, a UK tax.

They were initially used by hedge fund and institutional investors to hedge their exposure to stocks on the London Stock Exchange in a cost-effective way.

In the late 1990s CFDs were first introduced to retail investors. They were popular with a number of UK companies, whose offerings were typically characterized by innovative online trading platforms that made it easy to see live prices and trade in real time.

It was around the year 2000 that retail investors realized that the real benefit of trading CFDs was not the exemption from stamp tax but the ability to trade on leverage on any underlying instrument. This was the start of the growth phase in the use of CFDs. The CFD providers quickly responded and expanded their product offering from London Stock Exchange (LSE) shares only to include indexes, global stocks, commodities, treasuries and currencies. Trading index CFDs, such as those based on the major global indexes (e.g. Dow Jones, NASDAQ, S&P 500, FTSE, DAX, and CAC) quickly became the most popular type of CFD to be traded.

Around 2001 a number of the CFD providers realized that CFDs have the same economic effect as financial spread betting except that the tax regime was different, making it in effect tax free for clients.

Up until this point CFDs were always traded over the counter (OTC); however, on 5 November 2007 the Australian Securities Exchange (ASX) listed exchange-traded CFDs on the top 50 Australian stocks, 8 FX pairs, key global indices and some commodities.

Example 6.1

The easiest way to show the use the leverage of a CFD is as follows: if you had $1750 to invest, and wished to purchase a stock at $35 and sell at $37, a standard trade would look as follows

BUY: 50 × $35 = $1750
SELL: 50 × $37 = $1850
PROFIT = $100 or 5.7 %
Using a CDS, the above example reads as follows:
BUY: 1000 × $35 = $1750 (5 % deposit) + $33,250 (95 % borrowed funds)
SELL: 1000 × $37 = $37,000
PROFIT = $2000 or 114 %

As you can see, the profit received after using leverage was far greater than without. It is important to note, that losses are also magnified when using leverage.

6.1.2 Risk

Due to the dynamic nature of the stock markets and increased leveraged possible with CFDs (up to 500 fold) it is important that traders regularly calculate their risk, position sizes and the return required to cover CFD loss and overnight financing costs to manage their moving portfolio and changing risk. With the leveraging of CFDs it is possible to lose a lot more money than your account size should positions go against you.

CFDs allow a trader to go short or long on any position using margin. There are always two types of margin (see the appendix) with a CFD trade

1. Initial margin, which is normally 5–30 % for shares and 1 % for indices and FX.
2. A daily margin, which is then "marked to market".

Variation margin is applied to positions if they move against a client. The daily margin can therefore have either a negative or positive effect on a CFD trader's cash balance. But initial margin will always be deducted from a customer's account and replaced once the trade is covered.

Another dimension of CFD risk is counterparty risk; a factor in most OTC-traded derivatives. Counterparty risk is associated with the financial stability or solvency of the counterparty to a contract. In the context of CFD contracts, if the counterparty to a contract fails to meet their financial obligations, the CFD may have little or no value regardless of the underlying instrument. Exchange-traded contracts traded through a clearinghouse have less or no counterparty risk.

6.2 Binary Options/Digital Options

Binary options, sometimes also known as "rebate" options, are vanilla put and call options conditioned by something else other than just the price and the expiration date. They refer to conditional scenarios that, if they come true, either validate or invalidate the option. The trader fixes the amount of the desired payout he/she will get if their conditional scenario proves to be right. The price of the option or premium represents a percentage of that payout.

Accordingly, digital options are less expensive than one-touch options with the same strike and expiration date. Digital premiums can be half the price of no-touch options premiums with the exact same strike price and expiration dates, but the trader has to weigh the advantage of lower cost against the risk price will settle even 1 pip below the target at expiration.

Traders often combine various option types to build their option trading strategies. By associating different option types, some traders manage to minimize the risk they are taking. Some even claim to have found infallible methods. Others see it as a simple hedging instrument and use it to secure their funds.

6.2.1 Cash-or-Nothing Options

There are two kind of digital option: *cash-or-nothing options* and *asset-or-nothing options*. These options have as most of the others, a strike price and a given time to maturity.

A cash-or-nothing digital option pays a fixed amount of cash if they expire in the money, no matter how deeply, otherwise nothing. They are also known as *binary options, all-or-nothing options* or *bet options*.

Cash-or-nothing digitals differ from conventional options since their payoff does not depend on the extent to which the option is in-the-money (ITM), only on whether it is ITM. So digitals are a bet on the market reaching a certain level, with a cash payout if the bet is won. Otherwise the payout is zero. In Fig. 6.1 we illustrate the profile of digital options. The step function represents the call option at maturity and the other curves the call- and the put options with the following parameters; $S = K = 70$, $T = 0.5$, $r = 2.0\%$ and $\sigma = 40$ %.

Digitals are not only traded as standalone products, they are also often found embedded in structured securities.

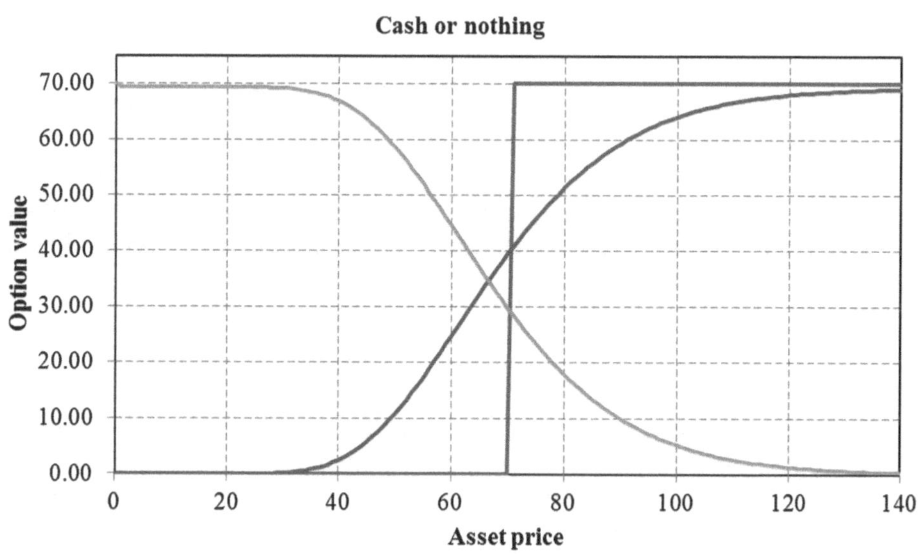

Fig. 6.1 The payout (profit) of digital cash-or-nothing options

6.2.2 Pricing

Mathematically, European-style digitals are easier to price than conventional options. The price for a cash-or-nothing option, is as we have seen given as one part of the Black–Scholes formula

$$P_{call} = e^{-rT}K \cdot N(d)$$
$$P_{put} = e^{-rT}K \cdot N(-d),$$

where

$$d = \frac{\ln\left(\frac{S}{K}\right) + \left(r + \frac{\sigma^2}{2}\right)T}{\sigma \cdot \sqrt{T}}.$$

The price of an asset-or-nothing option is given by

$$P_{call} = S \cdot N(d)$$

and

$$P_{put} = S \cdot N(-d).$$

What makes the digital option exotic is its risk profile, in particular the behaviour of its delta.

The stepped expiry payoff of a digital option means that the delta of an ATM option converges to infinity as it approaches expiry, while the delta of an OTM or ITM option converges to zero. A close-to-the-money digital approaching expiry becomes almost impossible to delta-hedge. Option traders are therefore careful to

- Limit the amounts of digitals they carry in their books
- Build significant safety margins into their pricing, to allow for the additional risks—a good example of the difference between theoretical pricing, based on a mathematical model, and actual pricing based on the potential costs of hedging the risks in practice

An exotic option is a contract whose "Greeks" behave differently from those of a conventional option. Whereas a conventional long call position is always theta-negative and vega-positive, the equivalent position in a digital may be positive or negative in theta and vega, depending on whether it is OTM or ITM. In practice, digitals are sometimes priced off (and hedged with) conventional call or put spreads. The expiry payoff of the digital call cannot be replicated exactly using conventional options.

Example 6.2

The solution of a digital call with strike $K = S(0)$ can be found as

$$\Phi(S_T) = \begin{cases} 1 & S_T > K = S(0) \\ 0 & \text{else} \end{cases} = 1_{I\{s>K\}}$$

$$\begin{aligned} C(t,s) &= e^{-r(T-t)} E^Q\left[1_{\{S_T>K\}}\right] \\ &= e^{-r(T-t)} E^Q\left[1_{\{S_0 \exp\{(r-\sigma^2/2)(T-t)+\sigma(W_T-W_t)\}>K\}}\right] \\ &= \frac{K}{\sqrt{2\pi}} \int_{-\infty}^{d_2} e^{-x^2/2} dx = K \cdot N[d_2]. \end{aligned}$$

6.2.3 Supershare Options

A *supershare option* is a type of binary option. In a common binary option the payout is a fixed amount if the underlying price is greater than (or less than) the strike.

In a supershare option, there is a lower and upper boundary. If the underlying at expiry is between these boundaries the payoff is

$$\text{Payoff} = \text{Underlying/Lower Boundary}$$

If the underlying security price is outside these boundaries the payoff is zero. The supershare option (introduced by Håkansson 1976), entitles the holder to a payoff of 0 if $X_L > S > X_H$ and S/X_L otherwise.

$$P = \frac{e^{-rT}}{X_L}[N(d_1) - N(d_2)]$$

where

$$d_1 = \frac{\ln\left(\frac{S}{X_L}\right) + \left(r + \frac{\sigma^2}{2}\right)T}{\sigma \cdot \sqrt{T}}; \qquad d_2 = \frac{\ln\left(\frac{S}{X_H}\right) + \left(r + \frac{\sigma^2}{2}\right)T}{\sigma \cdot \sqrt{T}}.$$

Simply put, the digital price is given by the present value of the fixed payout multiplied with the probability of reaching the strike.

6.2.4 Other Digital Options

There are two other kinds of digital options, *American digitals* and *asset or nothing*. The American-style digital, like all American options, has a market price that may not be lower than its intrinsic value, so the option's price rises even more steeply than the European equivalent as it approaches the strike. Moreover, the option has no time value when it is ITM and should be exercised immediately.

The asset-or-nothing digital call gives the holder the right to buy the underlying asset at a specified discount to the strike, while the asset-or-nothing digital put gives the holder the right to sell the underlying asset at a premium to the strike.

In Fig. 6.2 we illustrate the profile of an asset-or-nothing option with the following parameters; $S = K = 70$, $T = 0.5$, $r = 2.0$ % and $\sigma = 40$ %.

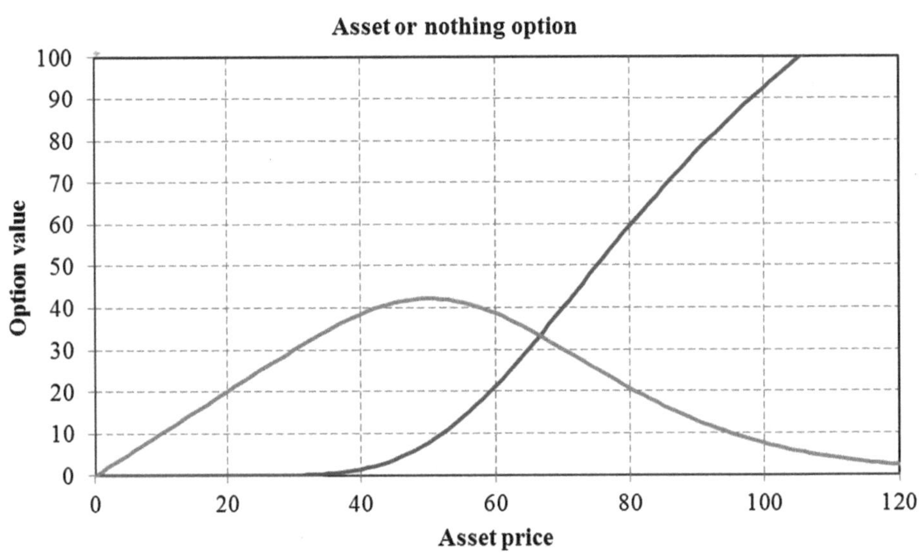

Fig. 6.2 The payout (profit) of an asset-or-nothing call and an asset or nothing put option. The bumpy curve represents the put option.

Reiner and Rubenstein derived a formula for American digitals in 1991. Sometimes they are called one-touch binary/digital or binary-at-hit. The value of a call option of the type one-touch-down digital is given by

$$P_{one\_touch\_down} = K \cdot \left[\left(\frac{H}{S}\right)^{\mu+\lambda} N(z) + \left(\frac{H}{S}\right)^{\mu-\lambda} N\left(z - 2 \cdot \lambda \cdot \sigma\sqrt{T}\right) \right]$$

and a one-touch-up digital by

$$P_{one\_touch\_up} = K \cdot \left[\left(\frac{H}{S}\right)^{\mu+\lambda} N(-z) + \left(\frac{H}{S}\right)^{\mu-\lambda} N\left(2 \cdot \lambda \cdot \sigma\sqrt{T} - z\right) \right]$$

where the barrier is given by H and

$$z = \frac{\ln\left(\frac{H}{S}\right)}{\sigma\sqrt{T}} + \lambda\sigma\sqrt{T}$$

$$\mu = \frac{r - \frac{1}{2}\sigma^2}{\sigma^2}$$

$$\lambda = \sqrt{\mu^2 + \frac{2r}{\sigma^2}}.$$

In Haug's book *The Complete Guide to Option Pricing Formulas* a huge family of digital barrier options are given as well as other exotics.

6.2.5 Gap Options

Gap options are a combination of a conventional option and a cash-or-nothing digital. A gap option is the right to buy, for a call, or the right to sell, for a put, an asset at time $T > 0$ for a price $G > 0$ if the asset exceeds, for the call, or falls below a price $X > 0$. It is straightforward to write down the price for a gap call as

$$C_g(S,T,G,X) = e^{-\bar{r}T} E^Q[S(T) - G | S(T) > X].$$

We can evaluate this conditional expectation directly without much trouble, but there is a better way. A gap call can be written as a portfolio with a usual call and a digital call. To see this we rewrite the above equation as

$$C_g(S,T,G,X) = e^{-\bar{r}T}\left(E^Q[S(T) - X|S(T) > X] - E^Q[G - X|S(T) > X]\right).$$

A put gap can be decomposed in a similar way as

$$P_g(S,T,G,X) = e^{-\bar{r}T}\left(E^Q[X - G|S(T) < X] - E^Q[X - S(T)|S(T) < X]\right).$$

Thus, we have

$$C_g(S,T,G,X) = C(S,T,X) - (G - X)C_d(S,T,X).$$

and

$$P_g(S,T,G,X) = (X - G)P_d(S,T,X) - P(S,T,X).$$

6.2.6 Collars

A collar is an option to buy an asset at strike price $X > 0$, but the total payoff is capped at $Z > X$. The claim at time T is given by min $((S(T) - X)^+, (S(T) - Z)^+)$, which equals $((S(T) - X)^+ - (S(T) - Z)^+)$. Thus, a collar can be priced as

$$C_c(S,T,X,Z) = C(S,T,X) - C(S,T,Z).$$

6.3 Barrier Options: Knock-out and Knock-in Options

The payoff of a conventional option depends only on the price of the underlying relative to the strike at the time of exercise, but there are so-called *path-dependent options* whose payoffs also depend on the history of the underlying price—that is, where the market has been before expiry. One important class of this type is the barrier option.

There are two general classes of so-called barrier options:; in-options and out-options. With in-options the buyer gets an option that becomes active if and when the underlying hit a given barrier value. If the underlying never reaches this value, the option will expire without a value.

An out-option is an option, which is active from the beginning, but becomes inactive, that is, expires immediately if the underlying hits the barrier value.

It is possible to combine both types. If we have a down-and-out-call option and a down-and-in-call option and the underlying hit the barrier, the down-and-out becomes inactive while the down-and-in becomes active. Therefore, this combination is an exact replication of a plain vanilla European call option.

As we will see below, it is possible to create many kind of barrier options on all kind of markets. All such barrier options can be either call or put options.

6.3.1 One-touch Options

When buying a *one-touch currency option* traders set that if the currency trades at a specified rate or trigger, then he/she will receive a profit whose amount he has decided upon. He thus knows in advance the extent of his potential profit and loss, the premium.

Let's say that the EUR/USD pair is trading at 1.2900. A trader could buy a 1.3000 one-touch option expiring in two days for 45 % of payout. In this case a trader would pay $45 and if price reached 1.3000 he would receive $100, or a 122 % return on his trade ($100 payout—$45 premium = $55; $55/$45 = 122 %).

Timing is especially critical with exotic options. You must know the exact time of expiration, and each broker may have different cut-off conventions. Typically, exotic options are timed against the New York cut-off, which is 10 a.m. ET. However, some brokers will set the cut-off time at 24:00 GMT (4 a.m. ET), so confirm the time before making a trade.

One-touch options are suited for conditions when you have a strong opinion about the direction of a currency pair and you are convinced the move will happen soon. A one-touch option with a far-away target (perhaps 200 pips away) and a very short time span (24–48 hours) will have a very high reward–risk ratio (typically 3:1 or less) precisely because the payout on such a trade will be rare.

6.3.2 No-touch Options

A *no-touch currency option* is profitable if the price of a currency pair does not reach the target by a specified time. For example, a 10-day no-touch option of GBP/USD at 1.9200 when the pound is trading at 1.9100 may be priced at 40 % of payout. This means you will pay $40 and receive $100 after 10 days if price does not decline to 1.9100.

A no-touch option offers better payout odds when the strike price is closer to the market price and the expiration date is farther away because the chances the currency will not touch the strike price diminish considerably the longer the trader has to wait.

One interesting property of the no-touch is the fact the underlying currency pair does not have to move in the trader's direction (that is, away from the strike price) in order to produce a profit. The currency pair simply has to stay relatively stationary in order for the trader to collect a payout.

6.3.3 Double One-touch Options

With this type of option, traders choose two triggers and set the profit they will make if either one is hit. Usually, double-one-touch options are used when traders expect highly volatile market conditions but don't know what direction the market will take. In this sense, double one touch options are similar to long straddle or strangle options.

The *double one-touch option* allows you to select two strike-price barriers and provides a payout if either one is touched. If the euro/US dollar (EUR/USD) spot was trading at 1.3000, you could buy a double one-touch with 1.2900 and 1.3100 strikes expiring 48 hours forward. If EUR/USD either rose to 1.3100 or declined to 1.2900, you would make a profit. The double one-touch is similar to a standard long strangle or straddle option trade in that it is a good tool to use when you have no strong opinion about direction but you expect volatility to explode.

6.3.4 Double No-touch Options

Double no-touch options are the opposite of the double one-touch options. Traders buy them when they expect a range-bound market with a relatively low volatility. In general, this type of option is profitable during the periods of consolidation that usually follow significant market moves.

This type of option is useful for a trader who believes that the price of an underlying asset will remain range bound over a certain period of time. Double no-touch options are growing in popularity among traders in the forex markets.

Large trend moves are often followed by periods of consolidation; the double no-touch can be a profitable trade to use in these cases. Assume the EUR/USD makes a strong up move from 1.2400 to 1.3400 over several weeks, but price then pauses and starts to weaken a bit. A trader could buy a double no-touch from 1.3200 to 1.3600 with expiration in a week. If the market remains within these boundaries, the trader will walk away with a profit.

One-touch and no-touch options are highly time-sensitive. A one-touch will be significantly cheaper the less time there is to expiration because the odds of reaching the target will be greatly reduced, while a no-touch will be priced in opposite fashion because the chances of not touching the target will diminish the more time is left on the contract.

However, the double one-touch and double no-touch options will have the same pricing parameters in terms of time but will vary greatly with respect to the width of the barriers. Double one-touch options, for example, will become progressively more expensive as the barriers narrow.

Recent pricing in double one-touch options in the US dollar/Japanese yen rate (USD/JPY) with 10 days to expiration and the spot rate trading at 104.75 were as follows: For strike barriers between 103.50 and 105.50 (meaning price had to hit either one of those points for the option to pay out), price was an eye-popping 95 % of payout, offering the trader only a potential 5-percent gain against a 95 % loss.

Expanding the boundaries to 102.50 and 106.50 reduced the premium to only 41 % of payout. Conversely, the double no-touch options would have the exact opposite properties, offering much higher payouts as the strike prices narrowed.

6.3.5 American Double No-touch FX Options

An *American double-no-touch option* will provide option buyers with an opportunity to earn a potential payoff if the upper barrier and the lower barrier have not been traded anytime during the observation period. Otherwise, such option will lapse. This option enables option buyer, who is willing to pay a premium upfront and has a range trading view over the currency pair anytime during the observation period, to earn a potential payoff.

6.3.6 American No-touch FX Option

An *American No-touch option* will provide option buyer with an opportunity to earn a potential payoff if the no-touch-up strike or no-touch-down strike has not been traded anytime during the observation period. Otherwise, such option will lapse. This option enables option buyer, who is willing to pay a premium upfront and believes that the pre-determined no-touch strike will not be traded anytime during the observation period, to earn a potential payoff.

6.3.7 American One-touch FX Option

An American one-touch option will provide option buyer with an opportunity to earn a potential payoff if the one-touch-up strike or one-touch-down strike has been traded anytime during the observation period. Otherwise, such option will lapse. This option enables option buyer, who is willing to pay a premium upfront and believes that the pre-determined one-touch strike will be traded anytime during the observation period, to earn a potential payoff.

6.3.8 European Digital FX Option

Unlike vanilla option, this type of option has a fixed payoff profile, instead of a linear payoff profile. Whether the investor is repaid with the fixed amount will depend on whether the pre-determined conditions have been satisfied before and at expiry.

6.3.9 American Double No-touch FX Option

An *American double-no-touch option* will provide option buyer with an opportunity to earn a potential payoff if the upper barrier and the lower barrier have not been traded anytime during the observation period. Otherwise, such option will lapse. This option enables option buyer, who is willing to pay a premium upfront and has a range trading view over the currency pair anytime during the observation period, to earn a potential payoff.

6.3.10 American No-touch FX Option

An *American no-touch option* will provide option buyer with an opportunity to earn a potential payoff if the no-touch-up strike or no-touch-down strike has not been traded anytime during the observation period. Otherwise, such option will lapse. This option enables option buyer, who is willing to pay a premium upfront and believes that the pre-determined no-touch strike will not be traded anytime during the observation period, to earn a potential payoff.

6.3.11 American One-touch FX Option

An American one-touch option will provide option buyer with an opportunity to earn a potential payoff if the one-touch-up strike or one-touch-down strike has been traded anytime during the observation period. Otherwise, such option will lapse. This option enables option buyer, who is willing to pay a premium upfront and believes that the pre-determined one-touch strike will be traded anytime during the observation period, to earn a potential payoff.

6.3.12 European Digital FX Option

Unlike vanilla option, this type of option has a fixed payoff profile, instead of a linear payoff profile. Whether the investor is repaid with the fixed amount will depend on whether the pre-determined conditions have been satisfied before and at expiry.

6.3.13 American Double No-touch Structured FX Option

This structured option enables investors who hold a range of trading views over the linked currency to earn an extra return if the exchange rate trades within the pre-set range during the observation period.

6.3.14 American Knock-in FX Option

An *American knock-in option* is an option that will become operative if the spot rate is larger than the upper in-strike, or if the spot rate is smaller than the lower in-strike any time during the observation period. Otherwise, the American knock-in option will lapse. This option enables "buyer", who believes the in-strike will be traded during the observation period, to buy the option at a cheaper cost; or "seller", whose view is in opposite, to earn premium from option sold.

6.3.15 American Knock-out FX Option

An *American knock-out option* is an option that will lapse if the spot rate is larger than the upper out-strike or if the spot rate is smaller than the lower out-strike anytime during the tenor of the option. Otherwise, the American knock-out option shall operate in the manner of a normal vanilla option. this option enables "buyer", who believes the out-strike will never be traded during the observation period, to buy the option at a cheaper cost; or "seller", whose view is in opposite, to earn premium from option sold.

6.3.16 Bonus Knock-out FX Option

A bonus knock-out option enables an investor who holds a bearish view over, say, USD to sell USD call against other major currencies and earn bonus Payout through incorporation of a bonus-strike. Once the bonus-strike has ever been traded during the observation period, the investor will receive the bonus Payout while the option will be terminated simultaneously.

6.3.17 European Knock-in FX Option

A *European knock-in option* is an option that will become operative if the fixing rate is larger than the upper in-strike, or if the fixing rate is smaller than the lower in-strike upon fixing time. Otherwise, the European knock-in option will never exist. This option enables "buyer", who believes the in-strike will be traded upon fixing time, to buy the option at a cheaper cost; or "seller", who believes the in-strike will not be traded upon fixing time, to earn premium from option sold.

6.3.18 European Knock-out FX Option

A *European knock-out option* is an option that will lapse if the fixing rate is larger than the upper out-strike or if the fixing rate is smaller than the lower out-strike upon fixing time. Otherwise, the European knock-out option shall operate in the manner of a normal vanilla option. This option enables "buyer", who believes the Out-Strike will not be traded upon fixing time, to buy the option at a cheaper cost; or "seller", who believes the out-strike will be traded upon fixing time, to earn premium from option sold.

6.3.19 Knock-in with Knock-out FX Option

It is a combination of both knock-in and knock-out option. The option will only be activated if the knock-in level is traded. However, the option can be terminated at any time if the knock-out level trades.

6.3.20 Window Barrier Knock-in FX Option

A *window barrier knock-in option* is a modified version of normal knock-in option, in which the knock-in mechanism is only valid for a specified period of the option tenor. The active period, during which the knock-in mechanism is operative, can be set at the front end (front knock in) or the rear end (rear knock in) of the option tenor. in case where the knock-in level has been triggered during the active period, the option shall become a normal vanilla option and remain in full force up to maturity whereby investor will be exposed to unlimited risks. This option is suitable for investor who wants to earn premium and believes that the knock-in strike will never be traded during the active period.

6.3.21 Window Barrier Knock-out FX Option

A *window barrier knock-out option* is a modified version of normal knock-out option, in which the knock-out mechanism is only valid for a specified period of the option tenor. The active period, during which the knock-out mechanism is operative, can be set at the front end (front knock out) or the rear end (rear knock out) of the option tenor. In case where the knock-out level has never been triggered during the active period, the option shall become a normal vanilla option and remain in full force up to maturity whereby investors will be exposed to unlimited risks. This option is suitable for investor who wants to earn premium and believes that the knock-out strike will be traded during the active period.

6.3.22 Analytical Formulas

Some barrier options have analytically solutions. An "*in*" barrier option becomes a plain vanilla option if the asset price has been below the barrier level H for a *down-and-in option* or if the asset price has been above H for an *up-and-in option*. An "*out*" barrier option is an option that equals a plain vanilla option as long as the asset price has always been above H for an *down-and-out option* or below H for an *up-and out option*. When some barrier options are knocked out, they pay a rebate K at maturity.

The payoff of an in-barrier in combination with an out-barrier of the same type is equivalent to a plain vanilla option and a cash payout equal to the rebate, K.

Rubenstein and Reimer summarized in 1991 the formulas below, where X is the strike and

$$\eta = \begin{cases} 1 & \text{if Down} \\ -1 & \text{if Up} \end{cases} \qquad \phi = \begin{cases} 1 & \text{if Call} \\ -1 & \text{if Put} \end{cases}$$

$$x_1 = \frac{\ln(S/X)}{\sigma\sqrt{T}} + (1+\mu)\sigma\sqrt{T}, \qquad x_2 = \frac{\ln(S/H)}{\sigma\sqrt{T}} + (1+\mu)\sigma\sqrt{T},$$

$$y_1 = \frac{\ln\left(H^2/(SX)\right)}{\sigma\sqrt{T}} + (1+\mu)\sigma\sqrt{T}, \qquad y_2 = \frac{\ln(H/S)}{\sigma\sqrt{T}} + (1+\mu)\sigma\sqrt{T},$$

$$z = \frac{\ln(H/S)}{\sigma\sqrt{T}} + \lambda\sigma\sqrt{T}$$

$$\mu = \frac{r - \sigma^2/2}{\sigma^2}, \qquad \lambda = \sqrt{\mu^2 + \frac{2r}{\sigma^2}}$$

$$A = \phi \cdot S \cdot N(\phi x_1) - \phi \cdot X \cdot e^{-rT} N\left(\phi x_1 - \phi\sigma\sqrt{T}\right)$$

$$B = \phi \cdot S \cdot N(\phi x_2) - \phi \cdot X \cdot e^{-rT} N\left(\phi x_2 - \phi\sigma\sqrt{T}\right)$$

$$C = \phi \cdot S \cdot \left(\frac{H}{S}\right)^{2(\mu+1)} N(\eta y_1) - \phi X e^{-rT} \left(\frac{H}{S}\right)^{2\mu} N\left(\eta y_1 - \eta\sigma\sqrt{T}\right)$$

$$D = \phi \cdot S \cdot \left(\frac{H}{S}\right)^{2(\mu+1)} N(\eta y_2) - \phi \cdot X \cdot e^{-rT} \left(\frac{H}{S}\right)^{2\mu} N\left(\eta y_2 - \eta\sigma\sqrt{T}\right)$$

$$E = Ke^{-rT}\left[N\left(\eta x_2 - \eta\sigma\sqrt{T}\right) - \left(\frac{H}{S}\right)^{2\mu} N\left(\eta y_2 - \eta\sigma\sqrt{T}\right)\right]$$

$$F = Ke^{-rT}\left[\left(\frac{H}{S}\right)^{\mu+\lambda} N(\eta z) - \left(\frac{H}{S}\right)^{\mu-\lambda} N\left(\eta z - 2\eta\lambda\sigma\sqrt{T}\right)\right].$$

The prices of call barriers are given by

Type	$X < H$	$X > H$	
Down-and-in	$S > H$	$A - B + D + E$	$C + E$
Up-and-in	$S < H$	$B - C + D + E$	$A + E$
Down-and-out	$S > H$	$B - D + F$	$A - C + F$
Up-and-out	$S < H$	$A - B + C - D + F$	F

The prices of Put barriers are given by

Type	$X < H$	$X > H$	
Down-and-in:	$S > H$	$A + E$	$B - C + D + E$
Up-and-in:	$S < H$	$C + E$	$A - B + D + E$
Down-and-out:	$S > H$	F	$A - B + C - D + F$
Up-and-out:	$S < H$	$A - C + F$	$B - D + F$

There are also many structural variations possible, for example:

- Some contracts have more than one barrier—e.g. a double knock-out knocks out if either a higher or a lower barrier is reached
- Some barrier options knock in or out depending on the performance of a different market. An example of this type is the *soft call provision* embedded in many euro convertible bonds, which gives the issuer the right to call the bond if the underlying shares reach a specified threshold level.

Some barrier contracts as we see above, also includes a *rebate clause.*

6.3.23 Some Applications of Barrier Options

Unless the rebates are very large, barrier options are typically cheaper than otherwise equivalent conventional options (since a barrier option can never perform better) and this is their main appeal to investors.

Barrier options are one of the most widely traded classes of exotic option. Knock-outs are more popular than knock-ins, perhaps because investors are reluctant to pay for something, which does not yet exist and may never exist. The most common flavours are *down-and-out calls* and *up-and-out puts.* These have the same payoff as the regular options, except that if the options go sufficiently OTM to hit the barrier they immediately expire worthless.

Example 6.3

Hedging with Up-and-out Puts

Consider the following situation: an equity investor remains fundamentally bullish, although short term he perceives some risk of a temporary market setback, if the forthcoming trade figures prove disappointing. The investor would like to protect his equity portfolio with conventional index options, but finds their premium cost prohibitive.

(continued)

Example 6.3 (continued)

Solution

Up-and-out puts might be a lower-cost alternative in this scenario: if the markets continued to rally the puts might cease to exist, but then the risk of a market setback might have receded too.

Example 6.4

A Touch Options
Consider the following contract

Type	Call spread
Strike	100
Touch level	120
Exercise date	3 months

This touch option, also known as a *capped* or *exploding spread* option, works like this:

- If the touch level is reached at any time during the life of the contract, the holder is entitled to the difference between the touch level and the strike, even if the underlying price subsequently pulls back. The locked-in spread may be paid immediately or deferred until the option's final expiry.
- If the touch level is not reached, then the option pays at expiry the difference between the underlying price and the strike (if positive), just like a conventional call spread.

This structure is a combination of two exotic options:

- An up-and-out call with a strike of 100 and an out level of 120
- An American style binary put, with a strike of 120 and a cash payout of 20.

6.3.24 Pricing Barrier Options

Barrier options may be priced using binomial models, Monte Carlo simulations or special versions of the standard Black–Scholes–Merton model, such as for the down-and-out put option above. Other variants can be found in Haug's book.

As with digital options, what is exotic about these contracts is the behaviour of their Greeks, especially delta. In Fig. 6.3 we illustrate the risk profile of an up-and-out call with a strike of 90 and an out level barrier of 100.

Initially, the price of the up-and-out call rises as the option moves into the money. However, as the out level is approached it begins to lose value. The

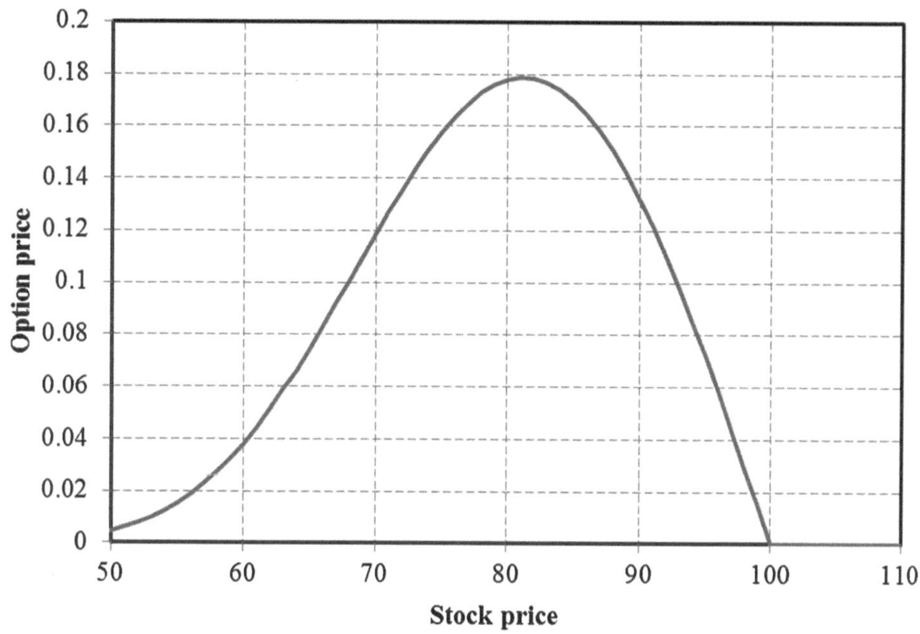

Fig. 6.3 The value profile of an up and out call option with strike price 90 and a barrier level 100.

position is delta-positive, if the call is OTM, but delta-negative closer to the out level. Moreover, the value of delta becomes very large close to the out level.

For a trader such potentially large swings in delta become impossible to hedge dynamically, so it is wise to

- Build a significant profit margin into the quoted option premium
- Never take on very large positions in such options

Whereas a conventional bought call option is always theta-negative and vega-positive, as you will see in the exercise in the next section, the equivalent position in this barrier call option may be positive or negative in theta and vega, depending on how close the underlying price is to the out level.

In Fig. 6.4 we show four different options, three different barrier options and a plain vanilla Black–Scholes call. The vanilla option has the same values as the up-and-in call option. The down-and-out call has the same value as the vanilla option at maturity.

The strike of the call options is 90 and the barrier level is 100. The option's price rises as it moves OTM and the in level be approached. In fact, until the in barrier is reached the position is delta-negative and the profile looks more like

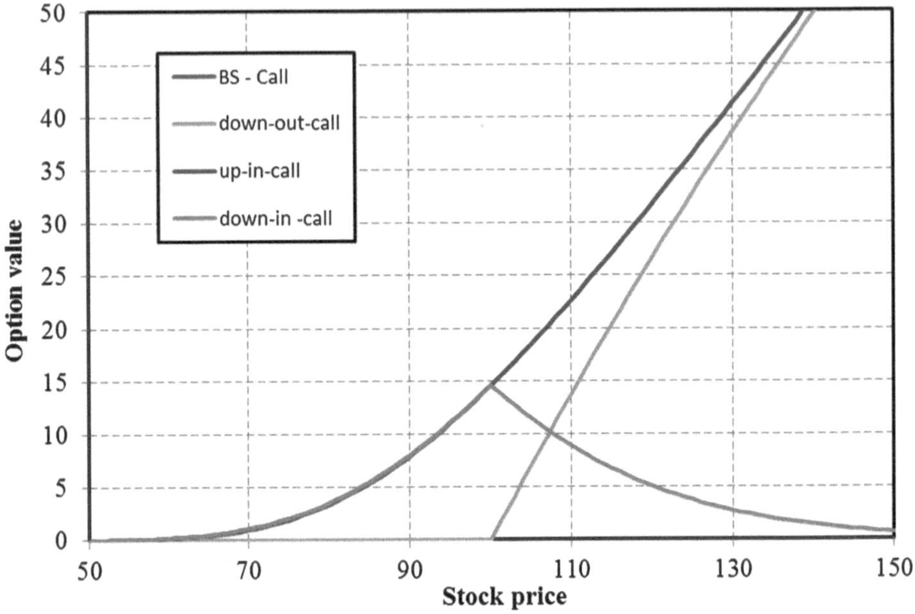

Fig. 6.4 The value profile of a pain vanilla call option and three different barrier call options

that of a put than a call. Once the barrier is hit, the option becomes a conventional call and its delta swings dramatically too positive.

Therefore, a trader delta-hedging a short position in this option starts off initially with a short position in the underlying but may have to turn the hedge very quickly into a long position, if the barrier is hit. Again, it is wise not to take on sizeable positions in such options as it may be very difficult, in practice, to turn the hedge around so quickly.

Not all barrier options have malignant barriers. For an up-and-in call, the "graduation" from a barrier option into a conventional one, as the in-level is hit, results in a much smaller change in delta, as both the strike and the in-barrier pull the option's price in the same direction. This is an example of a so-called *benign barrier*.

In Fig. 6.5 we show the corresponding put options with strike 90 and the barrier level at 100.

As well as the additional risks involved, which may be significant, there are two contractual issues that anyone trading barrier options must consider carefully.

- How frequently will the barrier be tested: hourly, daily or monthly? A barrier that is tested monthly is less risky than one that is tested hourly, as it

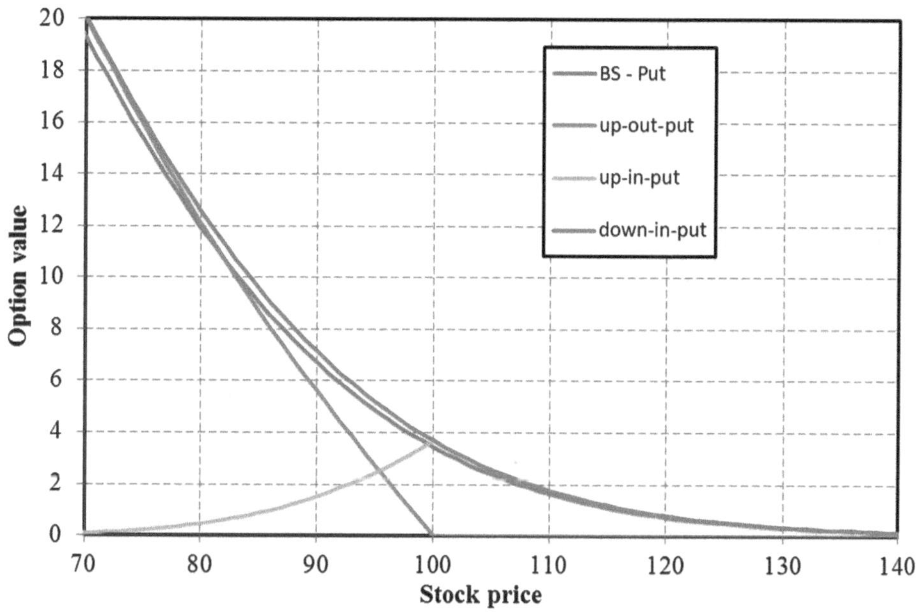

Fig. 6.5 The value profile of a pain vanilla put option and three different barrier put options

is possible that the underlying price may move through the barrier but retrace before the next look-up date.

- What is the reference market price against which the barrier will be tested: the price quoted by one market maker, an official market price or a broad market index? Obviously, testing the barrier against an official price or broad index makes the contract less vulnerable to temporary market distortions or abuse.

6.4 Lookback Options

A floating strike lookback call option gives the holder the right to buy the underlying security to the lowest observed value S_{min}, during the option lifetime.

$$P_{call} = SN(a_1) - S_{\min}e^{-rT}N(a_2)$$
$$+ Se^{-rT}\frac{\sigma^2}{2r}\left[\left(\frac{S}{S_{\min}}\right)^{-\frac{2r}{\sigma^2}} N\left(-a_1 + \frac{2r}{\sigma}\sqrt{T}\right) - e^{-rT}N(-a_1)\right]$$

$$a_1 = \frac{\ln\left(\frac{S}{S_{\min}}\right) + \left(r + \frac{\sigma^2}{2}\right)T}{\sigma \cdot \sqrt{T}}, \qquad a_2 = a_1 - \sigma \cdot \sqrt{T}.$$

Similarly, the holder of a floating strike lookback put options have the right to sell the underlying security to the highest observed price S_{max}, during the lifetime of the option.

$$P_{put} = S_{\max} e^{-rT} N(-b_2) - SN(-b_1)$$

$$+ Se^{-rT}\frac{\sigma^2}{2r}\left[-\left(\frac{S}{S_{\max}}\right)^{-\frac{2r}{\sigma^2}} N\left(b_1 - \frac{2r}{\sigma}\sqrt{T}\right) + e^{rT}N(b_1)\right]$$

$$b_1 = \frac{\ln\left(\frac{S}{S_{\max}}\right) + \left(r + \frac{\sigma^2}{2}\right)T}{\sigma \cdot \sqrt{T}}, \qquad b_2 = b_1 - \sigma \cdot \sqrt{T}.$$

A fixed strike lookback call option gives the holder the maximum difference between the price and the strike during a given period. Other types of lookback option can also be constructed (see Haug).

6.5 Asian Options

Asian options are especially popular on the currency and commodity markets. An average value option is less volatile than the underlying itself. Therefore, the price of an average-rate option is lower than a plain vanilla option. Options based on an average value are more stable and they are more difficult to be manipulated in price by the underlying.

Asian options come in two basic flavours:

- *Average price options:* at expiry the option pays the difference between the strike and an average of the underlying price achieved during a specified *averaging period* in the option's term
- *Average strike options:* the strike of the option is an average of the underlying price over the specified averaging period, and at expiry the option pays the difference between this strike and the underlying market price.

Of the two, the average price option is by far the most common.

Average price options are frequently found embedded in *principal-protected notes.* They offer another way of protecting the investors against a last-minute fall in the market, just before the option's expiry, which could disproportionately reduce any interim gains achieved.

A typical contract can have the following parameters:

Type	Call
Style	European
Underlying	An equity index
Strike	5000 (= ATM cash)
Expiry	5 years
Expiry payoff	The seller shall pay the difference between the strike and the arithmetic average of the index level achieved during the final 6 months of the option's term (the averaging period)

The benefit of this option for the buyer is that any last-minute market setback is averaged up. The drawback is that any last-minute market advance would likewise be averaged down.

Average price options also appeal to investors in thinly traded markets, since manipulation of the underlying price close to the expiry date will have little effect on the average.

The terms of an Asian option must specify

- **The averaging period**—e.g. the last six months, the entire term of the option or the closing prices at month-ends
- **The sampling frequency**—e.g. the daily, weekly or monthly closing prices during the averaging period
- **The averaging method**—e.g. a simple arithmetical average, a geometric average or some weighted average

6.5.1 A Mean Value Option

We will construct an exotic European option where the holder at the day of maturity T_2 receives

$$X = \frac{1}{T_2 - T_1} \int_{T_1}^{T_2} S(u)du,$$

where T_1 is a fix time $< T_2$. Calculate the arbitrage-free price. We know that

$$\begin{cases} dS(t) = r \cdot S(t) \cdot dt + \sigma \cdot S(t) \cdot dW \\ S(T) = s \end{cases}.$$

If we integrate this we will get

$$S(t) = s + r\int_0^t S(u) \cdot du + \sigma\int_0^t S(u) \cdot dW(u).$$

The price is given by

$$\Pi[X|\mathcal{F}] = e^{-r(T_2-t)}E_{t,s}^{Q}\left[\frac{1}{T_2 - T_1}\int_{T_1}^{T_2} S(u)du\right] = \frac{e^{-r(T_2-t)}}{T_2 - T_1}\int_{T_1}^{T_2} E_{t,s}^{Q}[S(u)]du.$$

We then calculate the expectation value of $S(t)$

$$E[S(t)] = s + r\int_0^t E[S(u)] \cdot du + 0.$$

Let $E[S(t)] = m$ and take the derivative with respect to time

$$\begin{cases} \dot{m}(t) = r \cdot m(t) \\ m(0) = s. \end{cases}$$

The solution is given by

$$m(t) = E[S(t)] = se^{rt}$$

This means that the price is given by

$$\Pi[X|\mathcal{F}] = \frac{s \cdot e^{-r(T_2-t)}}{T_2 - T_1}\int_{T_1}^{T_2} e^{r(u-t)}du = \frac{s/r}{T_2 - T_1} \cdot \left(1 - e^{-r(T_2-T_1)}\right).$$

6.5.2 Pricing Asian Options

There are three main approaches to pricing Asian options:

- European-style options based on geometric averages can be priced by adapting the analytical models. This is because if the underlying price is assumed to be log-normally distributed then its geometric average is also log-normal. The formulas below are given by Kemna and Vorst (1990):

$$P_{call} = S \cdot N(d_1) - Ke^{-rT}N(d_2)$$
$$P_{put} = Ke^{-rT}N(-d_2) - S \cdot N(-d_1)$$

where

$$d_1 = \frac{\ln\left(\frac{S}{K}\right) + \left(r + \frac{\sigma_A^2}{2}\right)T}{\sigma_A \cdot \sqrt{T}}, \qquad d_2 = d_1 - \sigma_A \cdot \sqrt{T}$$

$$b = \frac{1}{2}\left(r \quad \frac{\sigma^2}{6}\right)$$

and the adjusted volatility is given by

$$\sigma_A = \frac{\sigma}{\sqrt{3}}$$

- There is no equivalent solution for options based on arithmetic averages, because even if the underlying price is log-normally distributed, the arithmetic average is not. However, various analytic approximations have been developed which work reasonably well. A weak approximation by Turnbull and Wakeman is given in Haug.
- Any Asian option, no matter what its style or averaging method, may be priced by Monte Carlo simulation, but this is computationally much more intensive.

The volatility of an average is always less than that of the price itself, and the longer the averaging period the lower is its volatility.

Whatever pricing method is used, average price options come out very much cheaper than conventional ones. Comparing the price curves with different average curve one can observe that the option price sensitivity to spikes in the underlying market is reduced, hence also its price.

6.6 Chooser Options

A simple chooser option gives the holder the right to choose if the option will become a call or a put option after a certain time t_1. This is also known as U-choose option. The strike price K, is the same for both options and also the maturity T_2. In addition to all the standard terms of a conventional option, the chooser includes a clause that specifies the choose date—the date by which the buyer must tell the seller whether the option is to be a call or a put. After this date the option becomes a conventional call or put. A complex chooser option was introduced by Rubenstein (1991), where the strike was not the same for the call and the put options.

$$P = SN(d) - Ke^{-rT_2}N\left(d - \sigma\sqrt{T_2}\right) - SN(-y) + Ke^{-rT_2}N\left(-y - \sigma\sqrt{t_1}\right)$$

$$d = \frac{\ln\left(\frac{S}{K}\right) + \left(r + \frac{\sigma^2}{2}\right)T_2}{\sigma \cdot \sqrt{T_2}}, \qquad y = \frac{\ln\left(\frac{S}{K}\right) + rT_2 + \frac{\sigma^2}{2}t_1}{\sigma \cdot \sqrt{t_1}}$$

Like straddles, choosers are attractive to investors who expect higher price volatility but do not yet have a view on future market direction. However:

- With a straddle the investor retains the right to call or put the underlying right up to the expiry date
- With a chooser the investor loses one of these rights after the choose date

The chooser should therefore be cheaper than the straddle—and this is one of its attractions—unless of course the chooser date is the same as the expiry date!

Consider a call that expires at time T_1 with the strike price X_1 and a put that expires at T_2 with strike X_2. Both T_1 and T_2 are greater than T, the expiration time of the chooser option. At time T, the claim of the chooser option is max $\{C(S, T_1 - T, X_1), P(S, T_2 - T, X_2)\}$.

Consider the special case where $T_1 = T_2 = \tau$ and $X_1 = X_2 = X$, the solution is greatly simplified. Applying the put call parity, we can rewrite the claim at T as

$$\begin{aligned}\Phi(T) &= \max\left\{C(S, \tau - T, X), C\left(S, \tau - T, X\right) + Xe^{-r(\tau - T)} - S(T)\right\} \\ &= C(S, \tau - T, X) + \left[Xe^{-r(\tau-T)} - S(T)\right]^+.\end{aligned}$$

This is equivalent to the claim of a portfolio consisting of a call and a put. Thus the price of the chooser option is given by

$$CO(S,T,X) = C(S,\tau - T,X) + P\left(S,T,Xe^{-r(\tau-T)}\right).$$

For the general case where $T_1 \neq T_2$ or $X_1 \neq X_2$, the composition is a little bit more complicated. Let S^* be the solution S to $C(S, T_1 - T, X_1) = P(S, T_2 - T, X_2)$. When $S(T) > S^*$, the put becomes worthless and the value of the chooser option becomes that of the call. When $S(T) < S^*$, the call option becomes worthless and the value of the chooser option becomes that of the put. Hence, the claim of the chooser option at time T can be written as

$$\Phi(T) = [C(S,T_1 - T,X_1)]^+ + [P(S,T_2 - T,X_2)]^+,$$

which implies that the claim of a chooser option at time T can be replicated with a call on call and a call on put, each with a zero strike price. Thus, the price of a general chooser is

$$CO(S,T_1,T_2,X_1,X_2) = CC(S,T,T_1,0,X_1) + CP(S,T,T_2,0,X_2).$$

By our definition, the chooser meets the criteria of an exotic, in that the option type is not defined on the contract's effective date (when the premium becomes payable) but sometime later.

Example 6.5

An investment bank issues a simple chooser on the following terms

Style:	European
Strike:	100 (= ATM forward)
Expiry:	3 months
Choose date:	1 month

The investor has 1 month to decide whether the option is to be a call or a put. His/her decision will be based on which of the two options has a higher market value by the choose date. That will depend on whether the underlying price is higher or lower than 100 at that time.

The investor in a chooser gains nothing by choosing before the last possible chooses date, and therefore will delay his/her decision until then.

6.6.1 A Hedging Programme

If a bank is selling this chooser, they may hedge its risk as follows:

- Buy a conventional 3-month ATM European call
- Buy a 1-month ATM European put

Consider two possible scenarios on the last choose date

- **Scenario 1**: the underlying price rises above 100. Conventional European100 calls are therefore more valuable than the equivalent puts, so the investor chooses his/her option to be a call.

  At that point the 1 month put in the bank's hedge book expires OTM, leaving just the longer dated call, which mirrors exactly the investor's choice.
- **Scenario 2**: the underlying price falls below 100. Conventional 100 - European calls are therefore less valuable than the equivalent puts, so the investor chooses his/her option to be a put.

  At that point the 1 month put in the bank's hedge book expires ITM, creating a short position in the underlying. The combination short the underlying and long the calls gives the bank a synthetic long put position in its hedge book, which again mirrors exactly the investor's choice!

Thus, the simple chooser may be priced as a combination of a conventional call and a conventional put, both with the same strike but different expiry dates. It is cheaper than the straddle because one of the options has a shorter expiry than the other.

6.7 Forward Options

In a forward option, the strike is set at some specified future date, rather than on the effective date (i.e. when the premium is payable). This is also known as a *forward-start options* or *delayed-start options*. The actual strike will not be known until the future *effective date*, but the contract does specify what the strike will be in relation to the underlying market at the time—e.g. ATM, or 10 % OTM. In other words, the contract specifies the option's future *parity ratio* (= spot/strike) at the outset.

Example 6.6

Cliquets Options or Ratchets
Forward starting options are found in structures known as *resetting options, moving strike options* or *cliquets*. Cliquet is French for ratchet and, as the name suggests, the structure locks in any gains achieved as a result of favorable market movements by resetting the strike up or down at regular intervals in line with the market. Like ladder structures and Asian options, cliquets are sometimes embedded in principal-protected notes to protect investors against possible last-minute market.

Initially, such an option is struck at a current index level, but in the next year, when the market is finishes down at another level, the strike is moved to this new level. In year 2 we look at the market closing price and calculate the profit or loss. We continue like this until the maturity of the contract.

Normally, the Cliquet performed significantly better than an equivalent conventional call in the same scenario.

The Cliquet structure is made up of:

- A conventional European ATM call
- A series of forward-start ATM calls, each of which becomes effective on an anniversary date and expires a year later

Payment on the Cliquet, if any, is typically made at each reset date, but it may be deferred until the maturity of the whole structure.

6.7.1 Pricing Forward Starting Options

The buyer of a forward option pays a premium today for an option whose strike will be determined at some future date. How do you price an option whose strike you do not yet know?

Conceptually, there is a trick to pricing such options, which relies on the fact that the option's parity ratio is specified in advance.

For a given parity ratio, an option's premium is proportional to the underlying spot price (and strike).

In other words, other things being equal, if you double both the spot and the strike while keeping the parity ratio constant, then the option price doubles as well. This makes the profit/loss on a forward option position proportional to the underlying price and suggests a simple programme for hedging its risks until the effective date. As the example below illustrates, once you know how to hedge the derivative then you know how to price it!

Usually, forward-start options are at-the-money when issued. More generally, it is assumed that the strike price of a forward-start option expiring at time τ equals $\alpha S(T)$, where $\alpha > 0$ is a constant and $T < [0, \tau)$ is the issue time of the option. This special property allows us to write the Black–Scholes formula for such a European call at time T as

$$C(S(T), \tau - T, \alpha S(T)) = S(T)\Lambda$$

where

$$\Lambda = N\left(\frac{-\ln\alpha + (r + \sigma^2/2)(\tau - T)}{\sigma\sqrt{\tau - T}}\right) - \alpha e^{-r(\tau - T)} N\left(\frac{-\ln\alpha + (r - \sigma^2/2)(\tau - T)}{\sigma\sqrt{\tau - T}}\right)$$

Since Λ is a constant independent of the asset price, the pricing problem is obviously equivalent to the pricing of a futures contract at time t. The Future-Spot price parity implies

$$C_f(S, T, \tau, X) = \Lambda \cdot E^Q[S(T) | \mathcal{F}_0] = \exp(rT) \cdot \Lambda.$$

European forward-start puts can be priced in the same way.

Example 6.7

A client asks you to quote for the following contract

Type	Call
Style	European
Strike	ATM
Expiry	1 year
Effective date	6 months

At the time, the breakeven price for a conventional 1 year ATM call (effective spot) is 3.50, for an implied volatility of 10%.

Solution:

Assuming that in 6 months' time the 1-year ATM call will still be trading at 10 % implied, then the forward option may also be quoted at 3.50 on a breakeven basis. The hedging programme is as follows: if the client buys the contract at the quoted price, then the trader hedges her risks by applying the entire premium received to fund a long position in the underlying.

The Outcome:

Scenario 1: the underlying price rose 50 % so on the effective date the option's strike is set at 150. At this point, if the 1-year ATM call trades with 10% volatility, then its premium price will also be 50 % higher—i.e., 5.25.

(continued)

Example 6.7 (continued)

In this scenario the client made a profit of 1.75 on the trade (= 5.25 – 3.50) and the trader's option position will show a corresponding loss. However, the trader spent 3.50 to buy the underlying when it was trading at 100 and now the hedge is worth 50 % more—a gain of 1.50 that covers exactly the losses made on the option!

Scenario 2: the underlying price fell 50 % so on the effective date the option's strike is set at 50. If the 1-year ATM call trades with 10 % volatility, then its premium price would also be 50 % lower—i.e., 1.75.

In this scenario the client made a loss of 1.75 on the trade (= 1.75 – 3.50) and the trader's option position will show a corresponding gain. However, the trader spent 3.50 to buy the underlying when it was trading at 100 and now the hedge is worth 50 % less—a loss of 1.50 that matches exactly the gains made on the option!

On the effective date the option's strike is set and the trader replaces this static hedge with a conventional dynamic delta-hedging programme.

In 1990 Rubenstein presented the following formulas for forward start options with time to maturity T, that starts at-the-money or proportionally in- or out-of-the-money after a known elapsed time t in the future. The strike is set equal to a positive constant α times the asset price S after the known time t. If α is less than unity, the call (put) will start 1—α % in-the-money (out-of-the-money); if α is unity, the option will start it-the-money; and if α is larger than unity, the call (put) will start α—1 percentage out-of-the money (in-the-money). The forward start option pricing formula where b is the cost of carry rate is then given by

$$P_{call} = S\left[N(d_1) - \alpha e^{-r(T-t)}N(d_2)\right]$$

$$P_{put} = S \cdot \left[\alpha e^{-r(T-t)}N(-d_2) - N(-d_1)\right]$$

where

$$d_1 = \frac{\ln\left(\frac{1}{\alpha}\right) + \left(r + \frac{\sigma^2}{2}\right)(T-t)}{\sigma \cdot \sqrt{T-t}}, \qquad d_2 = d_1 - \sigma \cdot \sqrt{T-t}.$$

6.7.2 Ratchet Options

A ratchet option, sometimes called a moving strike option or Cliquet option, consists of a series of forward starting options where the strike price for the next exercise date is set equal to a positive constant times the asset price as of the previous exercise date. For instance, a one-year ratchet call option with quarterly payments will normally have four payments (exercise dates) equal to the difference between the asset price and the strike price fixed at the previous exercise date. The strike price of the first option is usually set equal to the asset price of today. A ratchet option can be priced as the sum of forward starting options.

$$P_{call} = \sum_{i=1}^{n} S \cdot \left[N(d_1) - \alpha e^{-r(T_i - t_i)} N(d_2) \right]$$

where n is the number of settlements, t_i is the time to the forward start or strike fixing, and T_i is the time to maturity of the forward starting option. A ratchet put is similar to a sum of forward starting puts.

6.7.3 Estimating Forward Volatility

Conceptually, a forward option may be priced as if it was a conventional (spot-start) contract. What makes the forward option exotic is the fact that the trader does not know, on the trade date, whether the option she is pricing will be trading with the same implied volatility as the conventional option does today. In our example we assumed that this was indeed the case, but the chances are that it will not. So the main risk with forward options is getting the assumed volatility wrong; there may be significant vega risk in pricing these contracts.

It is possible to estimate the forward-forward volatilities implied in the volatility curve, just as we can derive forward yields from the yield curve.

Example 6.8 illustrates how to estimate the forward volatility implied in a pair of market implied volatilities for conventional options.

Example 6.8

Suppose we observe the following two points on the volatility curve for ATM European options

6 months (σ_{0x6})	10 %
18 months (σ_{0x18})	12 %

What is the implied 6 × 18 months' volatility ($\sigma_{6\times18}$)?

(continued)

Example 6.8 (continued)

We can apply the same technique that is used for calculating portfolio risk by noting that a 0 × 18 position in the underlying asset is in fact a combination of two elements:

A 0 × 6 month position
A 6 × 18 month position in the same asset

The flat return (*R*) on this position over the period 0 × 18 is the sum of the annualized returns on the individual positions, each one pro-rated by its holding period

$$R = 1.5 \times R_{0\times18} = 0.5 \times R_{0\times6} + 1.0 \times R_{6\times18}$$

where

$R_{0\times18}$ = Annualized return over the period 0 × 18
$R_{0\times6}$ = Annualized return on 0 × 6 position
$R_{6\times18}$ = Annualized return on 6 × 18 position.

Therefore

$$R_{0\times18} = \frac{0.5}{1.5}R_{0\times6} + \frac{1.0}{1.5}R_{6\times18}.$$

This now looks like a standard two-asset portfolio, where the return on the portfolio is a weighted sum of the returns on the individual assets. The risk on such a portfolio is related to the risks of its constituent assets by the expression

$$\sigma^2_{0\times18} = \left(\frac{0.5}{1.5}\sigma_{0\times6}\right)^2 + 2\cdot\frac{0.5}{1.5}\cdot\frac{1.0}{1.5}\sigma_{0\times6,6\times18} + \left(\frac{1.0}{1.5}\sigma_{6\times18}\right)^2,$$

where

$\sigma_{0\times6,6\times18}$ = Covariance of returns between the two'assets'.

Assuming, as the option pricing models do, that the return on the underlying is log-normally distributed, this implies that $\sigma_{0\times6,6\times18} = 0$ and we calculate $\sigma_{6\times18} = 17.292\ \%$.

It is also possible, to an extent, to lock into a forward volatility level through a package of conventional options. The trader in our example above could hedge her vega risk with a combination of:

An ATM calendar spread:

(continued)

Example 6.8 (continued)

- Long an 18-month ATM call
- Short a 6-month ATM call

An ATM butterfly spread:

- Long a 6-month ATM straddle
- Short a 6-month strangle

The strikes of the strangle are set so that the net position is as far as possible neutral in all the Greeks until the expiry of the short-dated options. Unless the underlying market has moved significantly during this time, any losses on the forward-start option, arising from getting its volatility wrong, should be largely offset by gains in the remaining long-dated call, and vice versa.

Neither of these techniques for estimating (or hedging) forward volatility is fail-safe, so the trader must be careful to price sufficient profit margin into this product to cover the additional vega risk.

6.8 Compound Options: Options on Options

The buyer of a compound option pays an initial premium for the right to pay a second set premium by a certain future date for the ownership of a call or a put with an agreed strike and expiry. These options are also known as *instalment options*.

Compound options are options on options—that is, the 'underlying' is another option. There are four types:

- Call on a call
- Call on a put
- Put on a call
- Put on a put

A model to price options on options was first given by Geske (1977). The model was enlarged and discussed by Hodges (1979), Selby (1987), and Rubenstein (1991).

Compound options give buyers two main benefits

- More geared exposure to changes in the underlying asset price (or to changes in volatility) than conventional options
- The flexibility to pay for the rights to the underlying asset "in instalments"

Example 6.9

An investment client is considering buying a 12-month ATM call on an asset which is currently trading at 100, but finds the cost of a conventional option, at 3.61, too high in the current market.

Solution:

The bank offers the client the following contract:

Type	Call on a call
Strike of underlying option	100.00
Expiry of underlying option	1 year from today
Strike of compound option	2.68
Expiry of compound option	6 months
Price	1.84

Analysis:

The contract gives the buyer the right to purchase a 6-month 100 call in 6 months' time for an eventual premium of 2.68. Thus, the client pays:

1.84 to buy the call on the call.

- A further 2.68 in 6 months, if the call on the call is exercised, i.e. if the 6-month 100 call trades higher than 2.68 at the time.
- A further 100.00 in 12 months, if the underlying call is exercised, i.e. if the underlying asset trades higher than 100 at the time.

With the compound option, the right to acquire the underlying asset at 100 in 12 months' time could cost the client up to 4.52 (= 1.84 + 2.68), which is potentially more than the cost of the conventional 12-month call. However, the second instalment could cost less than 2.68, if the underlying call trades cheaper at the time, so the net cost could turn out less than 3.61.

Compound options are also used in the context of contract tenders or corporate acquisitions, where the outcome of the bid is uncertain. Banks providing the bidder with floating rate financing facilities often require the company to place an interest rate cap on this debt, to ensure that if the bid is successful the company will be able to maintain adequate interest cover on its additional debt.

However, for the bidder the purchase of a cap may prove to be an unnecessary expense, if the bid were to fail. A more cash-efficient solution is to buy a caption—a call on a cap—for a fraction of the cost.

6.8.1 Pricing Compound Options

Compound options may be priced using modified versions of the analytic, where the volatility of the underlying option price is derived from that of the underlying asset price. A model for pricing options on options was first published by Geske (1977). It was later extended and discussed by Geske (1979), Hodges and Selby (1987), Rubinstein (1991) and others.

6.8.2 A Call on a Call

The payoff is given by: $[BS(S, K_1, T_2) - K_2, 0]^+$, where K_1 is the strike price of the underlying option, K_2 the strike price of the option on the option, and BS (S, K, T) is the Black–Scholes call option formula with strike K and time to maturity T.

$$P_{call} = S \cdot M(z_1, y_1, \rho) - K_1 e^{-rT_2} M(z_2, y_2, \rho) - K_2 e^{-rt} N(y_2)$$

where

$$y_1 = \frac{\ln\left(\frac{S}{I}\right) + \left(r + \frac{\sigma^2}{2}\right)t_1}{\sigma \cdot \sqrt{t_1}}, \qquad y_2 = y_1 - \sigma \cdot \sqrt{t_1}$$

$$z_1 = \frac{\ln\left(\frac{S}{K_1}\right) + \left(r + \frac{\sigma^2}{2}\right)T_2}{\sigma \cdot \sqrt{T_2}}, \qquad z_2 = z_1 - \sigma \cdot \sqrt{T_2}$$

$$\rho = \sqrt{\frac{t_1}{T_2}}$$

where T_2 is the time to maturity on the underlying option, and t_1 is the time to maturity on the option on the option. M is the cumulative bivariate normal distribution and I a critical value given by solving: $BS(I, K_1, T_2 - t_1) = X_2$. Similar formulas are given in Haug for put on call, call on put and put on put.

6.9 Multi-asset Options

Multi-asset options are options whose payoffs at exercise depend on the performance of more than one underlying market.

There are two major varieties of multi-asset option:

- *Basket options*: the underlying is a portfolio of different assets, typically in the same class, in pre-defined amounts. Equity index options are basket options, but many OTC basket options are specially designed for investors seeking exposure to specific market sectors—e.g. a basket of technology stocks.
- *Outperformance options*: also known as an *exchange option* or a *spread option*, this gives the holder the right to receive one asset (A_1) in exchange for another asset (A_2) according to a specified *conversion ratio*, or the right to receive the amount by which the price of one asset exceeds that of another. A typical payoff might be

$$\mathit{Exercise\,payoff} = \max\{A_1 - A_2, 0\},$$

where A_1 and A_2 are the market values of the two underlying assets on the exercise date.
The payoff depends on the amount by which A_1 exceeds A_2, if positive, irrespective of whether both asset prices have moved up or down. This is equivalent to an option to exchange the lower-valued asset for the higher-valued one, or to go long the higher-valued asset and pay for it by going short the lower-valued one—that is, an option to enter into a spread position.

- *Better of options*: also known as a *rainbow option*, this entitles the holder to receive the gain on the best performing of two or more assets. A typical payoff might be

$$\Phi(T) = \max\{A_1, A_2, A_3, \cdots A_n, 0\}$$

where A_1 ... An are the gains in the prices of the n assets represented in an *n-colored rainbow*, typically expressed in percentage terms.
Here the payoff is based on the performance of the best-performing asset, if positive, rather than on its performance relative to the other assets, as in the spread option.

- *Correlation option* – the option pays the difference between the price of an asset (A_1) and a strike (K_1), ***provided*** the price of a related asset (A_2) is higher (or lower) than a certain level (K_2).

$$\Phi_{\text{call}}(T) = \max\{A_1 - K_1, 0\} \text{ if } A_2 > K_2, \text{otherwise } 0$$
$$\Phi_{\text{put}}(T)^{\circ} = \max\{K_1 - A_1, 0\} \text{ if } A_2 > K_2, \text{otherwise } 0$$

6.9.1 Pricing Multi-asset Options

All the options described in this section may be priced using derivations of the analytical models, or by Monte Carlo simulation. A number of multi-asset options are given in Haug.

6.10 Basket Options

With a basket option, the underlying is a portfolio of assets. Where the basket options are widely traded—e.g. index options—the volatility of the basket is already implied in the quoted premium prices, but where this is not the case it must be derived from the volatilities and correlations of the constituent assets using the standard portfolio risk.

6.10.1 Rainbow Options

Rainbow options refer to a family of options on the minimum or the maximum of two or more risky assets. Consider two assets

$$\begin{cases} dS_1(t) = S_1(t)\mu_1(t)dt + S_1(t)\sigma_1(t)dW_1(t) \\ dS_2(t) = S_2(t)\mu_2(t)dt + S_2(t)\sigma_2(t)\rho(t)dW_1(t) + S_2(t)\sigma_2(t)\sqrt{1-\rho(t)^2}dW_2(t) \end{cases}$$

where σ_1, σ_2 and ρ are all deterministic processes, while processes μ_1 and μ_2 are predictable. The two standard Brownian motions W_1 and W_2 are independent, so the correlation coefficient between $d\ln S_1(t)$ and $d\ln S_2(t)$ is $\rho(t)$. The two assets also generate continuous yields at deterministic rate processes θ_1 and θ_2, respectively. A call on the minimum of these two assets for a strike price $X > 0$ can be priced as

$$
\begin{aligned}
C_{\min}(S_1, S_2, T, X) &= e^{-\bar{r}T} E^Q\left[\left(\min(S_1(T), S_2(T)) - X\right)^+\right] \\
&= e^{-\bar{r}T} E^Q[S_1(T) \,|\, S_2(T) > X \quad \text{and} \quad S_2(T) > S_1(T)] \\
&\quad + e^{-\bar{r}T} E^Q[S_2(T) \,|\, s_1(T) > X \quad \text{and} \quad S_1(T) > S_2(T)] \\
&\quad - e^{-\bar{r}T} X \cdot E^Q[1 \,|\, S_1(T) > X \quad \text{or} \quad S_2(T) > X].
\end{aligned}
$$

The arbitrage-free condition is equivalent to the existence of an equivalent measure Q, so that

$$
\begin{cases}
S_1(T) = S_1(0)\exp\left\{ rT - \int_0^T \theta_1(s)ds - \frac{1}{2}\int_0^T \sigma_1(s)^2 ds + \int_0^T \sigma_1(s)dW_1(s) \right\} \\
S_2(T) = S_2(0)\exp\left\{ rT - \int_0^T \theta_2(s)ds - \frac{1}{2}\int_0^T \sigma_2(s)^2 ds \right\} \\
\qquad \times \exp\left\{ \int_0^T \sigma_2(s)\rho(s)dW_1(s) + \int_0^T \sigma_2(s)\sqrt{1-\rho(s)^2}dW_2(s) \right\}.
\end{cases}
$$

The covariance matrix for [ln $S_1(T)$ ln $S_2(T)$]' is

$$
\begin{bmatrix} \sigma_1^2 & \rho\sigma_1\sigma_2 \\ \rho\sigma_1\sigma_2 & \sigma_2^2 \end{bmatrix}.
$$

Noticing that the variance of ln $S_2(t)$ – ln $S_1(t)$ is

$$
\sigma^2 = \sigma_1^2 + \sigma_2^2 - 2\rho\sigma_1\sigma_2,
$$

that the correlation coefficient between ln $S_1(t)$ and ln $S_2(t)$ – ln $S_1(t)$ is $(\rho\sigma_2 - \sigma_1)/\sigma$, and that the correlation coefficient between ln $S_2(t)$ and ln $S_1(t)$ – ln $S_2(t)$ is $(\rho\sigma_1 - \sigma_2)/\sigma$, we evaluate the conditional expectation:

$$C_{\min}(S_1,S_2,T,X) = S_1\exp\left(-\int_0^T \theta_1(s)ds\right)N_{biv}\left(h_1,h_3,\frac{\rho\sigma_2-\sigma_1}{\overline{\sigma}}\right)$$
$$+ S_2\exp\left(-\int_0^T \theta_2(s)ds\right)N_{biv}\left(h_2,h_4,\frac{\rho\sigma_1-\sigma_2}{\sigma}\right)$$
$$- e^{-rT}X\cdot N_{biv}\left(h_1-\sigma_1\sqrt{T},h_3-\sigma_2\sqrt{T},\rho\right)$$

where

$$h_1 = \frac{\ln(S_1/X) - \int_0^T \theta_1(s)ds + (r+\sigma_1^2/2)T}{\sigma_1\sqrt{T}}$$

$$h_2 = \frac{\ln(S_2/X) - \int_0^T \theta_2(s)ds + (r+\sigma_2^2/2)T}{\sigma_2\sqrt{T}}$$

$$h_3 = \frac{\ln(S_2/S_1) + \int_0^T (\theta_1(s)-\theta_2(s))ds + (r-\sigma^2/2)T}{\sigma\sqrt{T}}$$

$$h_4 = \frac{\ln(S_1/S_2) + \int_0^T (\theta_2(s)-\theta_1(s))ds + (r-\sigma^2/2)T}{\sigma\sqrt{T}}.$$

The key is to realize $C_{min}(S_1, S_2, T, 0) = \min(S_1, S_2)$ because the asset prices are always positive. Now let us turn attention to options on the maximum. Its claim at T can be written as

$$\begin{aligned}K(T) &= (\max(S_1(T),S_2(T))-X)^+\\ &= [(S_1(T)-X)+(S_2(T)-X)-(\min(S_1(T),S_2(T))-X)]^+\\ &= (S_1(T)-X)^+ + (S_2(T)-X)^+ - (\min(S_1(T),S_2(T))-X)^+.\end{aligned}$$

The last equality follows because the third term must cancel out one of the first two terms. Hence, a call on the maximum is equivalent to a long position in two regular calls and a short position in a call on minimum. Its price can be valued as

$$C_{\max}(S_1, S_2, T, X) = C(S_1, T, X) + C(S_2, T, X) - C_{\min}(S_1, S_2, T, X).$$

Similarly, the claim of a put on maximum at time T is

$$\begin{aligned} K(T) &= (X - \max(S_1(T), S_2(T)))^+ \\ &= [(X - S_1(T)) + (X - S_2(T)) - (X - \min(S_1(T), S_2(T)))]^+ \\ &= (X - S_1(T))^+ + (X - S_2(T))^+ - (X - \min(S_1(T), S_2(T)))^+. \end{aligned}$$

Hence, a put on the maximum is equivalent to a long position in two regular puts and a short position in a put on minimum. Its price can be valued as

$$P_{\max}(S_1, S_2, T, X) = P(S_1, T, X) + P(S_2, T, X) - P_{\min}(S_1, S_2, T, X).$$

Rainbow options are typically

- More expensive than conventional ATM options on any one of the assets represented, because with a rainbow there is a higher probability that at least one of the assets will perform well
- More expensive than a basket option on the same assets, because the performance of the worst-performing assets is included in the value of the basket, whereas it is excluded from the rainbow's payoff
- Cheaper than a portfolio of ATM calls on each of the assets, because only the best-performing asset is considered in the rainbow; the other ones are eliminated from the payoff calculation, even if they perform well

The higher the correlation between the assets the *lower* is the price of the rainbow option.

High positive correlation means that if the best-performing asset performed well, then the other assets are also likely to have performed well. Since only the best-performing asset is considered in the payoff calculation, the opportunity cost of exercising into this asset is higher. Put a different way, with negative correlation it is more likely that at least one of the assets will perform well, whereas with positive correlation it is more likely that none of the assets will perform.

Example 6.10

A 20/20 Option: Rainbow options are particularly useful as asset allocation tools, ensuring that the investor always ends up in the best-performing asset or sector.

Situation:

A fund manager believes that US and some European equity markets have considerable upside potential, although there are also risks of possible setbacks.

The manager would like to profit from this anticipated scenario but the rules of the fund do not allow him to gear up the fund by placing bets on all markets at the same time. An investment bank offers the investor the following contract

Expiry	12 months
Expiry payoff	0.75 x MAX { %Rise on S&P, %Rise on DAX, %Rise on FTSE, 0 }

Analysis:

This is a 3-coloured rainbow with a strike of zero. It is similar in spirit to a lookback, in that it allows the holder to capture best performance with hindsight. The difference is that with the rainbow the choice is about the best-performing asset class, whereas with the lookback it is about the best market timing. As with lookbacks, you can expect this option to be quite expensive.

6.11 Correlation Options

Correlation options are cheaper than conventional options, as the requirement that both assets must pass through the strike impose an additional risk to the holder. Indeed, this is one of their main appeals to investors.

The higher the correlation between the assets the lower is the price of the rainbow option.

6.12 Exchange Options

Compare the payoff of a conventional call with that of an exchange option

- When an investor exercises a conventional call she receives the underlying asset (A_1) in exchange for payment of the strike, K which has a fixed cash value

$$\Phi_{\text{call}}(T) = \max\{A_1 - K, 0\}$$

- When the investor exercises an exchange option she receives one of the assets (A_1) in exchange for payment in another asset (A_2):

$$\Phi_{\text{Exchange}}(T) = \max\{A_1 - A_2, 0\}$$

So the strike in an exchange option (A_2) is also a variable, with an uncertain future value.

Here we are dealing with an option whose strike is uncertain. We can now use the same method as we did with forward options

For a given parity ratio, the options premium is proportional to the underlying spot price and strike, K.

In other words, other things being equal, if you double both the spot and the strike while keeping the parity ratio constant, then the option price doubles as well. This makes it possible to transform the standard option pricing formula as follows

$$\Pi_{\text{call}}(t) = F(A_1, K, T, \sigma, r_{yield}, r_{funding}) = K\,F(A_1/K, T, \sigma, r_{yield}, r_{funding})$$

Applying this transformation to an exchange option, where $K = A_2$, we have:

$$\Pi_{\text{exchange}}(t) = \Pi(t, A_2)\,\Pi_{\text{call}}(t, A_1/A_2\; K = 1)$$

Now we can use a standard option pricing model to price a call on the variable A_1/A_2 with a strike of 1. The volatility of A_1/A_2 is the same as that of the original spread position—long A_1 and short A_2—and is estimated from the volatilities and correlation of these two assets.

6.13 Currency-linked Options

Currency-linked options are options on foreign currency assets whose payoffs and risks are linked in some way to the performance of the rate of exchange, as well as to the performance of the underlying assets.

There are three main types of currency-linked contract:

- *Options on foreign currency assets struck in local currency* where the strike is fixed in the local currency and at expiry the option pays the difference between the strike and the value of the foreign asset converted into the local currency at the spot FX rate prevailing at the time. If we take as an example a call on EUR-denominated assets payable in USD

$$\Phi_{\text{call}}(T) = \max\{EUR\,Asset\,price \times Spot\,FX\,rate - Strike, 0\}$$

where *Spot FX* rate is in USD per EUR 1.

- *Currency-protected options* also known as *quantity-adjusted options*, *quantos* or *fixed exchange rate options*, these are options on foreign currency assets which pay the intrinsic value in the local currency at a rate of exchange that is fixed on the trade date. Again, if we take as an example a call on EUR assets payable in USD

$$\begin{aligned}\Phi_{\text{call}}(T) &= \max\{EUR\,Asset\,price - EUR\,Strike, 0\} \times Fixed\,FX\,rate \\ &= \max\{EUR\,Asset\,price \times Fixed\,FX\,rate - Strike, 0\}\end{aligned}$$

- *Asset-linked FX options* an FX option whose contract value is linked to the price of a foreign currency asset at expiry. Taking a EUR call / USD put payable in USD as an example:

$$\Phi_{\text{call}}(T) = EUR\,Asset\,price\,\mathrm{x}\max\{Spot\,FX\,rate - FX\,Strike, 0\}$$

Quantos are especially attractive to international investors who prefer to separate their asset allocation decisions from their currency decisions, effectively making currency a separate asset class.

Example 6.11

A US investor would like to buy a call on the Nikkei but is concerned that the gains made on the Japanese equities could be severely reduced by a weaker yen. This FX exposure could be hedged with forward FX contracts, but the hedge would be imperfect because the exact amount of FX cover required will depend on the performance of the underlying equities, which is uncertain.

An investment bank offers the client the following contract

Type	Call
Style	European
Underlying	Nikkei 225 index
Strike	21,250 (= ATM cash)
Expiry	3 months
Payoff	USD 100 per index point

Outcome: At the option's expiry the Nikkei is at 21,750 and the option expires 500 index points in the money. The investor receives a payout of USD 50,000 (= 500 x 100) *irrespective of the USD/JPY exchange rate at the time.*

Analysis: The seller of the quanto absorbs the currency risk, in addition to its exposure to the underlying market (in this case short a call), and this additional risk must be factored into the price.

For pricing, a number of models are given in Haug.

6.13.1 Fixed Domestic Strike Options

An option on a foreign asset, struck in the local currency is essentially an option on a basket consisting of

- A specified amount of the foreign asset
- An FX position in the same amount (long the foreign currency)

Therefore, to price this option we can apply the same technique that we apply to pricing any basket option (multi-asset option). The payoff will depend on

- The performance of the foreign asset
- The rate of exchange prevailing when the option expires
- The correlation between the two

The higher the correlation between the exchange rate and the price of the foreign asset, the higher is the cost of this option.

6.14 Pay-later Options

A pay-later option is the right to buy (for a call) or sell (for a put) an asset at time $T > 0$ for a strike price $X > 0$ with the following features:

- The premium for this option is paid only on the exercise,
- The option must be exercised if the asset price is above (for a call) or below (for a put) X.

The price of a pay-later call can be written as

$$\begin{aligned} C_{pc}(S,T,X) &= e^{-\bar{r}T}E^{Q}[(S(T)-X-X_c)\,|\,S(T)>X] \\ &= e^{-\bar{r}T}E^{Q}[(S(T)-X)\,|\,S(T)>X]-X_c e^{-\bar{r}T}E^{Q}[1\,|\,S(T)>X] \\ &= C(S,T,X)-X_c\cdot C_d(S,T,X) \end{aligned}$$

where X_c is the premium of the option. Thus, a pay-later call is a combination of a long position in a usual call and a short position in a digital call. Since it is costless to get such an option, its price must be zero. This results in the exact value

$$X_c = \frac{C(S,T,X)}{C_d(S,T,X)}$$

which remains constant throughout the life of the option. Accordingly, the price of a pay-later put is

$$P_p(S,T,X) = P(S,T,X) - X_p \cdot P_d(S,T,X)$$

where the value of X_p can be solved by setting $P_p(S, T, X) = 0$:

$$X_p = \frac{P(S,T,X)}{P_d(S,T,X)}.$$

6.15 Extensible Options

An extensible option grants the holder the right to extend the option to a later expiration time with a new strike price. Consider a call with a strike price $X_1 > 0$ and maturity $T_1 > 0$ when the holder can extend the option to time $T_2 > 0$ with a new strike price $X_2 > 0$ by paying a premium $A > 0$. The claim of this option at time T_1 is

$$\begin{aligned} K(T_1) &= \max\left(0, C\left(S(T_1), T_2 - T_1, X_2\right) - A, S(T_1) - X_1\right) \\ &= \max\left(\left(C\left(S(T_1), T_2 - T_1, X_2\right) - A\right)^+, \left(S(T_1) - X_1\right)^+\right). \end{aligned}$$

This is a compound rainbow option. The first asset is a compound option while the second is a standard call. It is straightforward to price this call as

$$C_c(S,T_1,T_2,X_1,X_2,A) = C_{\max}\left(CC(S,T_1,T_2,X_2,A), C\left(S,T_1,X_1\right), T_1, 0\right)$$

A put with a strike price X_1 and maturity T_1 when the holder can extend the option to time T_2 with a new strike price X_2 by paying a premium A can similarly be priced as

$$P_c(S,T_1,T_2,X_1,X_2,A) = P_{\max}\big(PP(S,T_1,T_2,X_2,A), P\big(S,T_1,X_1\big),T_1,0\big).$$

6.16 Quantos

Quantos are a family of contingent claims whose payoff are defined with respect to the value of some foreign asset in their own currency, but denominated in the domestic currency. The price of a foreign asset follows a diffusion process

$$dS(t) = S(t)\mu_S(t)dt + S(t)\sigma_S(t)dW_1(t).$$

The exchange rate follows another diffusion process

$$dC(t) = C(t)\mu_C(t)dt + C(t)\sigma_C(t)dW_1(t) + C(t)\sqrt{1-\rho^2(t)}\sigma_C(t)dW_2(t),$$

where W_1 and W_2 are two independent standard Brownian motions, σ_s, σ_c and ρ are all deterministic processes, while μ_s and μ_c are both predictable. Thus, the correlation between $d\ln C(t)$ and $d\ln S(t)$ is $\rho(t)$. Moreover, the asset S is associated with a proportional dividend process θ. Let r and r_f be the domestic and the foreign deterministic interest rate processes, respectively.

In the domestic market, foreign assets are not directly tradable, but the foreign currency and the foreign asset value in the domestic market are. To eliminate the arbitrage opportunity

$$\begin{bmatrix} Y_1(t) \\ Y_2(t) \end{bmatrix} = \exp\left(-\int_0^T r(s)ds\right)\begin{bmatrix} \exp\left(\int_0^T r_f(s)ds\right)C(t) \\ \exp\left(\int_0^T \theta(s)ds\right)S(t)C(t) \end{bmatrix}$$

has to be martingale under the risk-neutral measure Q. Starting with

$$\begin{aligned} dY_1(t) &= Y_1(t)\big(\mu_C(t) - r(t) + r_f(t)\big)dt + Y_1(t)\sigma_C(t)dW_1(t) \\ &\quad + Y_1(t)\sqrt{1-\rho^2(t)}\sigma_C(t)dW_2(t) \end{aligned}$$

$$\begin{aligned} dY_2(t) &= Y_2(t)(\mu_C(t)+\mu_S(t)-r(t)+\theta(t))dt + Y_2(t)(\sigma_S(t)+\rho(t)\sigma_C(t))dW_1(t) \\ &\quad + Y_2(t)\sqrt{1-\rho^2(t)}\sigma_C(t)dW_2(t) \end{aligned}$$

we have

$$dC(t) = C(t)\big(r(t) - r_f(t)\big)dt + C(t)\rho(t)\sigma_C(t)d\widetilde{W}_1(t) + C(t)\sqrt{1-\rho^2(t)}\sigma_C(t)d\widetilde{W}_2(t)$$

$$dS(t) = S(t)(r(t) - \theta(t) - \rho(t)\sigma_S(t)\sigma_C(t))dt + S(t)\sigma_S(t)d\widetilde{W}_1(t)$$

under the martingale measure Q. The foreign asset at time T under the measure Q becomes

$$S(T) = S(0)\exp\left(-\int_0^T \theta(s)ds + \int_0^T \sigma_S(s)dW_1(s)\right) \times \exp\left(\int_0^T r_f(s)ds - \int_0^T \rho(s)\sigma_S(s)\sigma_C(s)ds - \frac{1}{2}\int_0^T \sigma_S^2(s)ds\right).$$

Without loss of generality, it is assumed that $C(0) = 1$. It follows

$$S(0) = \frac{1}{B(T)}E^Q[C(T)S(T)|\mathcal{F}_0].$$

Since interest rates are deterministic, forward constraints are equivalent to futures contracts. Compared with the future spot price parity for domestic securities, a quanto forward or futures price can be written as

$$f = F = \exp\left(-\int_0^T \theta(s)ds\right)E^Q[S(T)|\mathcal{F}_0] = S\exp\left((r_f - \sigma_{CS}^2)T - \int_0^T \theta(s)ds\right),$$

where r_f is the foreign interest rate. Depending on the sign of the covariance, the quanto futures price can be either greater or less than the standard futures price. It is correlated to the exchange process because the replicating portfolio involves the foreign currency and the foreign asset.

Next we turn into a digital call. Its price can be written as

$$C_d(S,C,T,X) = e^{-rT}\Pr[S(T) > X] = e^{-rT}N(d)$$

$$d = \frac{\ln(S/X) - \int_0^T \theta(s)ds + (r_f - \sigma_{CS}^2 - \sigma_S^2/2)T}{\sigma_S\sqrt{T}}.$$

A quanto call is no more complicated. It can be priced as

$$C_q(S,C,T,X) = e^{-\bar{r}T}E^Q\left[(S(T)-X)^+\right]$$
$$= S\exp\left(-\int_0^T \theta(s)ds\right) \times e^{\left(\bar{r}_f-\bar{r}-\bar{\sigma}_{CS}^2\right)T}N\left(d-\bar{\sigma}_S\sqrt{T}\right) - Xe^{-\bar{r}T}N(d).$$

As we seen, the quanto option is designed to eliminate the FX risk to the buyer, so its payoff depends only on the performance of the foreign asset. However, the counterparty selling the quanto has to absorb the currency risk, so in pricing this option they must take into account not only the volatility of the exchange rate but also the correlation between it and the foreign asset price.

The higher the correlation between the exchange rate and the price of the foreign asset:

- The lower is the cost of the quanto call
- The higher is the cost of the quanto put

If there is a tendency for the EUR to strengthen every time the asset price rises (i.e. positive correlation) then the volatility of EUR asset price × fixed FX is lower than the volatility of EUR asset price × spot FX:

- The buyer of the quanto call foregoes the potential currency gain (if the underlying asset price rises) so the quanto call is cheaper
- The buyer of the quanto put is protected against currency losses (if the underlying price falls and the FX rate weakens) so the quanto put is more expensive

The opposite is the case for the investor buying these options when the correlation between the FX rate and the asset price is negative: the quanto call is more expensive and the quanto put is cheaper.

The seller of a quanto option must dynamically hedge her FX risk by carrying an outright FX position.

Whether the hedging FX position for the quanto seller should be long or short the foreign currency depends on the type of option sold (i.e. call or put).

As we have seen, currency-linked options are directly or indirectly subject to correlation risk. Pricing these options therefore requires careful estimation of the correlation between the underlying index and the rate of exchange. As with any other multi-asset option, these correlations can be quite unpredictable and this is what makes currency-linked options exotic.

6.17 Structured Products

Structured capital market products, including ever-new ones, have become increasingly complex in recent years. For market participants to be able to evaluate and control the risks involved, they must be well grounded in the intricacies of these innovative instruments as well as in adequate valuation techniques.

Structured products refer to combinations of individual financial instruments, such as bonds, stocks and derivatives. At first glance, most of these composite products are very similar to plain vanilla coupon bonds.

Structured products tend to involve periodical "interest payments" and redemption at maturity. What sets them apart from bonds is that both interest payments and redemption amounts depend in a rather complicated fashion on the movement of stock prices, indices, exchange rates or future interest rates.

Since structured products are made up of simpler components, we usually break them down into their integral parts when we need to value them or assess their risk profile and any hedging strategies. This should facilitate the analysis and pricing of the individual components. While this is indeed true in many cases, replication need not automatically entail a considerable simplification.

We will here give a few structured products to give an idea of how to treat them in different situations. For the purpose of valuation, structured products are generally replicated with simpler instruments. The portfolio of these simpler products must have the same payoff profile as the structured product and, given the (assumed) absence of arbitrage opportunities in financial markets, must thus also have the same market value. The merits of this approach are that, first, simple valuation rules can be used to calculate fair market prices for the simpler products. Second, risk control is more efficient since the replicated parts either are directly tradable or may be hedged more easily.

It is not possible to break all products down into simple components. In cases where the structured product has to be depicted as a combination of instruments which are themselves complex in nature and thus difficult to value and to hedge on the capital market, numerical procedures have to be employed in order to value the products and assess the risks involved.

6.17.1 Capped Call Option

In principal, the redemption amount for capital-guaranteed bonds with embedded call options can be infinitely high. The issues described in this section place a cap, expressed as a percentage of the instrument's face value, on the redemption amount. The bearer only participates in the relative performance of the underlying asset up to a certain maximum value. A capped call is the combination of a long position in a call option C_1 with a low strike price X_1 and a short position in a call option C_2 with a higher strike price X_2: As soon as the value of the underlying asset reaches the upper limit X_2, the holder of option C_2 will exercise his/her right and claim any further increase in the value of the underlying asset.

The issuer promises a redemption amount proportionate to the change in the underlying asset price. In cases where the price of the underlying asset decreases, the issuer guarantees a minimum redemption amount. At the same time, the issuer limits the investor's participation in the instruments performance by stipulating an upper limit (i.e., the cap).

Example 6.12

A guarantee certificate

Maturity: 15 December 2016 to 13 December 2020 (4 years)

Redemption rate: The redemption rate expressed as a percentage of the face value is proportionate to the change in the underlying asset price (S_T/S_0) expressed as

$$T = N \cdot (100\% + \min(9\%; \max(0\%; (S_T - S_0)/S_0)))$$

where

S_0 Closing price of XY stock on 15 December 2016

S_T Closing price of XY stock on 13 December 2020

Issue price: 100 %

Denomination: EUR 1000

If the price of the underlying asset increases between the issue date and the maturity date, then the investor will participate up to a rate of $a = 9$ %.

In order to replicate these products, it is necessary to express the formula for calculating the redemption amount in a different manner.

$$\begin{aligned}\Pi &= N \cdot \left\{1 + \min\left[a, \max\left(\frac{S_T - S_0}{S_0}, 0\right)\right]\right\} \\ &= N + N \cdot \max\left(\frac{S_T - S_0}{S_0}, 0\right) - N \cdot \max\left(\frac{S_T - S_0}{S_0} - a, 0\right) \\ &= N + \frac{N}{S_0} \cdot \max(S_T - S_0, 0) - \frac{N}{S_0} \cdot \max(S_T - (1 - a) \cdot S_0, 0).\end{aligned}$$

The capital-guaranteed products with an embedded European capped call option can be broken down into a portfolio consisting of a zero coupon bond

(continued)

Example 6.12 (continued)

with face value N, a long position of a European call option with a strike price of S_0, and a short position in a European call option with a strike price of $(1 + a)\cdot S_0$.

The face values of the zero coupon bonds is the coupon payments and the guaranteed redemption amount of the bond. Note that cash flows typically do not take place until the maturity date. The strike price of the long option is S_0, while that of the short option is $(1 + a)\cdot S_0$:

Assuming a principal of 1000, we can replicate this instrument with the purchase of a zero coupon bond which reaches maturity on 13 December 2016, and has a face value of EUR 1000, the purchase of $1000/S_0$ European call options on the underlying asset with a strike price of S_0, expiring on 13 December 2018 the sale of $1000/ S_0$ European call options on the underlying asset with a strike price of $1.09\cdot 2$, expiring on 13 December 2020.

6.17.2 Currency Basket Bonds with Caps and Floors

Currency basket bonds are zero coupon bonds whose redemption amount is notionally paid out in several currencies. The redemption amount is spread evenly across the reference currencies and paid out at the respective exchange rate applicable on the redemption date.

Defined caps and floors generally apply to the entire redemption amount, not to the individual currencies.

Example 6.13

A Quattro Bond

Maturity	22 October 2017, to 21 April 2022 (4.5 years)
Total principal	EUR 14,534,567
Base currency	EUR
Coupons	None
Redemption	22 April 2022
Redemption rate	see below
Denomination	EUR 726.73

Redemption: In order to calculate the redemption rate, 25 % of a bank bond is translated into EUR, 25 % into SEK, 25 % into GBP and 25 % into NOK at the Vienna exchange official middle exchange rate on 20 October 2017. Toward the end of the bonds term, the amounts in each currency are multiplied by a redemption factor of 1.2631 and the resulting amounts are translated into EUR at the Vienna exchange's official middle exchange rate on 18 April 2022. The redemption rate must be between 110 % and 139 % of the bonds face value.

(continued)

Example 6.13 (continued)

The Quattro bonds redemption amount depends on a portfolio of currencies whose development cannot be considered independently of one another. As the interest on the principal depends on the joint development of the various currencies and the cap and floor is defined for the entire portfolio, the instrument can only be valued using basket options.

The formula calculating the return on the principal invested can be rearranged as follows

$$r = \max(\min(0.25 \cdot 1.2631 \cdot (1 + r_{SEK} + r_{GBP} + r_{NOK}), 1.39), 1.1)$$

$$r = \frac{1.2631 \cdot 3}{4} + \frac{1.2631 \cdot 3}{4} \cdot$$

$$\max\left(\min\left(\frac{r_{SEK}}{3} + \frac{r_{GBP}}{3} + \frac{r_{NOK}}{3}, \frac{1.39 \cdot 4}{1.2631} - \frac{1}{3}\right), \frac{1.1 \cdot 4}{1.2631} - \frac{1}{3}\right)$$

At maturity, the bond is redeemed at a rate of at least 110 %. This minimum redemption amount can be replicated with a zero coupon bond with a face value equaling 110 % of the bond face value.

If the exchange rates develop in the bearer's favour, he/she will then participate in the positive performance of the three reference currencies. This payoff profile is equivalent to a call option on the currency basket with a strike price of

$$\frac{1.1 \cdot 4}{1.2631} - \frac{1}{3} = 0.828$$

As the underlying asset of the option, the currency basket has a standardized initial value of 1; its value at time T is calculated using the formula

$$1 + \frac{1}{3}(r_{SEK} + r_{GBP} + r_{NOK})$$

as the arithmetic mean of the changes in the component exchange rates in the currency basket.

The redemption rate cap of 139 % limits the payoff of the call on the currency basket. This cap is equivalent to the sale of a call option on the currency basket with a strike price of

$$\frac{1.39 \cdot 4}{1.2631 \cdot 3} - \frac{1}{3} = 1.134$$

The face value of the zero coupon bond is equal to the minimum redemption amount.

(continued)

Example 6.13 (continued)

Assuming a face value of EUR 100, we can replicate this instrument with the purchase of a zero coupon bond with a face value of 110 and a 4.5-year maturity, the purchase of k call options on a currency basket consisting of 1/3 SEK, GBP and NOK, strike price 0.828, expiration in 4.5 years and the sale of k call options on a currency basket consisting of 1/3 SEK, GBP and NOK, strike price 1.134, expiration in 4.5 years where

$$k = \frac{1.2631 \cdot 3}{4} \cdot 100$$

The zero coupon bonds are valued using the relevant spot interest rate. Options on currency baskets present the problem of that a sum (or the average) of log-normally distributed time series is not itself log-normally distributed. Basket options can be valued with Monte Carlo simulation assuming correlated Brownian motions.

6.17.3 A Quanto Asian Multi-basket Digital Option

We will here present an example of a structured product. Such a product is created by financial engineers. The Asian multi-digital consists of several Asian digital options with strike prices X_i equal to the initial underlying prices $S_i(t_0)$. The payoff is digital and depends on the number of Asian prices being above their strike prices.

The price is given by

$$\Pi(0) = e^{-rT} E^Q[\Phi(T)]$$

where the contingent claim with maturity T is given by

$$\Phi(T) = \begin{cases} 1 & \text{if} \quad \sum_{i=1}^{d} H_i \geq K \\ 0 & \text{else} \end{cases}$$

where

$$H_i = \begin{cases} 1 & \text{if} \quad A_i \geq S_i(t_0) \\ 0 & \text{else} \end{cases}$$

$$A_i = \frac{1}{N}\sum_{j=1}^{N} S_i(t_j)$$

What we have is a *basket* of d assets, S_i, $i = 1, 2, \ldots d$. We start by calculate the average underlying price (Asian price) A_i for each asset for N reset days, t_j, $j = 1, 2, \ldots N$. The average price makes the instruments *Asian*. We then make a test if the assets reach the strike. If asset i reach the strike, ($H_i = 1$, the digital payoff of asset i) it will pay 1 unit of cash. This makes the instrument *digital*. At the end, we count how many of these assets reach the strike. If the number of such assets is greater than K, the option will pay 1 cash unit. That's make the option *multi-digital*.

To find the price we calculate the discounted expected payoff with respect to the martingale measure Q.

An instrument as this has to be evaluated by Monte Carlo simulations. The same situation appears for all instruments that cannot be replicated into known instruments with a closed form solution.

6.18 Summary of Exotic Instruments

The following table summarizes the exotic options covered. For details, see appropriate sections.

Name	Structure	Variations/combinations
Chooser	Buyer can decide whether it is a call or a put by a certain date	Simple chooser Complex chooser
Digital	Buyer receives a fixed cash payout on exercise	Cash-or-nothing Asset-or-nothing
Barrier	Option either knocks out or knocks in if a certain market level is reached	Double barrier Ladder Touch spread
Lookback	Buyer receives the best performance achieved by the market during the option's life	Lookback price Lookback strike
Asian	Buyer receives difference between strike and an average of market prices (Asian price), or strike is calculated as an average of market prices (Asian strike)	Arithmetic averages Geometric averages Different averaging periods
Forward	Strike is set on a specified future date	Cliquet
Basket	Pays difference between a strike and the price of a basket of financial instruments	Equity index options OTC baskets Options on foreign assets struck in local currency

(continued)

Name	Structure	Variations/combinations
Rainbow	Payment based on the best-performing of any number of assets	Two or more colors
Exchange	Buyer has the right to exchange units of one asset for units of another asset	Convertible bonds Spread options
Correlation	If the option is ITM, payout is contingent on the price of another asset also having reached a certain level	
Quanto	Option on a foreign currency asset, where payout is in the local currency at a fixed rate of exchange	
Compound	Buyer has the right to exercise into another pre-defined option contract	

6.19 Something About Weather Derivatives

To give an example of completely different kinds of derivative, we will look at so-called weather derivatives. The weather has an enormous impact on businesses, for example, energy producing companies, farmers, travel agents and wine producers. The weather affects each of these industries in different ways. For example, a warm winter will reduce the profits of an energy producer, whereas a warm summer will produce better grapes wine producers. In some situations, "bad" weather can result in companies making a loss. For this reason, a number of companies have started trading derivatives to hedge against losses due to weather events, just as one hedge against any price changes on an asset by acquiring an option. Essentially, this makes a weather derivative rather similar to weather insurance.

A weather derivative is exactly as it says—a derivative written on the weather. These can be either swaps, futures or options (most commonly in the form of swaps between two companies). So how can the weather be treated as an underlying asset? Well, the weather is not strictly speaking a physical asset, but just something to base the pricing of the derivatives upon.

The first transactions on weather derivatives took place in the US in winter 1997, after the strongest El Niño event on record. (The El Niño phenomena have worldwide implications on the weather.) The significant coverage of this phenomenon by the American press meant companies decided to hedge against losses due to unseasonable weather. European companies soon followed, however, the weather derivatives market in the USA is still worth some 10 times more and that of its European counterpart. This makes the euro market rather illiquid.

Initially, it was the energy sector that began trading in the weather derivatives market. By September 1999, with the increase of investors outside the energy sector, the Chicago Mercantile Exchange (CME) started an electronic market for the trading of weather derivatives.

A weather derivative can be written on many different parameters, such as, temperature, rain and snowfall measured at some mutually agreed weather station (e.g. Heathrow, London). Energy producer even have models for the quality of the snow, to calculate how much water it will give when it melt. The most common parameter is the temperature.

To develop a temperature index we start by defining first the temperature on day i,

$$T_i = \frac{T_i^{\max} + T_i^{\min}}{2}$$

where $T_i^{\max}$ and $T_i^{\min}$ are the maximum and minimum temperatures (in degrees Celsius) recorded at a particular station, respectively. This will allow us to define the *heating degree-days*, HDD_i , and the *cooling degree-days*, CDD_i ,

$$HDD_i = \max\{18 - T_i, 0\} \cdots \text{and}$$

and

$$CDD_i = \max\{T_i - 18, 0\}$$

respectively. Here 18 ° C is here used as a reference point, since it is believed that if the temperature is above (or below) 18 ° C people will turn on their air conditioning (or heating) and cool down (or heat up) their homes, thus a cooling (or heating) day. Most temperature based derivatives as based on the accumulation of *HDDs* or *CDDs* over some period, usually one month or a winter/summer period. Then the contracts can be written on the form (from the above expressions):

$$S = k\max\{Hn - K, 0\}$$

where k is the amount of money paid out per degree day index (a proportionality constant), known as the *tick size*.

This makes the pricing of a weather derivative rather like a weather forecasting game, the party with the best forecast better off since they can price their derivatives more appropriately.

6.19.1 Modelling Temperature

Modelling the temperature's statistical behaviour in terms of a stochastic process is an essential for temperature-derivative pricing. However, methods used to predict the temperature vary widely.

So why choose weather derivatives over insurance contracts?

The main difference between a weather derivative and contract insurance is that the holder of an insurance contract has to prove that he/she has suffered financial loss as a result of the weather in order to be compensated. (This is mainly for extreme events such as typhoons and hurricanes). Weather derivative payouts depends solely on the outcome of the weather, regardless of how it affects the holder's profits. Therefore, the weather derivatives market is rapidly growing.

When modelling the temperature in whether derivatives the process is described by a fractional Brownian motion (FBM). These processes are known as fractional Ornstein–Uhlenback-processes. Since the temperature varies periodically like $\sin(\omega t + \varphi)$ where t denotes the time, measured in days, and $\omega = 2\pi/365$ the period of oscillation neglecting leap years. The phase function ω is necessary if we let $t = 1, 2, \ldots$be the first, second, and so on, days of the year—since the seasonal cycle is out of phase with the western calendar. Another trend that is often seen on some datasets of temperature is a slight annual increase in mean temperature. Reasons for this can be the global warming trend, or, urbanization near big cities, with cities growing in size causing a net warming of the surroundings. This trend is much weaker than that of the seasonal cycle, so, to first approximation on an approximate this to a polynomial expression dominated by the linear term. Using this and the seasonal cycle, one can write the mean temperature, as a function of time:

$$T_t^m = A + Bt + C\sin(\omega t + \varphi)$$

where the parameters A, B, C and φ are constants. So far we have just a deterministic model for the mean temperature. However, we know that temperature is a stochastic process. Naturally, as one does very often in financial mathematics, a Wiener process $(W_t, t \geq 0)$, is put forward as a first approximation to the stochastic part of the model. In Fig. 6.6 we show the daily changes of the temperature in Stockholm, Sweden since 1 January 1756[1]. We see that it agrees pretty well to a sum of two Gaussian distributions.

[1] See http://www.smhi.se/klimatdata/meteorologi/temperatur/stockholms-temperaturserie-1.2847. From 1756 to 1875 the thermometer was hung in the free air outside a north-facing window on the second floor

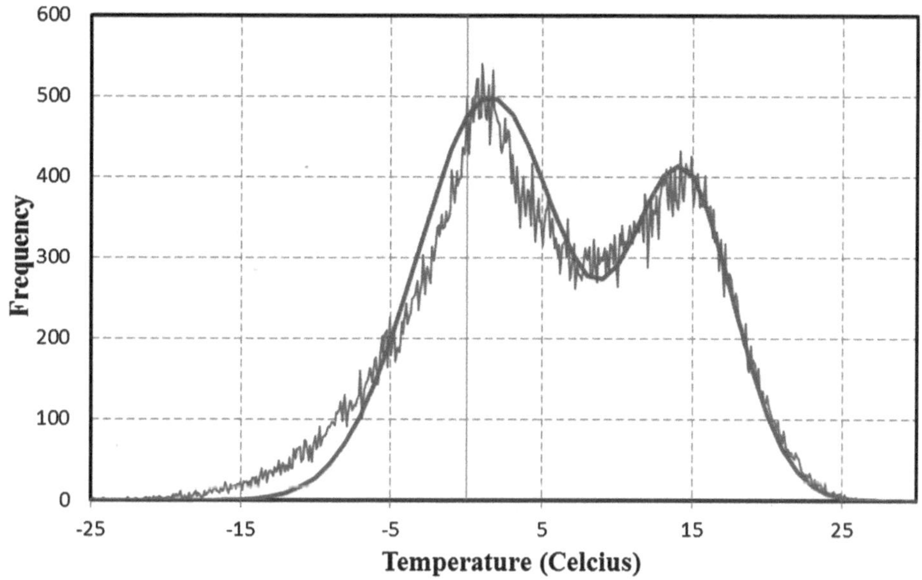

Fig. 6.6 Histogram of daily Stockholm temperature fluctuations 1756–2015

Thus, the "noise" term in our stochastic differential equation (SDE) will be of the form $\sigma_t W_t$, where σ_t is the time varying standard deviation. Sometimes the standard deviation is found to be approximately constant for each month, so σ_t is chosen from a set given by each month of the year. This deals with our stochastic term. Now the deterministic part – already found for the mean daily temperature T_t^m. We present the following form for the change in temperature

$$dT_t = a\left(T_t^m - T_t\right)dt + \sigma_t dW_t$$

The $\left(T_t^m - T_t\right)$ term comes from the mean reversion property of temperature. i.e. cannot deviate away from the mean temperature on long time scales. Constant a (real and positive) determines the speed of mean reversion. A better physical explanation for this may be found in the thermodynamic conditions of the system (atmosphere), though this is not trivial to see. This

of the old astronomical observatory building in Stockholm. From 1876 to 1960 the thermometer was placed outside a north-facing window on the first floor of the old astronomical observatory building in Stockholm. From 1961 to summer 2006 the thermometer was placed in a SMHI-screen (Stevenson-type screen) about ten meters north-east of the former position. Since summer 2006, a platinum resistance thermometer in a modern cylindrical screen close to the SMHI-screen replaced the mercury thermometer in the SMHI-screen.

equation describes an Ornstein-Uhlenback processes, and is found to agree well with empirical data. The only problem with our SDE so far is that the long-term mean is not the same as T_t^m (this can be shown when solving the SDE). By adding the differential of the mean T_t^m,

$$\frac{dT_t^m}{dt} = B + C\cos(\varphi + \omega t)$$

to the deterministic drift term the solution of the SDE gives our desired long-term mean T_t^m. Thus,

$$dT_t = \left\{\frac{dT_t^m}{dt} + a\left(T_t^m - T_t\right)\right\}dt + \sigma_t dW_t \quad t \geq 0.$$

The solution is

$$T_t = \left(T_t - T_0^m\right)e^{-a(t-t_0)} + T_t^m + \int_{t_0}^{t} e^{-a(t-\tau)}\sigma_\tau dW_\tau$$

where

$$T_t^m = A + Bt + C\sin(\omega t + \varphi)$$

We have now developed a physically representable stochastic model of temperature, which can be used to price temperature derivatives.

Recently, modelling wind has become of importance since it is predicted that within five years 10 % of the UK energy will be produced by this renewable source. The problem with measuring wind is that, unlike temperature, local variations are very large.

7

Pricing Using Deflators

7.1 Introduction to Deflators

We will now discuss the use of deflators for the valuation of financial contracts that can be used on many kind of contract, insurance policies and pension plans. Many such contracts contain option features. For example, several life insurance policies contain rate of return guarantees. Pension funds typically aim for full indexation of the benefits to price inflation, but in scenarios where inflation is extremely high or the funding ratio is low, indexation can be reduced or skipped altogether. This type of payout is difficult to value with standard present value calculations, as it is not obvious which discount rate to use. However, the payout of the contract depends on some underlying variable like the stock price or the inflation rate. We can therefore see such contracts as contingent claims, or derivatives.

Starting with the path-breaking work of Black and Scholes (1973), a large literature on the valuation of derivatives has emerged. The key observation in the Black and Scholes analysis is that the derivative can be hedged by a position in the underlying instrument. The resulting cash position is risk-free and therefore should earn the risk-free rate of interest. Working out this argument formally leads to the risk-neutral valuation (RNV) method used in the previous sections. The RNV method calculates the price of the derivative as the expected payoff of the derivative in a risk-neutral world, discounted at the risk-free rate. In that risk-neutral world, the expected return on the stock is set equal to the risk-free rate.

J.R.M. Röman, *Analytical Finance: Volume I*,
DOI 10.1007/978-3-319-34027-2_7

RNV is a very convenient and powerful method for the valuation of derivatives where the underlying value is a traded asset. It is less trivial to apply the method when the underlying determinant if the contingent claim is not a financial instrument, though. An important example of such a claim is an indexed pension benefit, whose payout depends on the price level or the inflation rate. The price level is not the price of a traded asset, so we cannot do as if the "return" on the price level equals the risk-free rate. For such claims, a valuation method known as deflators is useful.

We will now explain the concept of deflators by a few very simple examples, starting with the stock price model of Black and Scholes (BS). We will show how to construct a deflator for the BS model and how to use it to calculate the value of stock options. Then we will turn to inflation contingent claims like pension contracts, and show how to construct a deflator for such claims.

7.1.1 The Black–Scholes Deflator

Consider a simple one-period model (think of the period as one year). The price of the stock at the beginning of the period is S_0, and S denotes the stochastic price at the end of the period. In the Black-Scholes model, stock prices follow a lognormal distribution. Formally, the returns on the stock are generated by

$$\ln\left(\frac{S}{S_0}\right) = \left(\mu - \frac{1}{2}\sigma^2\right)t + \sigma \cdot \varepsilon\sqrt{t},$$

where ε is a standard normal random variable with mean zero and variance equal to one. The parameter μ equals the expected return, $E[S/S_0] = \mu$, and σ is the standard deviation of returns. A parameter that will play an important role in this section is the market price of risk or the sharp ratio (sometimes sharp quote) of the stock, defined as the risk premium (expected return minus the risk-free interest rate, r), divided by the standard deviation of the return

$$\lambda = \frac{\mu - r}{\sigma}.$$

For the valuation of derivatives of the stock price, risk-neutral valuation is the most common approach in the options literature, but here we will present the valuation of derivatives using the deflator method.

The deflator is a stochastic discount factor, that is, a discount factor that varies with the random variables driving the stock returns. We have already almost derived the expression of the Black–Scholes deflator when we used the Girsanov transformation between the observable market probabilities, P, and risk-neutral probability measure, Q. The Girsanov kernel $g(t)$ that took us from P to Q was the market price of volatility risk, that is, $g(t) = \lambda$.

We then used that $dQ(t) = L(t)dP$, where

$$\begin{cases} dL(t) = g(t)L(t)dW(t) \\ L(0) = 1 \end{cases}$$

With the known solution to $L(t)$ (using Itô on $\ln(L)$)

$$L(t) = \exp\left\{ \int_0^t g(s)dX(s) - \frac{1}{2}\int_0^t g^2(s)ds \right\}$$

we get

$$\begin{aligned} L(t) &= \exp\left\{ \int_0^t \frac{r-\mu}{\sigma} dW(s) - \frac{1}{2}\int_0^t \left(\frac{r-\mu}{\sigma}\right)^2 ds \right\} \\ &= \exp\left\{ \frac{r-\mu}{\sigma} W(t) - \frac{1}{2}\left(\frac{r-\mu}{\sigma}\right)^2 t \right\} \qquad = \exp\left\{ -\lambda \cdot W(t) - \frac{1}{2}\lambda^2 t \right\} \end{aligned}$$

which we can express like

$$L(t) = \left\{ -\lambda \cdot \varepsilon\sqrt{t} - \frac{1}{2}\lambda^2 t \right\}.$$

This means that the discount factor e^{-rt} in Q can, in P be expressed as

$$D = \exp\left(-\left(r + \frac{1}{2}\lambda^2 \right) t - \lambda \cdot \varepsilon\sqrt{t} \right).$$

This is our definition of the Black–Scholes deflator. As we can see, $1/D$ is a stochastic process with a normal distribution

$$N\left[\frac{r+\lambda^2}{2}, \lambda^2(T-t) \right].$$

If we do not use continuous compounding, we can approximate the deflator (using a second order Taylor approximation) by

$$D = \left(\frac{1}{1+r}\right)\left(\frac{1}{1+\lambda\cdot\varepsilon+\lambda^2}\right).$$

The deflator is the product of the risk-free discount factor and a stochastic term, which depends on the shocks to the stock price.

The deflator can be used to calculate the value of derivatives of the stock price as follows. Denote the payoff of the derivative by X, which will be a function of the stock price at the end of the period $X = f(S)$. The value of the derivative is then given by the expectation of the product of the deflator D and the payoff X:

$$X_0 = E[DX] = E[Df(S)].$$

For an interpretation, let's assume that the risk premium on the stocks is positive. The deflator then takes low values in states with a high stock return (high ε), and vice versa. One could also say that in states with low stock returns, the implicit discount rate for payoffs is low. Payoffs in "bad" states of the world, namely states where the stock price is low, will therefore have a relatively high value. Derivatives that mainly pays off in "bad" states of the world, such as put options, will therefore be relatively expensive compared to their expected payoff. On the other hand, assets that pay off when the stock returns are high, such as call options, will be less valuable.

We now discuss two important properties of the deflator. First, the deflator's expectation is equal to the risk-free discount rate

$$\begin{aligned} E[D] &= E\left[\exp\left(-rt - \frac{1}{2}\lambda^2 t - \lambda\varepsilon\sqrt{t}\right)\right] = \exp(-rt)E\left[\exp\left(-\frac{1}{2}\lambda^2 t - \lambda\varepsilon\sqrt{t}\right)\right] \\ &= E\left[\exp\left(-\frac{1}{2}\lambda^2 t\right)\right]E\left[\exp(-\lambda\varepsilon\sqrt{t})\right]\exp(-rt) \\ &= \exp\left(-\frac{1}{2}\lambda^2 t\right)\exp\left(\frac{1}{2}\lambda^2 t\right)\exp(-rt) \\ &= \exp(-rt) \end{aligned}$$

where we have used the theorem: For $X \sim N(m, \sigma^2)$ and $\gamma \in R$ we have

$$E\left[e^{-\gamma X}\right] = \exp\left\{-\gamma m + \frac{1}{2}\gamma^2\sigma^2\right\}$$

that is, since ε is N[0, 1] we get

$$E\left[e^{-\gamma\varepsilon\sqrt{t}}\right] = \exp\left\{\frac{1}{2}\gamma^2 t\right\}.$$

A risk-less cash flow, say F dollars, will therefore be valued by the standard present value formula

$$PV(F) = E[DF] = E[D]F = \exp(-rt)F.$$

Another interesting special case is the stock itself. Working out the return equation we find the stock price at the end of the year,

$$S = S_0\exp\left(\left(\mu - \frac{1}{2}\sigma^2\right)t + \sigma \cdot \varepsilon\sqrt{t}\right).$$

It is then easy to show that the deflated value of the end-of-period stock price equals the current price, $E[DS] = S_0$.

To prove this, we do the following calculation

$$\begin{aligned}
E[DS] &= E\left[\exp\left(-rt - \lambda\varepsilon\sqrt{t} - \frac{1}{2}\lambda^2 t\right)S_0 \exp\left(\left(\mu - \frac{1}{2}\sigma^2\right)t + \sigma\varepsilon\sqrt{t}\right)\right] \\
&= S_0 E\left[\exp\left(\left((\mu - r) - \frac{1}{2}\{\sigma^2 + \lambda^2\}\right)t + \{\sigma - \lambda\}\varepsilon\sqrt{t}\right)\right] \\
&= S_0 \exp\left\{\left((\mu - r) - \frac{1}{2}\{\sigma^2 + \lambda^2\} + \frac{1}{2}\{\sigma - \lambda\}^2\right)t\right\} \\
&= S_0 \exp\{(\mu - r - \sigma\lambda)t\} = S_0
\end{aligned}$$

This shows that the deflator prices the (stochastic) stock payoff itself correctly. Another important example is a call option with exercise price K. The value of this option is given by

$$C_0 = E[D\cdot\max(S - K, 0)].$$

Working out this expectation can be done analytically, and will give the famous Black–Scholes option pricing formula. For less trivial payoffs, the expectation $E[DX]$ can be approximated using Monte Carlo simulation.

7.1.2 Deflators for Inflation Linked Claims

We now turn to the construction of a deflator for price index (or inflation) linked claims. Although the price level or inflation are not assets traded on a financial market, one can use index linked bonds (ILB's) to hedge claims that depend on changes in the price level. Specifically, let there be an (zero coupon) ILB, with a notional value. The payoff of this bond is $\exp(\pi)$, where π is the inflation. Assume that inflation is stochastic with distribution

$$\pi = \pi^e + \sigma_\pi \eta$$

where π^e is expected inflation, assumed to be known in advance, and $\sigma_\pi \eta$ is the unexpected inflation. The random shock η has a normal distribution with mean zero and variance one. The zero-time price of this bond is equal to exp $(-r)$ where r is the real interest rate, assumed to be known and constant. Hence, the nominal return on ILB is $r+\pi$, and its expected return is $r + \pi^e + 1/2\ \sigma_\pi^2$. The market price of inflation risk is defined as the Sharpe ratio of the ILB return

$$\lambda_\pi = \frac{r + \pi^e + \frac{1}{2}\sigma_\pi^2 - r}{\sigma_\pi},$$

where r is the nominal risk-free interest rate. The deflator is now defined in the same way as before

$$D = \exp\left(-\left(r + \frac{1}{2}\lambda_\pi^2\right)t - \lambda_\pi \eta \sqrt{t}\right).$$

The valuation formula for inflation linked claims with payoff $X = g(\pi)$ is

$$X_0 = E[DX] = E[Dg(\pi)].$$

So far, this repeats the analysis of the Black–Scholes deflator, but with the index-linked bond (ILB) as the underlying asset. There is one crucial difference, though. Whereas it is natural to assume that stocks have a positive risk premium and hence a positive Sharpe ratio, the expected return on ILBs is quite low, and can be lower than the (nominal) risk-free interest rate, that is, $r > r + \pi^e + 1/2\ \sigma_\pi^2$. This will happen when investors are prepared to pay a premium to hedge against inflation risk. This will drive up prices of ILBs and hence give low real interest rates. The market price of inflation risk is therefore negative, $\lambda_\pi < 0$. This means that payoffs in high inflation states will be relatively valuable, and payoffs in low inflation states are less valuable.

7.1.3 Extensions

In the above discussed what deflators are and how they can be used for the valuation of contingent claims such as options and pension benefits. We discussed two specific examples, a deflator for stock price derivatives and a deflator for inflation contingent claims. A third important example are interest rate dependent claims, like portfolios of fixed income securities or the liabilities of insurance companies with embedded interest rate options.

Two extensions of the examples are relevant in practice. First, the examples above construct the deflator from the Sharpe ratio of traded asset prices. For risks that are not traded on financial markets, one can still construct a deflator of the form

$$D = \exp\left(-\left(r + \frac{1}{2}\lambda_\xi^2\right)t - \lambda_\xi \xi \sqrt{t}\right),$$

where ξ is the shock to the risk factor. The difference with the previous case is that the value of λ_ξ cannot be derived from the expected return on traded assets, but has to be fixed exogenously.

A second extension of the examples is the valuation of more complex payoff structures that depend on multiple risk factors simultaneously. An example is a pension benefit that depends both on inflation and on the returns on the pension fund's investments. It is possible, and actually quite easy, to build a deflator for multiple risks. Essentially, this can be done by multiplying the deflator for each individual risk factor. For example, consider a payoff that is contingent on both the stock price and the inflation rate, $X = h(S, \pi)$. The present value of this claim can be calculated as $X_0 = E[D_S D_\pi X]$, where D_S and D_π are the deflators for stock contingent claims and inflation contingent claims, respectively.

7.1.4 Monte Carlo Simulation

The valuation of complex contingent claims cannot be done with analytical methods. Typically, Monte Carlo simulation is used. For the Black–Scholes deflator, this works in the following steps:

1. Generate N values $\varepsilon^{(i)}$ from the standard normal distribution, and calculate the associated values of the stock price
   $S^{(i)} = S_0 \exp\left(\left(\mu - \frac{1}{2}\sigma^2\right)t + \sigma\varepsilon^{(i)}\sqrt{t}\right)$

the derivative payoff
$X^{(i)} = f(S^{(i)})$,
and the deflator
$D^{(i)} = \exp(-(r + \frac{1}{2}\lambda^2) + \lambda\varepsilon^{(i)}\sqrt{t})$.
2. Average the deflated payoff of the derivative over all N simulations
$\widehat{X}_0 = \frac{1}{N}\sum_{i=1}^{N} D^{(i)}X^{(i)}$.

For many simulations (large values of N), this estimate converges to the true value of the contingent claim.

7.1.5 Deflators and State Prices

As we have seen before, the fair value (present value) of a (simple) contract cannot be expressed as the discounted expected value of the future cash-flow

$$\text{PV} = \frac{1}{(1+r)^T} E[\text{CF}(T)].$$

There are two main techniques to find the value of the cash flow above:

- A change of probability measure.
- A change in the discounting factor.

The change in the probability measure is the risk-neutral method, where we calculate the expectation value under a risk-neutral measure

$$\text{PV} = \frac{1}{(1+r)^T} E^Q[\text{CF}(T)].$$

You then have to find the risk-free probability measure Q. See the binomial model.

The second method, where we make a change in the discounting factor, we can still use the real probability measure P but the discounting factor has to be changed and becomes stochastic

$$\text{PV} = E^P[D(T) \cdot \text{CF}(T)].$$

The stochastic discount function is the deflator. We will begin to study a simple, single period market model with one risk-free assed S_0 and d risky assets S_i defined on a probability space Ω

$$\begin{aligned} S_0(1) &= (1+r)S_0(0) \\ S_i(1) &= \{S_i(1,\omega_1), S_i(1,\omega_2), ..., S_i(1,\omega_N)\} \\ \Omega &= \{\omega_1, \omega_2, ..., \omega_N\} \text{ with } p_j = P(\{\omega_i\}) \end{aligned}$$

where r is the risk-free interest rate, $i = 1, 2, \ldots, d$ and $j = 1, 2, \ldots, N$.

First we introduce (define) the *state prices* Ψ_j:

Theorem 7.1 *If the market is free of arbitrage, there exist a random variable* Ψ *such as for any asset*

$$S_i(0) = \sum_{j=1}^{N} \Psi_j S_i(1,\omega_j) \quad \text{with} \quad \Psi_j = \Psi(\omega_j) > 0$$

For the risk-free asset ($i = 0$) *we then have*

$$\overline{\Psi} = \sum_{j=1}^{N} \Psi_j = \frac{1}{1+r}.$$

We can then create an artificial risk-neutral measure by

$$q_j = Q(\omega = \omega_j) = \frac{\Psi_j}{\overline{\Psi}} = (1+r)\Psi_j$$

with the properties

1. $0 \leq q_j \leq 1$
2. $\sum_{j=1}^{N} q_j = 1$
3. *For any asset, the mean return is given by the risk-free rate* $S_i(0) = \frac{1}{1+r}\sum_{j=1}^{N} q_j S(1,\omega_j)$

We now define the deflator.

Definition 7.2 The *deflator* is a random variable D

$$D_j = D(\omega_j) = \frac{\Psi_j}{p_j}$$

with the properties

1. $\sum_{j=1}^{N} p_j D_j = E[D] = \frac{1}{1+r}$
2. $S_i(0) = \sum_{j=1}^{N} p_j D_j S_i(1, \omega_j)$.

We can generalize this to any contingent claim, X on the market with cash flows $X(1, j)$ in scenario j as

1. Using state prices Ψj
$$X(0) = \sum_{j=1}^{N} \Psi_j \cdot X(1, \omega_j)$$
2. Using the risk-neutral measure Q
$$X(0) = \frac{1}{1+r} \sum_{j=1}^{N} q_j \cdot X(1, \omega_j) = \frac{1}{1+r} E^Q[X(1)]$$
3. Using the deflator D
$$X(0) = \sum_{j=1}^{N} p_j D_j X(1, \omega_j) = E^P[D \cdot X(1)]$$

State-price securities (also called Arrow–Debreu securities) are also used in the fixed income theory, where we use them to create binomial interest rate trees with forward induction, (see the Black–Derman–Toy model). A state-price security is a contract that will pay one unit of currency if a particular state is reached at a particular time in the future. In all models where we hope to replicate the market dynamics, we have to calibrate the model.

We will now consider a multi-period discrete model where $t = \{0, 1, \ldots, T\}$

$$\begin{aligned} S_0(t) &= (1+r)^t S_0(0) \\ S_i(t) &= \{S_i(t, \omega_1), S_i(t, \omega_2), \ldots, S_i(t, \omega_N)\} \\ \Omega &= \{\omega_1, \omega_2, \ldots, \omega_N\} \quad \text{with} \quad p_j = P(\{\omega_i\}) \end{aligned}$$

with state prices

$$S_i(0) = \sum_{j=1}^{N} \Psi_j(t) \cdot S_i(t, \omega_j) \quad \text{with} \quad \Psi_j(t) = \Psi(\omega_j, t) > 0$$

and deflator

$$D_j(t) = D(\omega_j, t) = \frac{\Psi_j(t)}{p_j}.$$

We can then price a replicable financial contract (contingent claim) X on this market by the generating successive stochastic cash flows: $\{cf(t,\omega)\}|t=1, \ldots, T; \omega \in \Omega\}$. The initial value can then be written as

$$X(0) = \sum_{t=1}^{T} \sum_{j=1}^{N} cf(t, \omega_j)\Psi_j(t)$$

or

$$X(0) = \sum_{t=1}^{T} \frac{1}{(1+r)^t} \sum_{j=1}^{N} q_j \cdot cf(t, \omega_j)$$

or

$$X(0) = \sum_{t=1}^{T} \sum_{j=1}^{N} p_j \cdot cf(t, \omega_j) D_j(t) = \sum_{t=1}^{T} E[D(t) \cdot cf(t)].$$

Example 7.3

An economy is currently in state A, and can evolve to future states over the next 2 years as shown in the picture below:

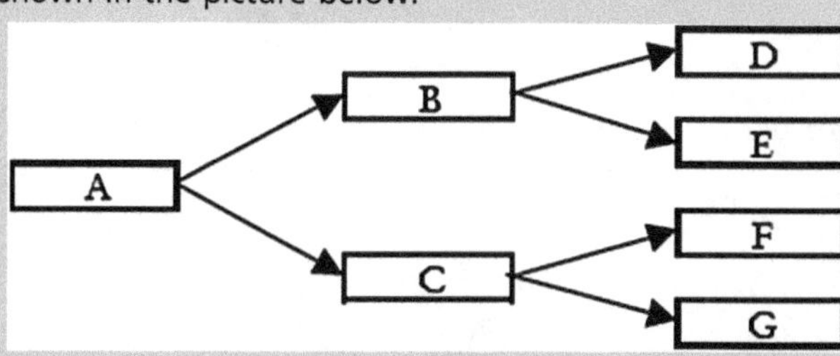

You are also given the corresponding price evolution for two securities in the economy.

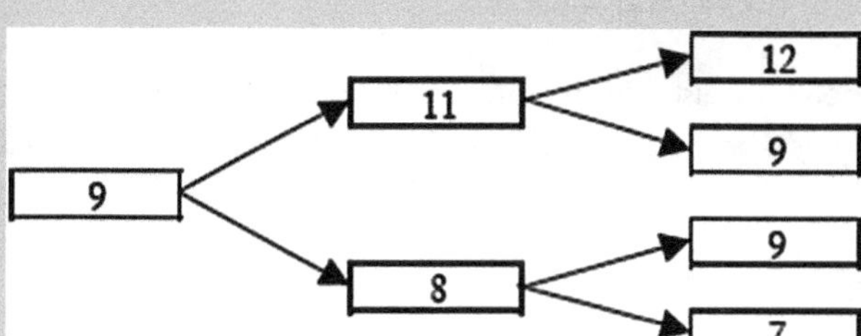

(continued)

Example 7.3 (continued)

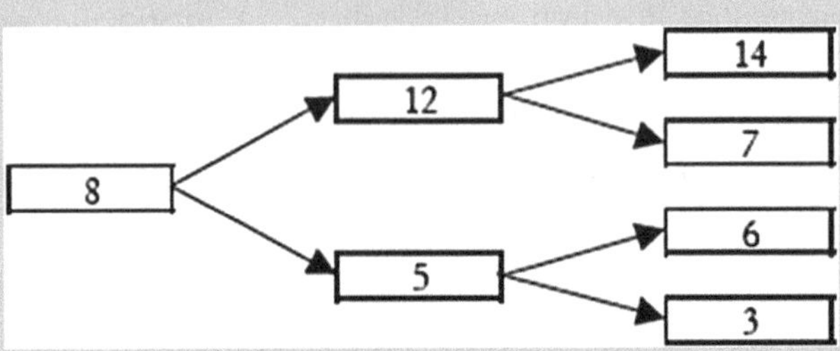

The annual effective risk-free rates of interest at the given states are

State	A	B	C
Rate	5.128 %	2.439 %	7.143 %

Using this information to:

(a) Define state price deflators and briefly discuss how to value a future cash Flow C_T at a time t in the future using deflators, where $t < T$.
(b) Describe advantages and disadvantages of risk-neutral valuation versus deflators.
(c) Calculate the state price vectors for states D, E, F and G.
(d) Calculate the state price deflators for states A, B, D, and E.
(e) Verify the value of 11 for Security 1 in state B using state price deflators.

Solution:

A state price deflator is defined as the ratio of the state price to the state probability, $D(s) = \psi(s)/p(s)$. The valuation of cash flows can be accomplished as: Value of C_T at time t *is*

$$\frac{E_t[D_T C_T]}{D_T}$$

Advantages of risk-neutral valuation

- State-independent discount factor, using the risk-free rate.
- Expected returns at the risk-free rate can be calculated for any asset.

(continued)

Example 7.3 (continued)

Disadvantages of risk-neutral valuation

- Difficult to understand conceptually (e.g., martingale measures).
- Complicated to implement multiple-currency models.

State price vectors
State A
The system

$$\begin{cases} 11\Psi_1 + 8\Psi_2 = 9 \\ 12\Psi_1 + 5\Psi_2 = 8 \end{cases} \text{gives} \quad \begin{cases} \Psi_1 = 0.4634 \\ \Psi_2 = 0.4878 \end{cases}$$

State B
The system

$$\begin{cases} 12\Psi_1 + 9\Psi_2 = 11 \\ 14\Psi_1 + 7\Psi_2 = 12 \end{cases} \text{gives} \quad \begin{cases} \Psi_1 = 0.7381 \\ \Psi_2 = 0.2381 \end{cases}$$

State C
The system

$$\begin{cases} 9\Psi_1 + 7\Psi_2 = 8 \\ 6\Psi_1 + 3\Psi_2 = 5 \end{cases} \text{gives} \quad \begin{cases} \Psi_1 = 0.7333 \\ \Psi_2 = 0.2000 \end{cases}$$

State-price vector for D = (0.4634)(0.7381) = 0.3420
State-price vector for E = (0.4634)(0.2381) = 0.1103
State-price vector for F = (0.4878)(0.7333) = 0.3577
State-price vector for G = (0.4878)(0.2000) = 0.0976
To solve for the deflators we first need to calculate probabilities at the various nodes

Deflators
State A
The system

$$\begin{cases} \Psi_1 = \dfrac{p_1}{(1+r)} \\ \Psi_2 = \dfrac{p_2}{(1+r)} \end{cases} \text{gives} \quad \begin{cases} p_1 = \Psi_1(1+r) = 0.4634 \cdot 1.05128 = 0.4872 \\ p_2 = \Psi_2(1+r) = 0.4878 \cdot 1.05128 = 0.5128 \end{cases}$$

(continued)

Example 7.3 (continued)

State B
The system

$$\begin{cases} p_1 = \Psi_1(1+r) = 0.7381 \cdot 1.02439 = 0.7561 \\ p_2 = \Psi_2(1+r) = 0.2381 \cdot 1.02439 = 0.7561 \end{cases}$$

State C
The system

$$\begin{cases} p_1 = \Psi_1(1+r) = 0.7333 \cdot 1.07143 = 0.7857 \\ p_2 = \Psi_2(1+r) = 0.2000 \cdot 1.07143 = 0.2143 \end{cases}$$

We now calculate the state price deflators as

$$\begin{aligned} D(A) &= 1 \text{ (this is a property of deflators)} \\ D(B) &= \frac{0.4634}{0.4872} = 0.9511 \\ D(D) &= \frac{0.3420}{(0.4872 \times 0.7561)} = 0.9284 \\ D(E) &= \frac{0.1103}{(0.4872 \times 0.2439)} = 0.9282 \end{aligned}$$

The value of security 1 in state B as calculated using the state price deflators

$$\text{Value} = \frac{0.9284 \cdot 0.7561 + 0.9282 \cdot 0.2439}{0.9511} = 11.00$$

8

Strategies with Options

8.1 Introduction

Before we begin to study strategies using derivatives, we will rehearse the concepts of options and forwards and study the market on the Swedish stock and derivative exchange.

A long position in an option means we have bought the option. In other words, we hold or own the option. A short or a written position means that we have sold the option. If we are long an option, we have the right, but not the obligation to by (for a call) or sell (for a put) the underlying stock to the given strike price. For a long position in an option we take no risk since we cannot lose more money than we invested. If we take a short position on the other hand, we take a risk since we have the obligation to sell (for a call) or buy (for a put) the underlying stock at the given strike price. If we don't own the stock in a short position we might need to buy the stock at a high price on the market to sell the stock to the holder of the option.

Options on stocks are usually of American type, which means they can be exercised at any time for the option lifetime. When exercised there will be a physical delivery of the underlying stock. Options on stock indices on the other hand, is said to be of European style. They can only be exercised at maturity of the contract. They are also cash settled since we cannot deliver an index, since this is just a fictive underlying.

A normal option contract on equities contains 100 stocks. To go into this contract the buyer pays an initial price (the option value) to the writer (seller) of the option. On an exchange, such as Nasdaq in Stockholm, the

J.R.M. Röman, *Analytical Finance: Volume I*,
DOI 10.1007/978-3-319-34027-2_8

clearinghouse is the counterparty to both the buyer and the seller. The clearinghouse then guarantees the delivery to the buyer. Therefore the buyer does not take a risk to enter an option contract. The writer of the option, on the other hand, takes a risk. It's always risky to go short in a position. The clearinghouse eliminates its own risk as far as possible by requiring the writer to pay an initial margin requirement. This margin requirement is recalculated by the clearinghouse every night. If the price of the underlying changes in a negative direction for the writer, he/she is told by the clearinghouse to increase the margin requirement. Nasdaq has its own system called RIVA (RIsk VAluation) to calculate this margin requirement. The writer can use other securities to secure his/her margins. Typically, 90 % of the value of a treasury bond can be used, or 80 % of the value of a corporate bond, or 50–70 % of the value of an equity. Moreover, cash on a bank account can be used as margin. Both the buyer and the seller pay a courtage to their broker or bank and a part of this is paid to the clearinghouse exchange.

The buyer of the call options has no obligation. The writer, on the other hand, has the obligation to sell the underlying asset to the buyer if the buyer wishes to exercise the option. If the writer does not own the underlying, he/she might have to buy the underlyings at a very high price and then sell them to the option holder at a low price. Hereby the writer takes the risk.

A contract of a put option is very similar. The buyer of a put option has the right, but not the obligation, to sell the underlying asset at a given price.

A long position means that you have bought a security. Then you are the holder and you hold the contract. A short position means that you have sold or written the security.

8.1.1 Characterization of Stock Options

A stock option (equity option) is characterized by an *identity* on the *underlying security* (e.g. a share in Volvo), a *contract size* (normally 100 shares), a *time to maturity* (6 month for short and 2 years for long) and a *strike price*. Series with new strikes are created each day if necessary. These series are created so that there is always a numbers of options, in-the-money (ITM), some out-of-the-money (OTM) and one at or close to at-the-money (ATM). Therefore, there is always at least a specific number of strike prices for each time to maturity. On the Swedish derivative exchange, the maturity is always on the third Friday in the month. At most there are five different maturities, three short and two long. All stock options on the Swedish exchange market are of American type. These options, as we will see, can be exercised at any time during the option lifetime.

8.1.2 Characterization of Forwards on Equities

A forward on a stock (on the Swedish exchange) is characterized by an *identity* on the *underlying security* (e.g. a share in Volvo), a *contract size* (normally 100 shares), a *time to maturity* (at most two years) and a *forward price*. The price to enter a forward contract is zero. In such a contract, both the buyer and the seller have an obligation to fulfil the trade at maturity. In this way both parties take a risk. On the Swedish derivative exchange the maturity is always on the third Friday in respective month.

Exchange traded forwards, as these in Sweden, are unique. Usually forwards are traded OTC, but, as we will see, thanks to the exchange traded forwards it is possible to build any strategy without going short or long in the underlying asset. Other exchanges usually trade futures. The difference between a forward and a future is that there are daily settlements in future contracts, that is, transfer of money between the two counterparties who have agreed on the contract, to buy or sell the underlying equity at the maturity of the contract.

8.1.3 Characterization of Options on the OMXS30 Index

An option on the OMXS30 index (the Swedish stock index at Nasdaq) is characterized by an *identity*, the 30 most actively traded assets on the OMX Stockholm Exchange, an *index level*, a *time to maturity* (6 months for short and 2 years for long) and a *strike price*. Since there is no deliverable underlying, these options are *cash settled*. This means that, if the option is ITM at maturity, a cash amount is calculated from the index level and transferred to the holder of the option. If the option is OTM it is worthless and no money will change owner. Since there is no deliverable underlying, these options are of European type.

Series with new strikes are created each day if necessary. These series are created so that there is always some option ITM, some OTM and one at or close to ATM. Therefore, there are always at least a specific number of strike prices for each time to maturity. On the Swedish derivative exchange the maturity is always on the fourth Friday in respective month. At most, there are five different maturities, three short and two long. All index options on the Swedish exchange market are of European type.

The assets in the index are weighted in such way that the most traded have more impact on the index level. Usually, Ericsson is the dominating asset in the index and the composition of the OMXS30 index is revised twice a year.

The value of index options is that they can be used to hedge a complex portfolio if it can be considered as well diversified and with a beta value similar to the OMXS30 index. They can also be used for speculation on the entire market.

As we know, European options cannot be exercised during their lifetime. However, a holder of an index option can sell the option at any time or buy the opposite position and in such way net his position.

OMX have a marketplace and a clearinghouse. The customers on this market are brokers, speculators, traders and market-makers. Market-makers have an obligation to set prices on both call- and put options and futures for some underlyings. The spread between the ask and bid prices must be in a certain level. Due to this obligation, the market-maker will be given lower transaction costs.

On other exchanges, similar contracts are traded as at Nasdaq in Sweden.

8.1.4 Standardized Names on Options and Forwards

Short names of options are built by the name, expiry and the strike price. Example: Let us use the Swedish stock Ericsson as the underlying. Then, ERIC6J45 is a call option with expiry in October (remember, the third Friday) 2016 with strike price 45 SEK (also 450 or 4.50 SEK is possible, but the current market price will tell us the strike). The letter J tells us that this is an October call. In the table below, we see how the letters are used. The corresponding RIC-name on Reuters for the same option is: LMEb450J6.ST

```
Month| J F M A M J J A S O N D
-----+-------------------------
Call | A B C D E F G H I J K L
Put  | M N O P Q R S T U V W X (also for forwards)
```

In addition, a corresponding name standard for forwards exists. Example: ERIC6B is a forward on Ericsson with expiry in February 2016. The corresponding RIC-code on Reuters is: LMEbG6.ST. Reuter uses the following rule for the months

```
Jan = F, Feb = G, Mar = H, Apr = J,
May = K, Jun = M, Jul = N, Aug = Q,
Sep = U, Oct = V, Nov = W, Dec = Z.
```

8.1.5 Alternatives When Trading Options

As a holder or a seller of options, you have alternative choices if the price of the underlying asset changes.

A holder (buyer) of an American option has the possibility to

1. Exercise the option, i.e., to buy (or sell) the underlying asset to the option strike price.
2. Net the position, i.e., to sell the option before maturity. This is the most common alternative.
3. Let the option die at maturity if the option is OTM.

The seller of an American option has the possibility to

1. Keep the option until maturity (and wait for the holder to exercise).
2. Net the position by buying a similar position to the current market price.

Remark! As a seller, you must consider the possibility to be exercised at any time if the option has a real value. Therefore, it is extremely important to follow the market and act thereafter.

8.1.6 Alternatives with Forwards

The seller of a forward can.

1. Deliver the underlying asset (a forward on a stock), or close the position (a forward on index) at maturity or
2. Net the position by buying the forward in the same series before maturity.

8.1.7 Ranking the Trades

Since many orders can exist at the same time in the order book on the exchange, orders have to be ranked. Orders with the highest rank will be exercised first. The orders are ranked as

1. Price; lowest ask- (to buy) and highest bid (to sell) price.
2. FIFO – first in first out.

8.1.8 The Reason for Buying Call Options

There are several reason for and advantages in trading options instead of the underlying asset:

1. To get a better lever, i.e., to make a better profit for every cash unit invested, compared with buying the underlying asset.
2. The risk is lower when buying the option compared with the underlying asset. If the price on the underlying decreases, you can only loose the invested capital, not the corresponding decrease of the underlying asset.
3. Avoiding tying up your capital. You can make other investments until you want to buy the underlying asset.
4. To plan to buy the underlying asset in the future.
5. If you want to sell the underlying asset but still want to earn money if the price increases.

8.1.9 The Reason for Selling Call Options

Reasons for selling call options can be:

1. To get a profit on a neutral or falling market.
2. To increase your profit on a neutral or a weak increasing market.
3. To be compensated on a decrease in the market.
4. To fix a certain ask price in the future.

8.1.10 The Reason for Buying Put Options

The reason for buying put options might be:

1. To get a profit on a falling market.
2. To protect an income from the underlying if the prices on the market will fall.
3. To lower the risk by owning the underlying asset.

8.1.11 The Reason for Selling Put Options

The reason for selling put options might be:

1. To get a profit on a neutral or weak market.
2. If you plan to buy the underlying asset in the future

8.1.12 The Reason for Buying Forwards or Futures

The reason for buying forwards and futures might be:

1. To tie up less capital than when buying the underlying asset.
2. To fix a price for the underlying asset in the future.

8.1.13 The Reason for Selling Forwards

The reason for selling forwards or futures contracts might be:

1. To get a profit on decreasing market.
2. To lock in a profit on the underlying asset.
3. To get a lower risk in a long position on the underlying asset.

8.1.14 Market Belief: Decision

If you want to learn to trade options and forwards/futures, you have to know about the existing base positions, which are illustrated in Figs. 8.1, 8.2 and 8.3. With these positions, you can create many different strategies by making combinations of different instruments. We will study these strategies in detail in the following sections in this chapter.

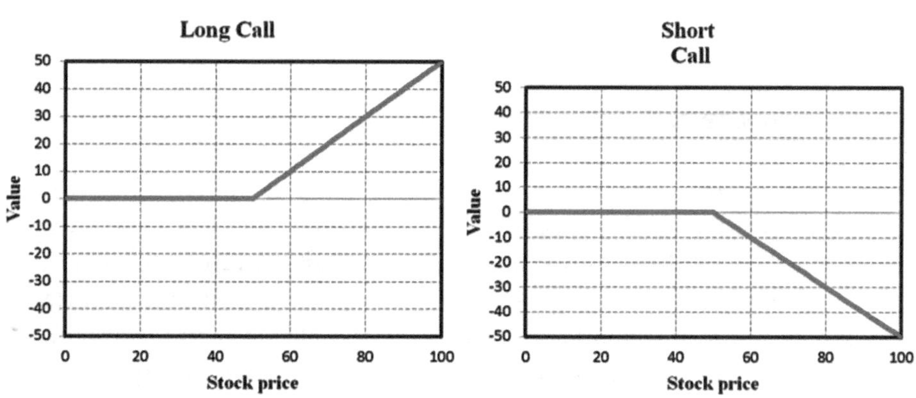

Fig. 8.1 The profit of a long and a short call option when the strike price = 50

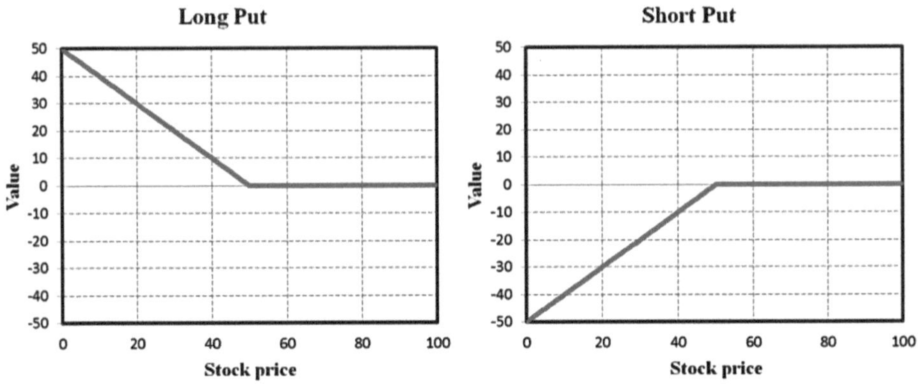

Fig. 8.2 The profit of a long and a short put option when the strike price = 50

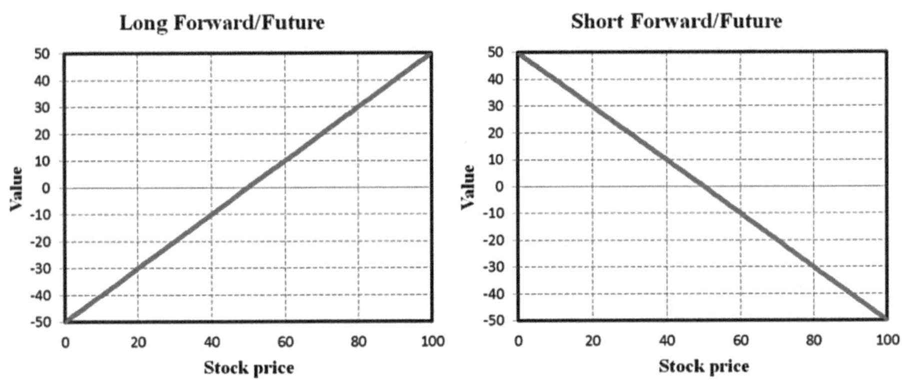

Fig. 8.3 The profit of a long and a short forward when the strike price = 50

8.1.15 Synthetic Contracts

To understand the trading strategies we first have to understand synthetic contracts. By using combination of the standardized contracts above, we can create so-called synthetic contracts. The profit–loss curve of such contracts shows the same features as single standardized contracts. With such contracts, you also have the possibility to create arbitrage, if one of the components is mis-priced. The price of a synthetic contract must be equal to the corresponding standard contract otherwise; you can buy one of them and sell the other to get a free lunch. You should have in mind to buy at a low and sell at a high price.

Most of the time you cannot make arbitrage with the use of synthetic contracts. The reason is that you have to pay transaction costs. However, a market-maker with lower transaction costs may have this option. If the possibility arises in the market, a market-maker will close this possibility in a very short time.

The synthetic combinations are

```
Bought Call    = Bought Forward + Bought Put
Sold   Call    = Sold   Forward + Sold   Put
Bought Forward = Bought Call    + Sold   Put
Bought Put     = Sold   Forward + Bought Call
Sold   Put     = Bought Forward + Sold   Call
Sold   Forward = Sold   Call    + Bought Put
```

We illustrate these contracts in Figs. 8.4 and 8.5.

8.1.16 CFD: An Alternative to Shares

As discussed in Sect. 6.1 a CFD or *contract for difference*, is a derivative with a stock or an index as the underlying instrument. A CFD is an agreement

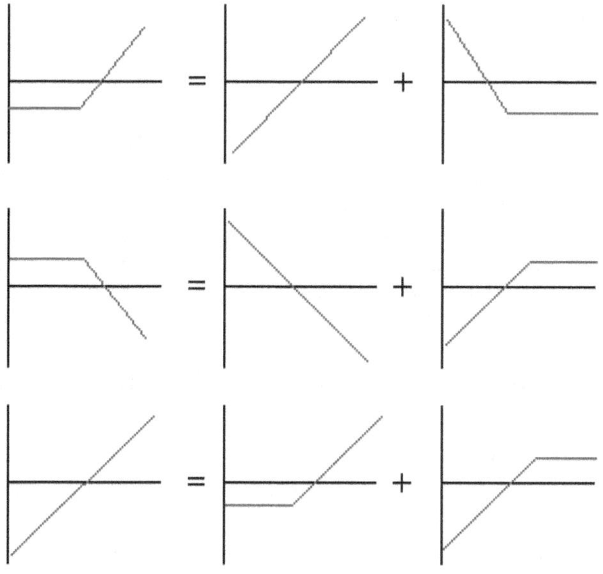

Fig. 8.4 Synthetic contracts where we show a synthetic long call, a synthetic short call and a long forward/future

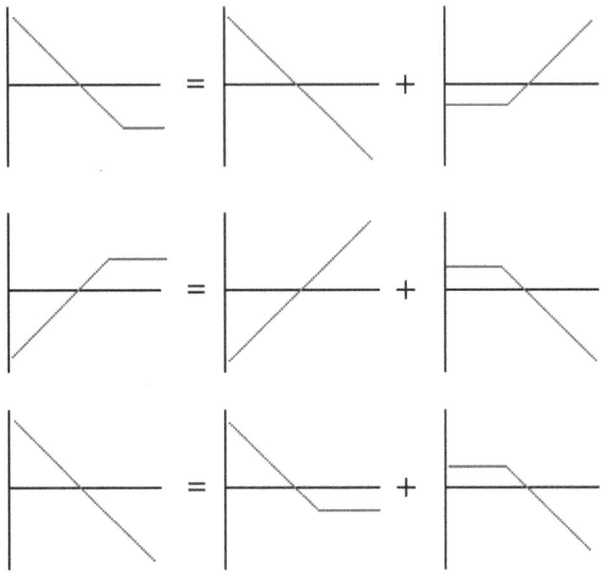

Fig. 8.5 Synthetic contracts where we show a synthetic long put, a synthetic short call put a short forward/future

between two parties to exchange the difference between the initial market price and the market price at the CFD maturity. Usually, one of the counterparties is a financial institute. CFD are sometimes used as an alternative to buy the stock.

Example

If the initial market price of an underlying stock is 360, the price of the CFD is almost 360. Then, if at maturity, the price of the stock is 380 you will be paid 20 (380–360). The big advantage of a CFD is that you can go short in such a trade and get a profit on a decrease without lending money, as would be the case if you short sale in the underlying stock instead.

CFDs, like future contracts, are also traded on commodities and in currencies.

Other advantages are the liquidity. You do not have to worry about this, as for options. Is also easier to buy a CFD than a stock if you do not want to own the underlying.

In some countries, the CFD has become very popular, where taxes on derivatives and stocks are different. The reason is that, when trading a CFD, no shares actually change ownership. But the tax situations differs between countries.

8.2 Strategies

By making combinations with different options, it is possible to create an infinite number of strategies. Your choice of strategy depends on how you, as an investor or speculator, believes that market conditions will change. In this section we will study some of the most common option strategies. With an option strategy, it is possible:

- To a lower initial cost, establish positions on the market.
- Get a better payoff on a given market belief.
- Get a lower risk with wrong market belief.
- To follow up a position on a wrong market belief.

When trading options, it is important to have a plain strategy from the beginning. Thereafter you have to continuously follow market changes to be able to follow up the strategy and to realize the profit when possible. A common mistake made by amateurs and beginners is to hold their positions for too long a time. Often, it is better to sell a call option to realize the profit, and then buy a new option at a higher strike price with later maturity (if you believe in a continuous increase in the underlying price).

We will use the following symbols to classify the strategies:

−1, 0, 1	The slope of the payoff curve as function of the underlying stock
*	If the position includes a long position in the underlying
+, −	The slope of the payoff if this is not piecewise linear

We will study four different market conditions: increasing (bullish), decreasing (bearish), neutral and volatile. The reason is that they are used on different market conditions.

One common myth is that the terms "bull market" and "bear market" are derived from the way those animals attack a foe, because bears attack by swiping their paws downward and bulls toss their horns upward. This may be a useful way to remember which is which, but it is not the true origin of the terms.

Long ago, "bear skin jobbers" were infamous for selling bears skins that they did not own—that is,the bears had not yet been caught. This term eventually was used to describe speculators who sell shares that they do not own but who hoped to buy them after a price drop and then deliver the shares to the owner. Obviously, these "bears" were hoping the market would go down.

Because bull and bear baiting were once popular sports, "bulls" came to be seen as the opposite of "bears". The bulls were those people who bought in the expectation that a stock price would rise, not fall.

Cartoonist Thomas Nast popularized the Bull and Bear as symbols for the market's movement. However, perhaps the final word on bulls and bears is the old Wall Street adage: bulls make money, bears make money, and pigs are slaughtered. Do not get greedy!

8.2.1 Basic Option Theory

In-, At- and Out-of-the-Money

A call option is ITM when the underlying price is higher than the option's exercise price, and is OTM when the underlying price is lower than the option's exercise price. A put option is ITM when the underlying price is lower than the option's exercise price, and is OTM when the underlying price is higher than the option's exercise price. An option is ATM when the underlying price is equal to the option's exercise price. In practice, the option with the exercise price nearest to the prevailing underlying price is called the ATM option.

Intrinsic and Time Value

The option price, or premium, can be considered as the sum of two specific elements: intrinsic value and time value—that is, real value = intrinsic value + time value.

The intrinsic value of an option is the amount an option holder can realize by exercising the option immediately. Intrinsic value is always positive or zero. An OTM option has zero intrinsic value.

Intrinsic value of ITM call option = underlying product price – strike price. Intrinsic value f ITM put option = strike price – underlying product price.

The time value of an option is the value over and above intrinsic value that the market places on the option. It can be considered as the value of the continuing exposure to the movement in the underlying product price that the option provides. The price that the market puts on this time value depends on a number of factors: time to expiry, volatility of the underlying product price, risk free interest rates and expected dividends (Figs. 8.6 and 8.7).

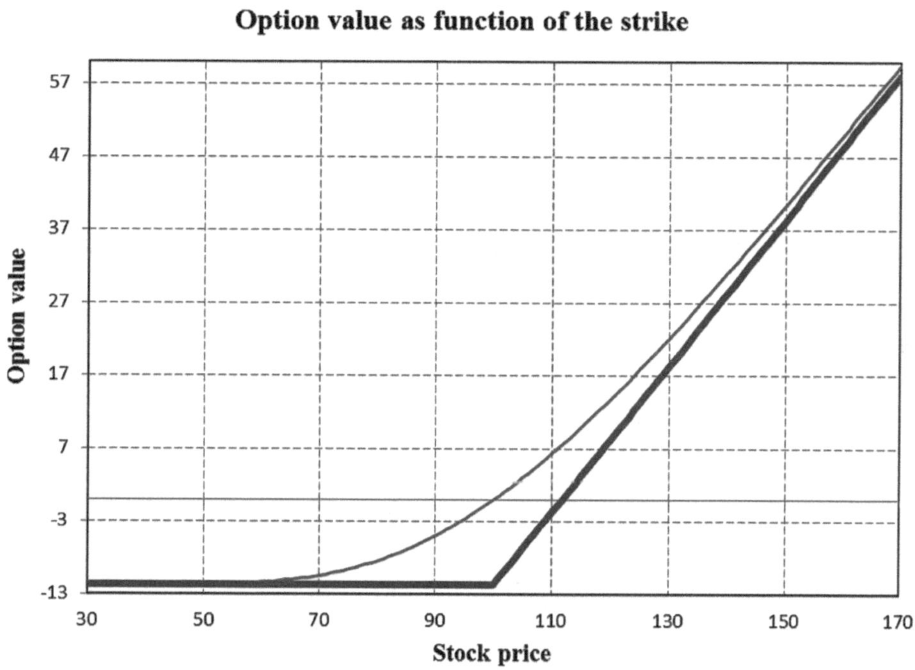

Fig. 8.6 The intrinsic value of a call option with strike 100 is represented by the thick line, the "hockey-stick", while the thin-lined curves represent the real value of the same option 6 months to maturity. The difference between the real and intrinsic value is the time value. Where the "hockey-stick" have a non-zero slope we are ITM. At 100 we are ATM and below 100 OTM. The negative value represents the premium payed for the option

Time to Maturity and the Time-Value

Time has value, since the longer the option has to go until expiry, the more opportunity there is for the underlying price to move to a level such that the option becomes ITM. Generally, the longer the time to expiry, the higher the option's time value. As expiry approaches, the value of an option tends to zero, and the rate of time decay accelerates.

Figure 8.8 shows how the time-value changes with time until the options maturity.

Volatility

The volatility of an option is a measure of the spread of the price movements of the underlying instrument. The more volatile the underlying instrument, the

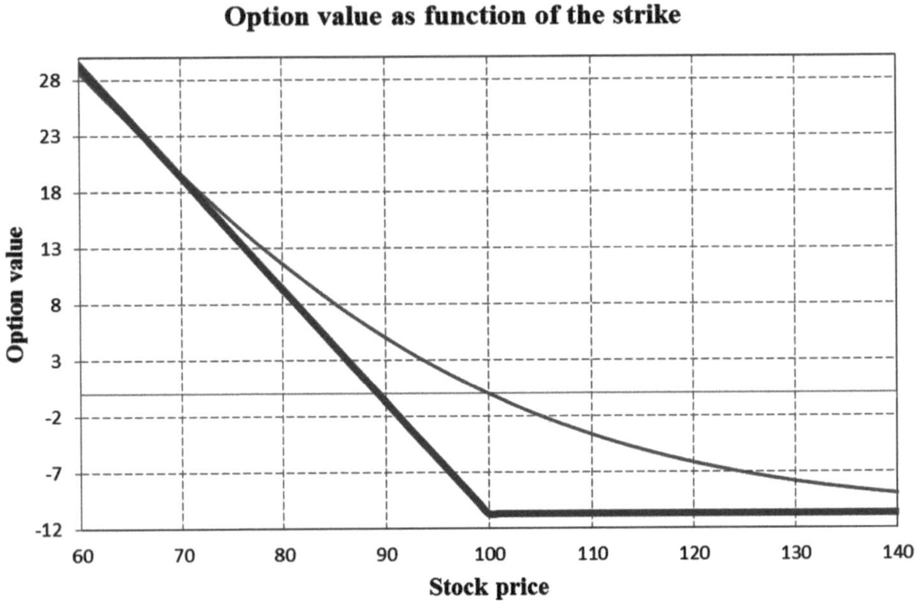

Fig. 8.7 Same as Fig. 8.6 but for a put option. Where the "hockey-stick" have a non-zero slope we are ITM. At 100 we are ATM and above 100 OTM

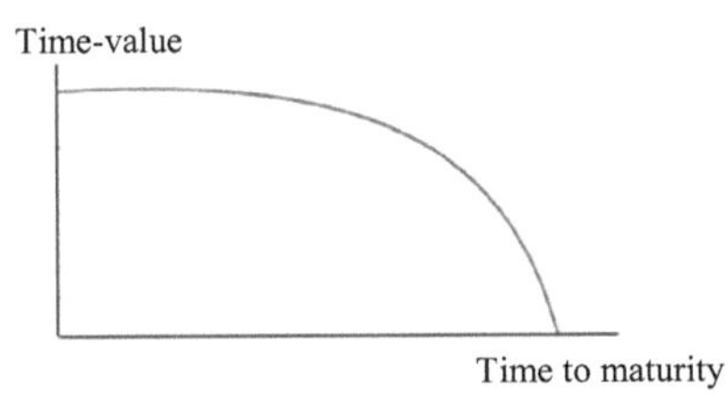

Fig. 8.8 The time-value decreases to zero at time to maturity

greater the time value of the option will be. This will mean greater uncertainty for the option seller who will charge a high premium to compensate. Option prices increase as volatility rises and decrease as volatility falls.

In Fig. 8.9 we show the effect of a volatility on a long straddle. Here, the strike is 100 in both the call and the put option, the risk-free interest rate is 2.0 %, time to maturity 3 month and the volatility 30 % and 50 %. Higher volatility gives the highest price. The thick line is the profit at maturity.

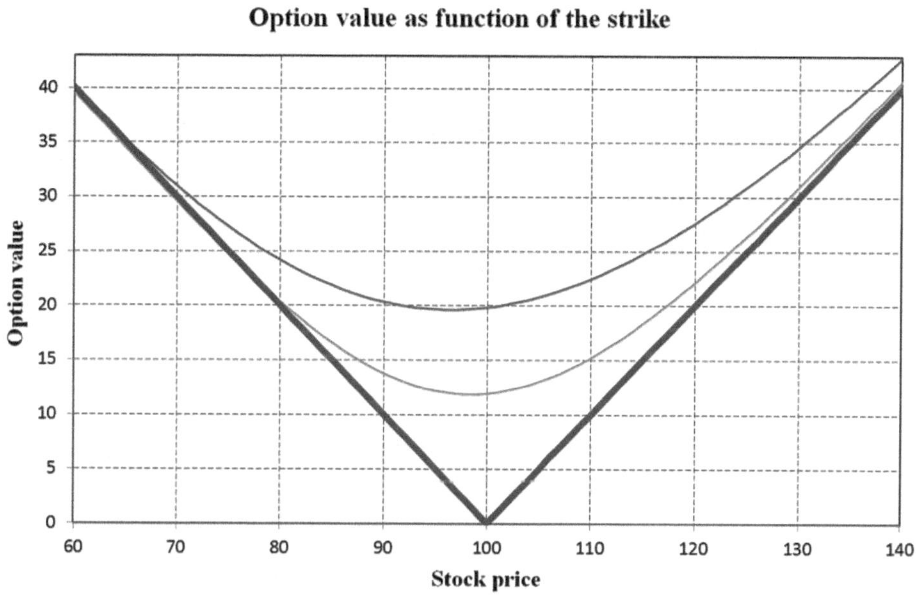

Fig. 8.9 The effect of the volatility for a long straddle build by a long call and a long put option with the same strike

Option Sensitivity

Throughout this chapter, the strategy examples refer to market sensitivities of the options involved. These sensitivities are commonly referred to as the "Greeks" and these are defined below.

Delta measures the change in the option price for a given change in the price of the underlying and thus enables exposure to the underlying to be determined. The delta is between 0 and +1 for calls and between 0 and 1 for puts, (thus a call option with a delta of 0.5 will increase in price by one tick for every two tick increase in the underlying).

Gamma measures the change in delta for a given change in the underlying. (E.g. if a call option has a delta of 0.5 and a gamma of 0.05, this indicates that the new delta will be 0.55 if the underlying price moves up by one full point and 0.45 if the underlying price moves down by one full point).

Theta measures the effect of time decay on an option. As time passes, options will lose time value and the theta indicates the extent of this decay. Both call and put options are wasting assets and therefore have a negative theta. Note that the decay of options is non-linear in that the rate of decay will accelerate as the option approaches expiry. As can be seen on the slope in Fig. 8.8, the theta will reach its highest value immediately before expiry.

Vega measures the effect that a change in implied volatility has on an option's price. Both calls and puts will tend to increase in value as volatility increases, as this raises the probability that the option will move ITM. Both calls and puts will thus possess a positive vega.

Rho measures the effect that a change in the risk-free interest rate has on an option's price.

Put Call Parity

Of particular importance with regard to arbitrage trades is the concept of put call parity. This is the relationship, which exists between calls and puts. It states that the value of a call (put) can be derived from the value of a put (call) with the same exercise price, maturity date and underlying price. Hence, for options on futures:

$$C = P + F - X$$

where:
$C =$ call price
$P =$ put price
$F =$ futures price
$X =$ exercise price

This assumes there are no carrying costs for options. A put call parity price for premium up front options can be found by slightly modifying this formula. Arbitrage trades are based on the relationships that exist between certain positions using options and futures. Referred to as synthetic positions, they are derived from put call parity and, by using this relationship, it is possible to perform arbitrage between synthetic positions and their outright equivalent.

8.3 A Decreasing Markets

A decreasing market is also called *baisse* or a *bearish* market. In this section we will discuss some of the strategies for situations when you believe that the underlying price will decrease.

8.3.1 Long Put [−1 0]

Market Belief
You believe on a decreasing market price of the stock or index.

Construction
Purchase a near-the-money or an ATM put option on the stock. The stock/index 5 days volatility should be low to neutral when compared with its 100 day volatility. You profit if the stock goes up beyond your break-even price. Probability should always be greater than 50 %. The more bearish you are the more OTM (lower strike) should be the option you buy.

Profit
The profit is almost unlimited on a default. The profit increases as stock falls. At expiration, break-even point will be option exercise price a less premium paid. For each point below break-even, profit increases by additional point.

Break-Even
The strike price of the option minus the initial premium.

Losses
Limited to the amount paid for option. Maximum loss is realized if the stock ends above option strike price. For each point below the strike, loss decreases by additional point.

Margin Requirement
None.

Comments
This position is a wasting asset. As time passes, value of position erodes toward expiration value. If the volatility increases, erosion slows, if the volatility decreases, erosion speeds up.

Trading Reasons

1. To get a profit on a decreasing market.
2. To protect a profit in the underlying stock.
3. To reduce the risk if owning the underlying stock.

Follow Up
On an increase:

- Create a positive or negative price-spread (see the price-spread below).
- Shift to a higher strike price. In that way you can realize the so far earned profit.
- Buy a forward. Then you take a risk, but you can buy the underlying for a low cost. This follow up requires being under guard.

On a decrease:

- Create a price-spread on a higher level.
- Create a ratio spread (see ratio spread below).
- Create a sloping synthetic position (see below).

8.3.2 Short (Written) Call [0 −1]

Market Belief
You believe the market will trend down or sideways for a period of 30 days or similar.

Construction
You sell (short) a far OTM call option. The strike price should be above the expected range of the stock during that time period. The stock/index 5 days volatility should be generally high or at least neutral when compared with its 100 day volatility. The probability of profit should be greater than 80 %.

Profit
Limited to the initial premium if the underlying price is below the strike price at maturity.

Break-Even
The option strike price minus the initial premium.

Losses
Unlimited if the underlying price increases to infinity.

Margin Requirement
Always required.

Comments
If the underlying price remains constant, the option value decreases with time due to the time value. You can also use this feature if you already own the underlying stock and you feel it will not move significantly for a period of time and want to earn some extra money by selling a covered call.

Trading Reasons

1. To get a profit on a falling or neutral market.
2. To increase the profit on a weakly increasing market.
3. To get compensation in a decrease in the underlying price.
4. To fix a satisfactory underlying price to sell the underlying.

Follow Up
On an increase:

- Create a price-spread, positive or negative.
- Create a time-spread.

On a decrease:

- Roll to a higher strike price – eventually, also issue a put option.
- Buy the underlying or a forward/future contract.
- Roll to a higher strike price with longer maturity.

8.3.3 Negative Price Spread [0 −1 0]

A negative price-spread is also called a *bear spread* or a *basis spread.*

Market Belief
You think the stock will go down somewhat or at least is a bit more likely to fall than to rise. This is a good position if you want to be in the stock but are unsure of bearish expectations. This is one of the most popular bearish strategy. This is also a conservative strategy when you believe more on a decrease than an increase.

Construction

1. A call option is bought with a higher strike and another call is sold with a lower strike, producing a net credit.
2. A put option is bought with a higher strike price and another put option is sold with the lower strike, producing a net debit.

In Fig. 8.12 we have constructed the positive price-spread using two call options. Both options have a maturity of 6 months, a volatility of 40 % and discounted with a risk-free interest rate of 2 %. We suppose that the initial stock price is 15 and we buy an option with strike price 20 and go short in another call option at strike 10. In Fig. 8.10 we illustrate the initial payoff and the payoff at maturity.

In Fig. 8.11 we see the payout at maturity of the same strategy as in Fig. 8.10.

In Fig. 8.12 we have constructed the positive price-spread using two put options. Both options have a maturity of 6 months, a volatility of 400 % and discounted with a risk-free interest rate of 2 %. We suppose that the initial stock price is 15 and we buy an option with strike price 18 and go short in another put option at strike 12.

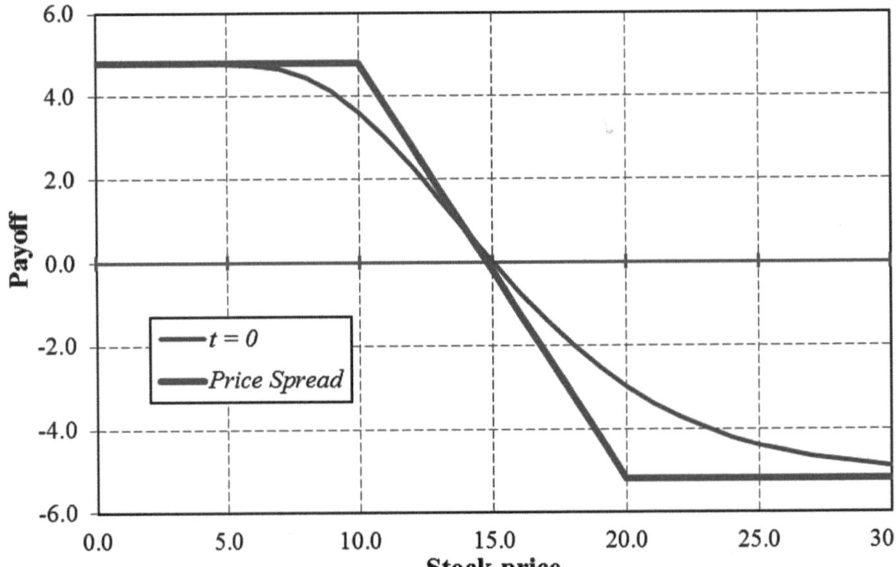

Fig. 8.10 A negative price-spread with call options. The thin line represent the option value when you buy the option (at time $t = 0$) and the fat line the profit at maturity

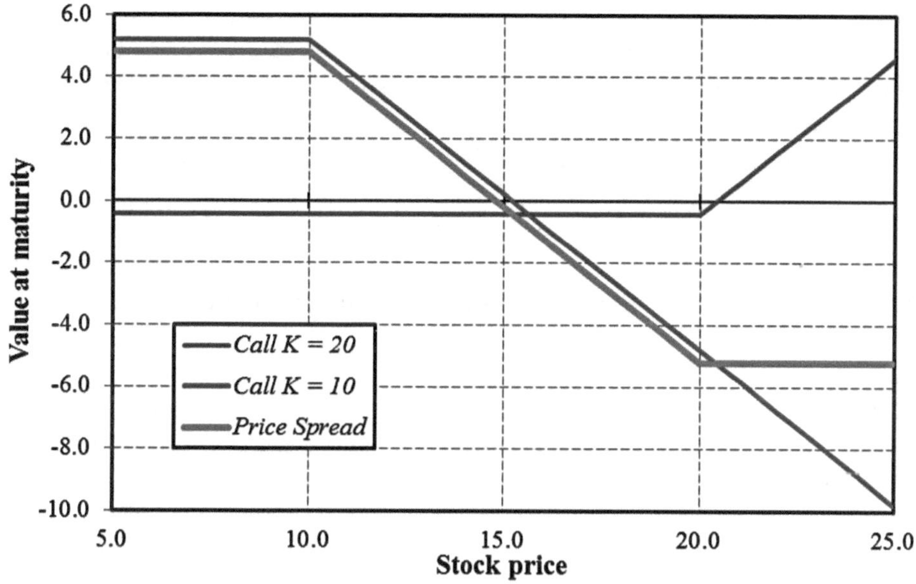

Fig. 8.11 The negative price-spread with call options at maturity

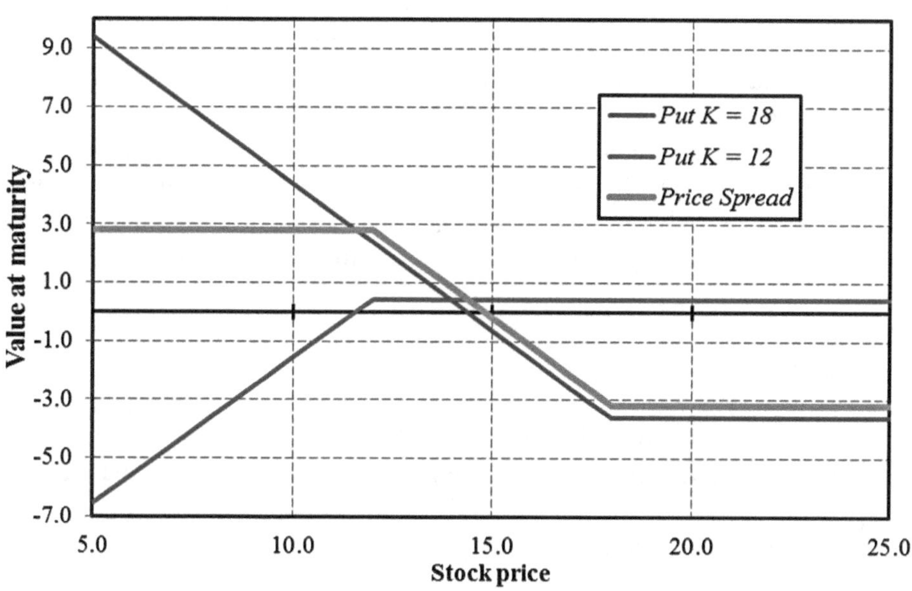

Fig. 8.12 A negative price-spread with put options

Profit

The total profit is limited, reaching maximum if stock ends at or below the lower strike at maturity. If a put spread used, the payoff is given as the difference between the strikes minus the initial debit. With the call spread in Fig. 8.10, the initial credit is given as $5.207 - 0.418 = 4.789$ where the values are calculated using Black–Scholes formula.

Break-Even

The strike price for the long option minus the initial premium.

Losses

Maximum, if stock at expiration is at or above the higher strike price. For a put spread, the maximum loss is the net initial debit. For a call spread, it is the difference between strikes minus the initial credit.

Margin Requirement

Here we have the possibility to offset the margin requirement.

Comments

The time value has only a small influence since the position is balanced. As we can see in the figures, the maximum loss is limited but also the profit. The maximum loss decreases on the cost of the maximum profit. If a negative price-spread succeeds and if you believe on a further decrease, the position can be "rolled" similarly as for a positive price-spread. On an increase, you can sell the long position. You can also issue more put options. This position requires less changes in the underlying price then a long put option and have a lower break-even. Normally, such a strategy has a maximum profit between 75 % and 150 %.

Trading Reasons

1. To give a higher probability to a profit than a long put option.
2. This strategy requires smaller changes in the underlying price than a long put option.
3. One can buy more contracts than on a naked put option.

Follow Up

On a decrease

- Roll the price-spread to a lower strike price

On an increase

- Issue more put options and create a ratio spread or a ladder.
- Issue a call option to compensate for the initial cost and create a three-legged position.

Due to a low initial cost, only a small change in the underlying is needed to get a profit. The maximum profit of is reached if the underlying price is below lower strike. If the position is taken together with a long position in the underlying, then the negative price-spread is a strategy to reduce the risk on a price decrease.

8.3.4 Negative Time-Spread [+−]

Market Belief
This spread is used when you are bearish on the stock/index over the next several months. The investor believes in a weak initial market, but with a strong decrease in the future. This strategy is also called a negative time-spread.

Construction
You buy a put option at near the money (or ATM) with a long maturity and sell put options with a lower strike with a shorter time to maturity.

Profit
You profit in two ways

1. The premium received for the sale of the put with near expiration and
2. From the downward movement of the stock price over the specified longer expiration.

Losses
Limited to the difference in strike price +/− the initial profit/cost.

Margin Requirement
Yes, but this is limited due to off-setting.

Comments
It is possible that the issued option will be exercised. Initially the sale of the put reduces the cost of the spread, but each month you sell another put generating

monthly income until the price of the stock gets close to the strike of the purchased put. Once the stock price has reached near the strike price of the long term purchased put, you continue to profit from the stock/index's price decrease without the need of selling more puts. Your potential profit is unlimited, but your risk is limited and will vary during the time the spread is in effect. Since the spread is continually adjusted over time, an accurate probability of profit cannot be determined prior to placing the trade. Margin requirements vary depending on the distance between strike prices. The rapid time decay of the closer expiration put sold and the slower time decay of the farther expiration put is your friend with this trade.

In Fig. 8.13 we illustrate a time-spread with two put options with strike prices at 15 for the sold option with maturity in a year and a bought put option at strike 13 with 6-month maturity. The initial stock price is 15. The risk-free interest rate is 2 % and the volatility 60.0 %.

Figure 8.14 illustrate the situation when we reach the first option maturity and still have a half of a year to the second option maturity.

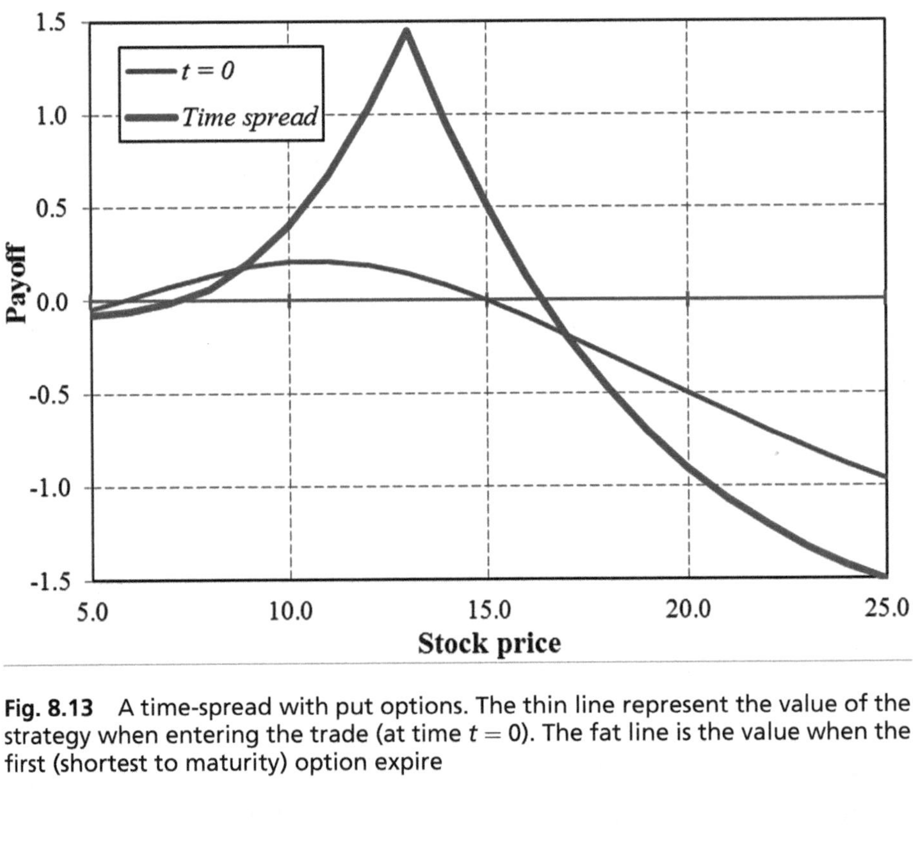

Fig. 8.13 A time-spread with put options. The thin line represent the value of the strategy when entering the trade (at time $t = 0$). The fat line is the value when the first (shortest to maturity) option expire

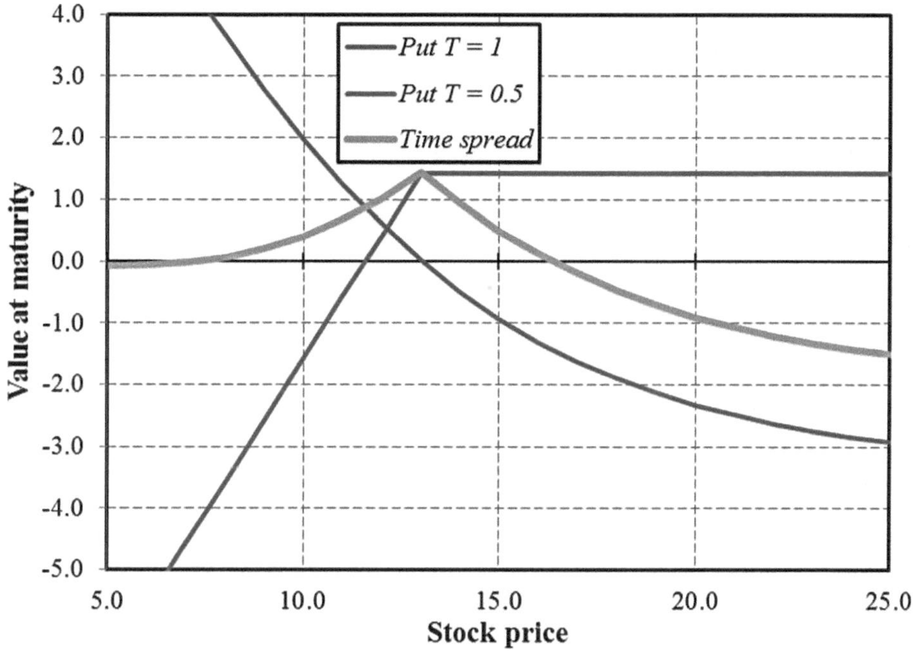

Fig. 8.14 The diagonal spread with put options at the fist maturity

A negative time-spread can also be made by call options. In Fig. 8.15 we illustrate a time-spread with two call options with strike prices at 17 for the bought option with maturity in a year and a sold put option at strike 14 with maturity six month. The initial stock price is 15. The risk-free interest rate is 2 % and the volatility 40 %.

Figure 8.16 illustrate the situation when we reach the first option maturity and still have a half of a year to the second option maturity.

8.3.5 Ratio-Spread with Put Options [1 −1 0]

Market Belief

This spread is used in the same way as a long put. You should be very bearish on the stock/index and the expected range of the stock during the particular time period should extend significantly beyond the break-even points of the position.

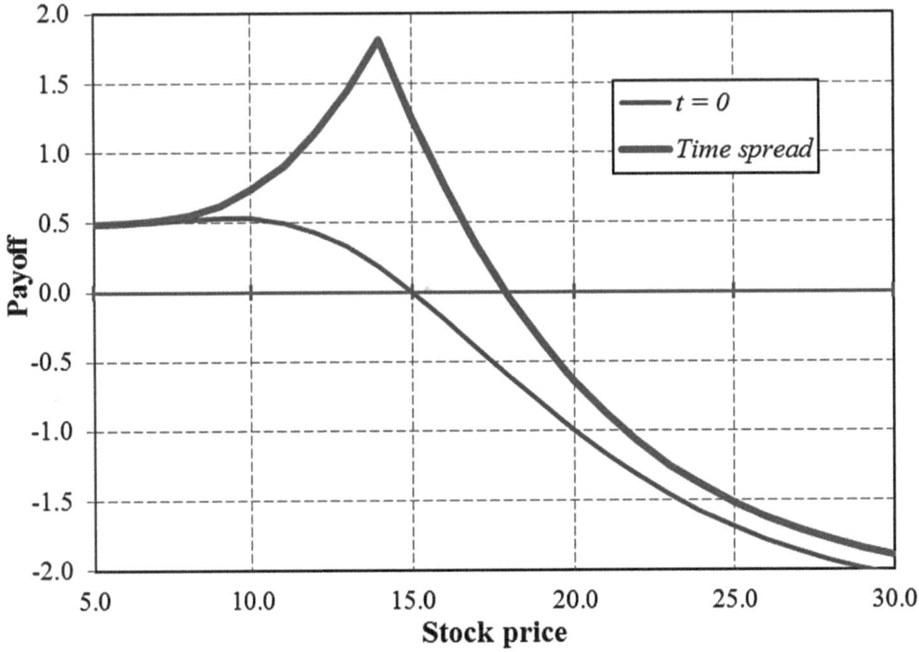

Fig. 8.15 A time-spread with call options. The thin line represent the value of the strategy when entering the trade (at time $t = 0$). The fat line is the value when the first (shortest to maturity) option expire

Construction

You buy one of the higher strike put options that are near the current price and sell two put at a lower strike price than those purchased. This ratio (buy 2; sell 1) reduces the cost of the two puts purchased, often to the point or a free trade.

Profit

Limited. The difference between the strike prices plus/minus the net profit/ cost for the options. The maximum profit is reached on the lower strike price.

Losses

Unlimited. The strategy gives losses on a big decrease in underlying price.

Margin Requirement

Always needed.

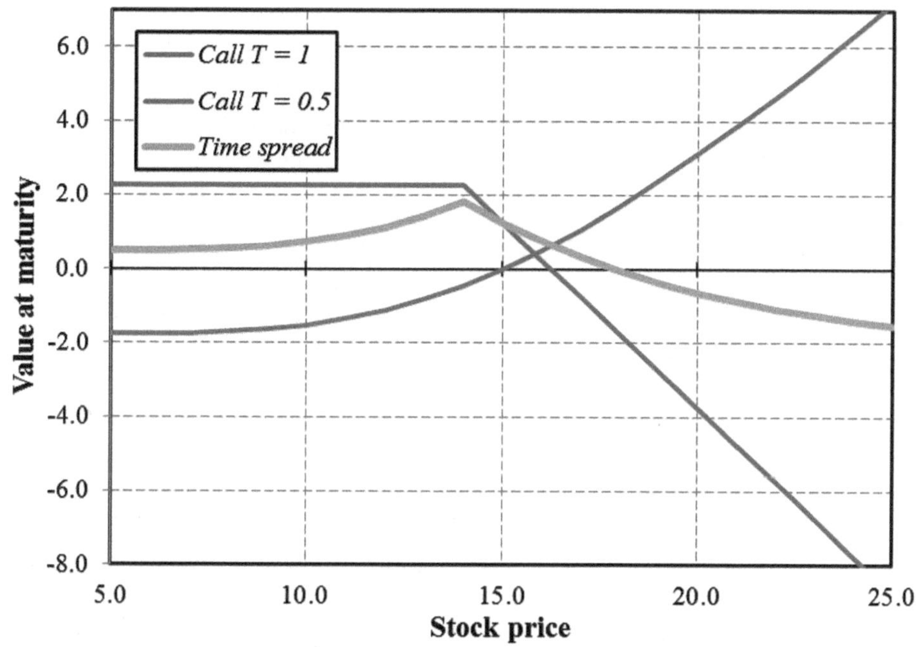

Fig. 8.16 The diagonal spread with call options at the fist maturity

Comments

Your break-even has two different points and you will lose money if the options expire at any point between the two break-even prices. You can also use a 3:2 ratio (buy 3; sell 2) which will reduce the cost further or increase the credit received, but the break-even points will be extended even further requiring an even greater move in the stock price for a profit. The probability of profit should be greater than 40 % and rarely will exceed 60 %.The strategy requires massive coverage.

In Fig. 8.17 we illustrate a put ratio spread with two put options. Here we buy one at strike 16 and sell two at strike 10. The initial stock price is 15, the risk-free interest rate 2 % and the volatility 40 %. Fig. 8.18 illustrate the same situation at maturity.

8.3.6 Negative Back-Spread [−1 1 0]

Back-spreads is a common name when you have more long options than short options. If you buy one more option, you will get a ladder or a stair which we will study in a later section.

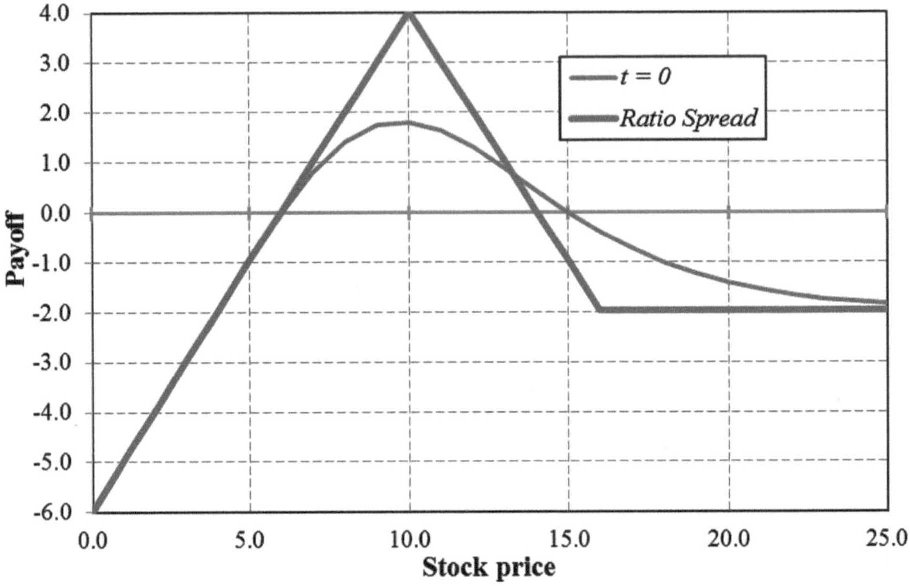

Fig. 8.17 A put ratio spread with put options. The thin line represents the value when the strategy is bought and the thick line the profit at maturity

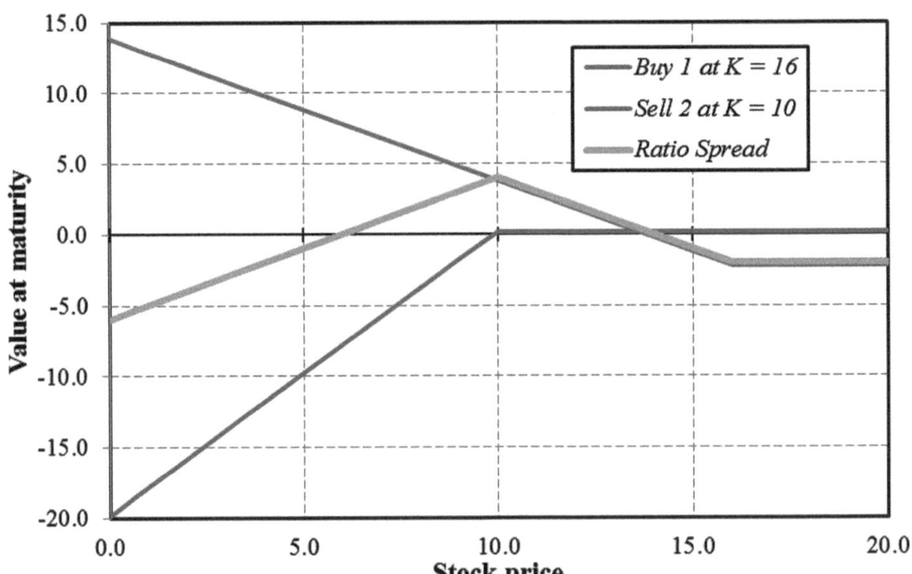

Fig. 8.18 The put ratio-spread with put options at maturity

Market Belief
The investor believes in a decrease of the underlying but wants a good protection for an increasing market.

Construction
Issue put options with a high strike and buy twice as many options with a lower strike at the same maturity.

Profit
Unlimited.

Losses
Limited to the difference between the strike prices plus possible costs or minus possible profits.

Break-Even
The lower strike price plus the difference between the strike prices plus possible losses or minus possible profits.

Margin Requirement
Very restricted.

Trade Reasons

1. To get a lower loss on an increase, compared with buying a put option.
2. To get a lower loss on an increase, compared with a three-leg position.

Follow Up
On an increase:

- Issue a put option to a lower strike price and buy back the issued option.
- Buy the underlying forward.

On a decrease:

- Issue put options on a lower strike price and call options on a higher strike.
- Issue put options on a lower strike price with later maturity.

In Fig. 8.19 we illustrate a negative back-spread with two put options. Here we buy two at strike 15 and sell one at strike 20. The initial stock price is

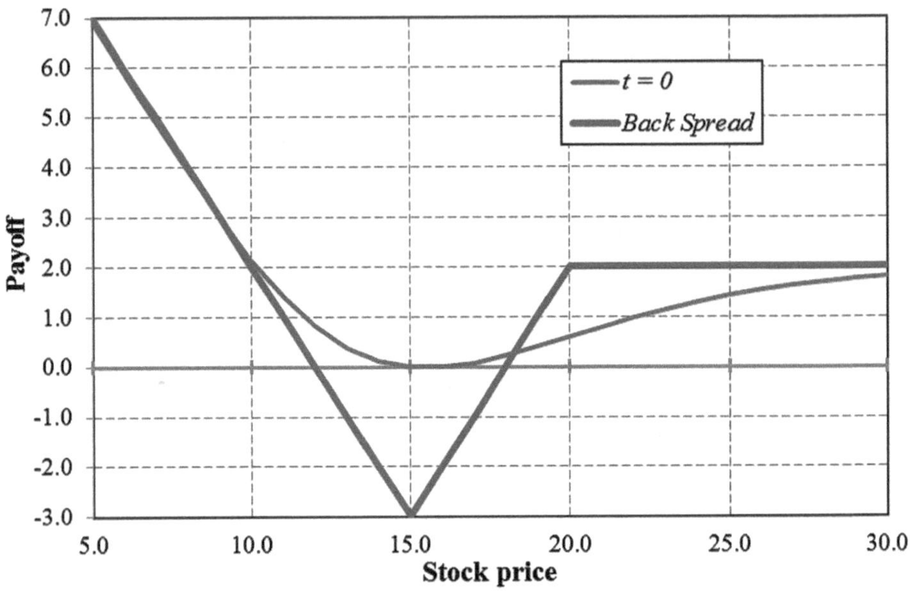

Fig. 8.19 A negative back-spread with put options. The thin line represents the value when the strategy is bought and the thick line the profit at maturity

15, the risk-free interest rate 2 % and the volatility 40 %. Figure 8.20 illustrates the same situation at maturity.

8.3.7 A Negative Three-Leg Position [−1 0 −1 0]

Market Belief

The investor believes in a strong decrease of the underlying, but wants at the same time a good protection on a decrease of the underlying price.

Construction

Buy a put option with a low strike price and a call option at a high strike. Then sell a call options with a strike in the middle (near at-the-money) of the other two strikes.

Profit

Unlimited.

Losses

Limited. The maximum loss is the difference between the strike prices on the call options plus a possible initial cost or minus a possible initial profit.

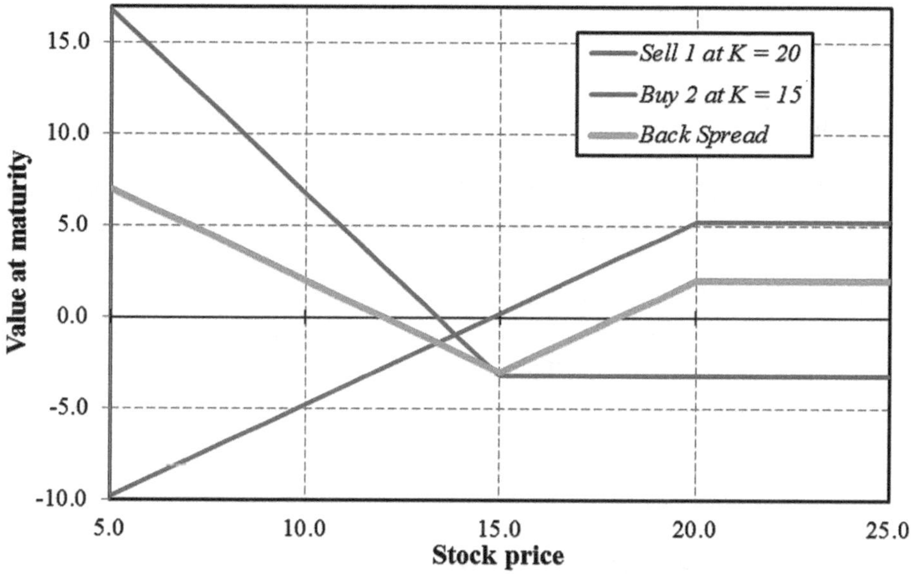

Fig. 8.20 The negative back-spread with put options at maturity

Break-Even

Upwards: The put option strike price minus the possible cost.

Downwards: The lower strike price on the call option plus the possible initial income.

Margin Requirement

Limited.

Trade Reasons

1. The position gives a lower cost than to buy a put option and have a lower Break-even.
2. The position gives no loss with a limited decrease as the back-spread above.

Follow Up

On profit:

- Issue put option on a lower level and use the income to close the negative price-spread.

- Roll the owned put option to a lower strike price and use the profit to close negative price-spread.

On losses:

- Sell the owned call option and roll the issued option to a higher strike price.
- Sell the owned put option and create a price-spread on a lower level.

In Fig. 8.21 we illustrate a put negative three-leg strategy where we have bought a put option at strike 10 and a call option at strike 20. Then we sell a call option at ATM, strike 15. The initial stock price is 15, the risk-free interest rate 2 % and the volatility 40 %. Figure 8.22 illustrate the same situation at maturity.

8.4 An Increasing Market

An increasing market is also called *Hausse* or a bullish market. In this section we will discuss some of the strategies for situations when you believe that the underlying price will increase.

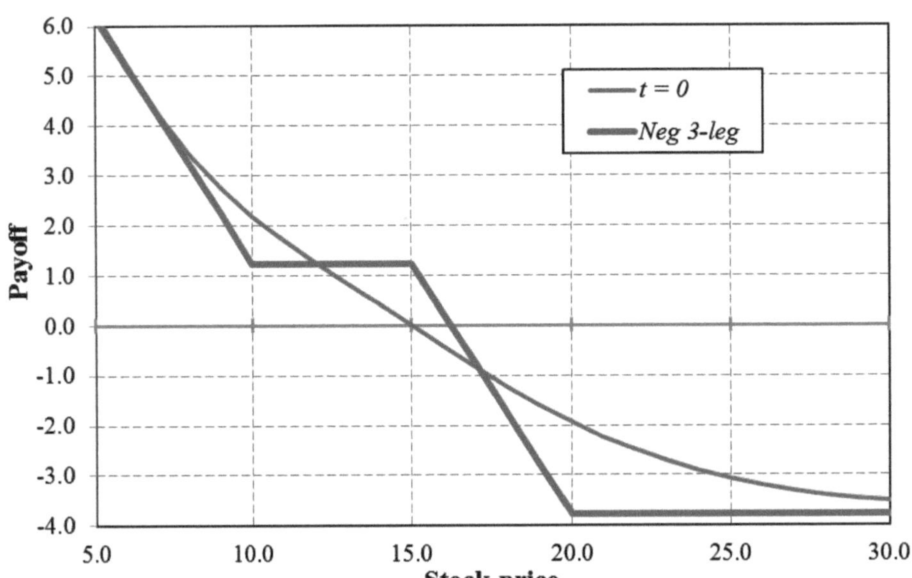

Fig. 8.21 A negative three-leg strategy with two call options and one put

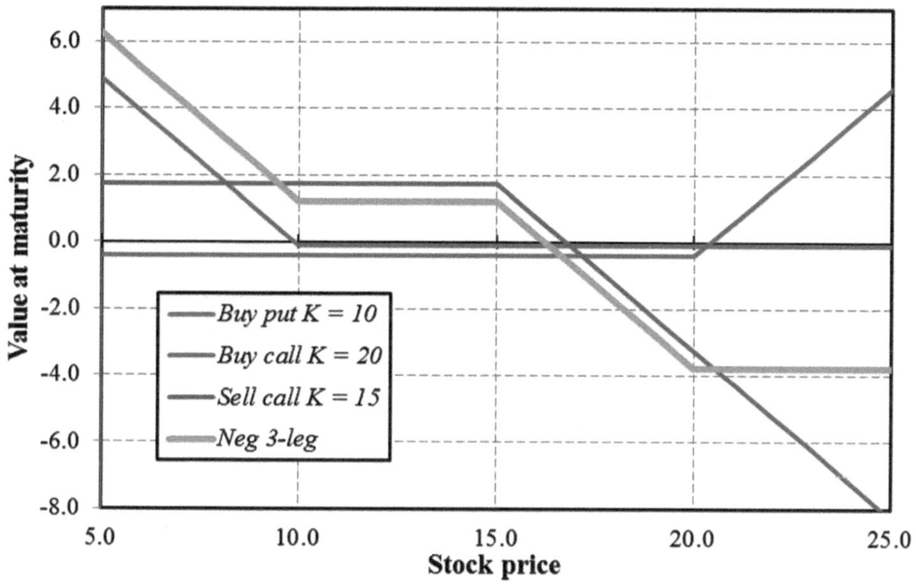

Fig. 8.22 The negative three-leg strategy at maturity

8.4.1 Long Call/Bought Call [0 1]

Market Belief
You are very bullish on the stock. The more bullish you are, the higher the strike should be. No other position gives you so much leveraged advantage with limited downside risk.

Construction
You purchase a near-the-money or ATM call option. The stock/index 5-day volatility should be low to neutral when compared with its 100 day volatility. Probability should always be greater than 50 %.

Profit
The profit increases as stock rises. At expiration, break-even point will be option strike a plus premium paid. For each point above break-even, profit increases by an additional point.

Break-Even
Strike price plus the premium.

Losses
The loss is limited to the premium paid. Maximum loss realized if the stock ends below A. For each point above A, loss decreases by additional point.

Margin Requirement
None.

Comments
This position is a wasting asset. As time passes, value of position erodes toward expiration value. If volatility increases, erosion slows; if volatility decreases, erosion speeds up.

Trade Reasons

1. To get a better lever.
2. To get a lower risk compared with buying the underlying.
3. To avoid tying up capital.
4. To insure a future bought of the underlying stock.
5. To sell stocks and still make a profit on a continuous increase of the underlying.

Follow Up
On an increase:

- Create a price-spread, positive or negative
- Shift to a higher strike price. Then we lock-in the earned profit.
- Sell a forward. You then take a risk, but you can buy the underlying to a favorable price.

On a decrease:

- Create a price-spread on lower level.
- Create a ratio spread.
- Create a sloping synthetic position.

8.4.2 Sold Put Option/Short Put [1 0]

Market Belief
You are certainly sure that the price will not fall.

Construction
Sell a put option and sell to a lower strike option if you are only somewhat convinced. Sell higher strike options if you are very confident the stock will stagnate or rise. If you doubt that the stock will stagnate, sell ATM options for maximum profit.

Profit
Limited to the premium received from sale. Maximum profit realized if stock settles at or above a.

Break-Even
At expiration, break-even point is strike price a less premium received.

Losses
Increases as stock falls. At expiration, losses increase by one point for each point stock is below break-even. Because the risk is open-ended, this position must be watched closely.

Margin Requirement
Always.

Comments
This position is a growing asset. As time passes, value of position increases as option loses its time value. Maximum rate of increasing profits occurs if the option is at-the-money.

Trade Reasons

1. To get a profit on a neutral and/or a weakly increasing market.
2. If you plan to buy the underlying.

Follow Up
On an increase

- Create a positive or negative price-spread.
- Create a time-spread (calendar spread).

On a decrease

- Roll the position to a lower strike price (maybe also issue a call option).
- Sell short (without ownership) the underlying or sell a forward contract.
- Roll the position to a lower strike price with later maturity.

8.4.3 Put Hedge, Protective Put or Synthetic Call [0 1]

Market Belief

You are very bullish on the stock. The more bullish you are, the higher the strike should be. No other position gives you so much leveraged advantage with limited downside risk. You can also protect an ownership for a possible decrease in price and to look-in the earned profit. At the same time, you can keep the underlying and get a continuous profit on a price increase.

Construction

You purchase a near-the-money or an ATM put option. The stock/index 5 days' volatility should be low to neutral when compared with its 100 day volatility. You profit if the stock goes down beyond your break-even price. Probability of profit should always be greater than 50 %.

Profit

Increases as stock rises. At expiration, break-even point will be option strike a plus premium paid. For each point above break-even, profit increases by an additional point.

Break-Even

The underlying price minus the initial premium

Losses

The losses are limited to the paid premium. Maximum loss realized if the stock ends below a. For each point above a, loss decreases by additional point

Margin Requirement

None.

Comments

This position is a wasting asset. As time passes, value of position erodes toward expiration value. If volatility increases, erosion slows; if volatility decreases, erosion speeds up. You can use delta to calculate the number of options needed to become delta-neutral. If you buy the options at-the-money, delta = 0.5. Then you should buy twice as many options to become delta-neutral. On price changes, you can issue call options with higher strike price.

Trade Reasons

To protect (hedge) the owned stocks. In that way you lock the profit, you have and decrease the risk by owning the underlying.

Follow Up
On profits

- Issue a call option on a higher level to compensate for the cost.
- Issue a call option and move the put option to a higher strike price to lock the earned profit.

On losses

- Issue a call option to compensate the cost. This is called a fence (= a positive price-spread). If you believe on a price turn, sell the put option and buy more stocks at the new lower price level.

In Fig. 8.23 we illustrate a protective put. Here we own the stock or buy the forward (future) and buy a put option at ATM strike, that is, at 15. The initial stock price is 15, the risk-free interest rate 2 % and the volatility 40 %. Fig. 8.24 illustrate the same situation at maturity.

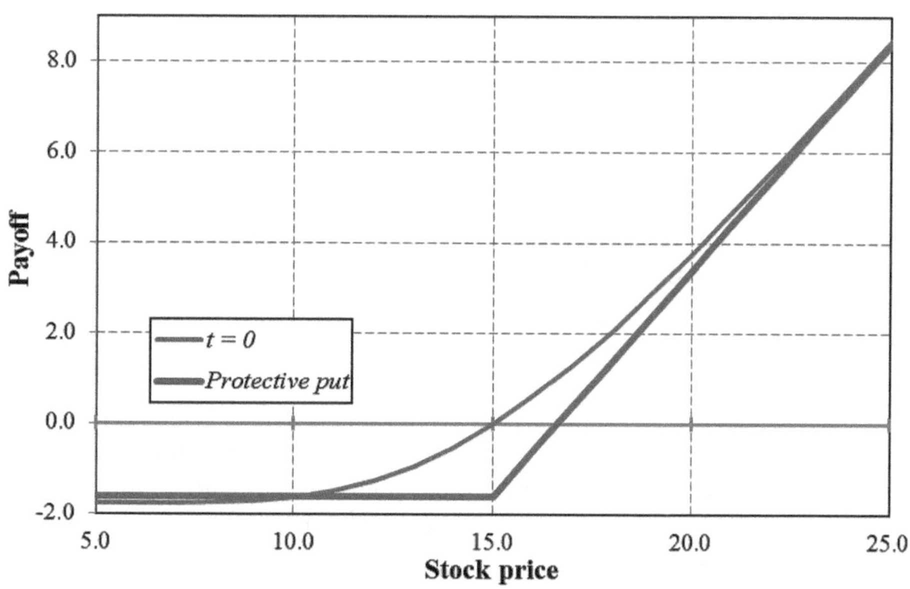

Fig. 8.23 A protective put where we also owns the underlying

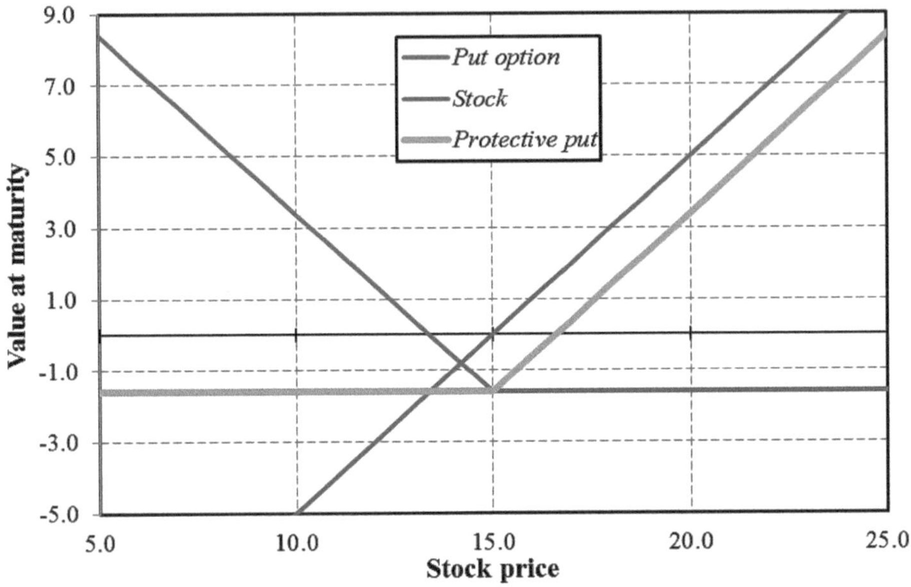

Fig. 8.24 The protective put at maturity, where we owns the underlying

8.4.4 Positive Price Spread/Bull Spread [0 1 0]

Market Belief

You think the stock will go up somewhat or at least is a bit more likely to rise than to fall. This is a good position if you want to be in the stock but are unsure of bullish expectations, and the most popular bullish strategy. It is also a conservative strategy to believe more in an increase than a decrease. (Limited positive.)

Construction

1. Call option is bought (usually an ATM) with a strike price of a and another call option sold with a higher strike producing a net debit.
2. Put option is bought with a lower strike and another put sold with a higher strike producing a net credit.

Profit

Limited, reaching maximum if stock ends at or above the higher strike b at expiration. If call spread used, difference between strikes minus initial debit. If put spread used, net initial credit. You get the maximum profit when the underlying price is above the higher strike price, b. The maximum profit varies between 75 % and 150 % on the initial cost.

Break-Even
Strike price of the bought option plus the premium.

Losses
Maximum loss if stock at expiration is at or below a. If call spread used, maximum loss is net initial debit. If put spread, difference between strikes minus initial credit. You get the maximum when the underlying price is below the lower strike price, a.

Margin Requirement
Possibility to off-set the margin requirement.

Comments
The time value has no influence since the position is in balance. As we can see in the Figures, the maximum loss is limited as well as the profit. The maximum loss is decreased on the cost of the maximum profit. At the same time, the total profit will be better than with a single call option. This is due to the premium for the sold call option. If a positive price-spread succeeds and if you believe on a further increase, it is possible to roll the position by buying back the issued put option and buy a new one with higher strike. On a decrease you can sell de ownership. Remark! always *buy cheaply and sell expensively.*

Trading Reasons

1. To get a maximum yield on our market belief.
2. To get a higher probability for a profit compared with buying a single call option.
3. It is easier to follow up this position than a single call option.

Follow Up
On an increase

- Roll the price-spread to a higher strike.

On a decrease

- Issue more call options and create a ratio spread or a ladder.
- Issue put options to compensate for the initial cost.

In Fig. 8.25 we illustrate a positive price-spread with call options. Here we buy a call option at 10 and sell another on at 20. The initial stock price is 15, the risk-free interest rate 2 % and the volatility 40 %. Figure 8.26 illustrate the same situation at maturity.

8.4.5 A Fence [0 1 0 *]

Market Belief
The investor is almost sure that the market will increase, but not how much. Therefore he wants to minimize the risk for a possible decrease.

Construction
Issue a call option and buy a put option. The investor owns the underlying.

Profit
The strike price of the call option minus the underlying price plus possible profit or minus the cost.

Break-Even
The stock price plus cost or minus the profit.

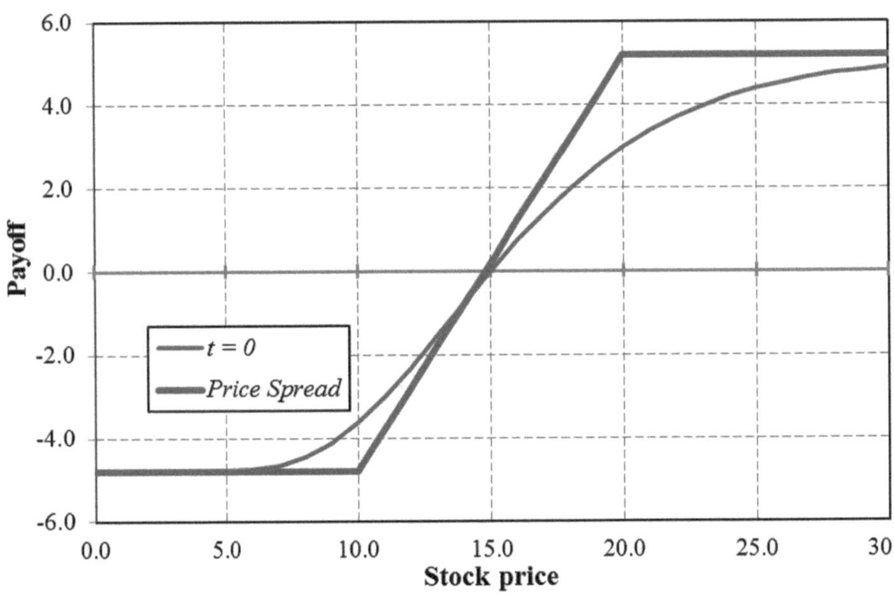

Fig. 8.25 A positive price-spread with call options

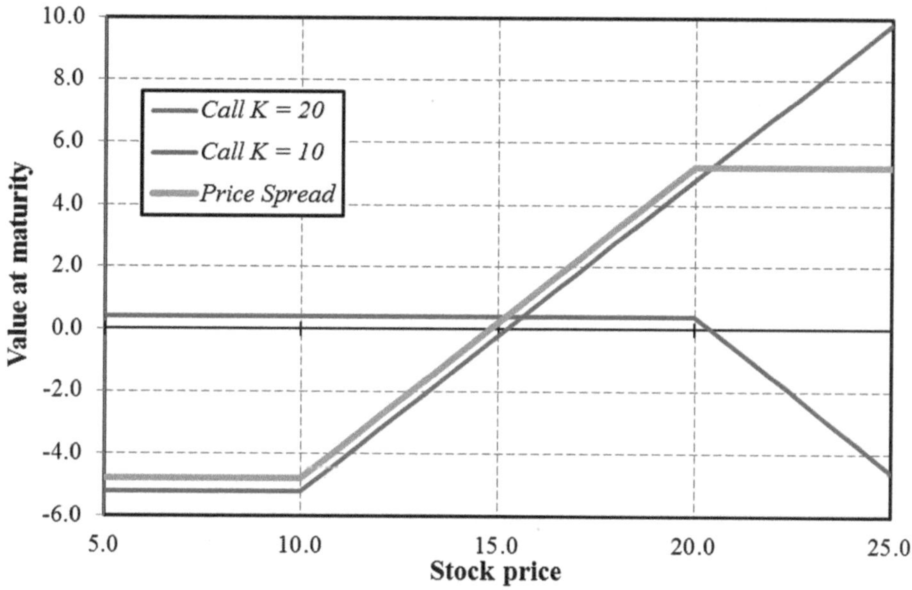

Fig. 8.26 The positive price-spread with call options at maturity

Losses
Limited; the stock price minus the strike price of the put option plus cost or minus the profit.

Margin Requirement
Possibility to off-set the margin requirement.

Trade Reasons

1. To lower the risk when you own the underlying.
2. To lock-in profits in the underlying.

Follow Up
On an increase:

- Sell the put option and roll the call option to a higher strike price.
- Roll the call option to a higher strike price and with later maturity.

On a decrease:

- Roll the issued call option to a lower strike price.

- If you believe on a turn on the market, sell the put option with profit and buy more of the underlying stocks to the new and lower price.

In Fig. 8.27 we illustrate a fench made by buying a put option with a strike 10 and a call with a strike 20 and at the same time holding the underlying stock or forward/future. The initial stock price is 15, the risk-free interest rate 2 % and the volatility 40 %. Figure 8.28 illustrate the same situation at maturity.

8.4.6 Positive Time Spread [- +]

Market Belief
This spread is used when you are bullish on the stock/index over the next few months.

Construction
You purchase a call option with a low strike and a long maturity and sell another call option with a higher strike with shorter time to maturity.

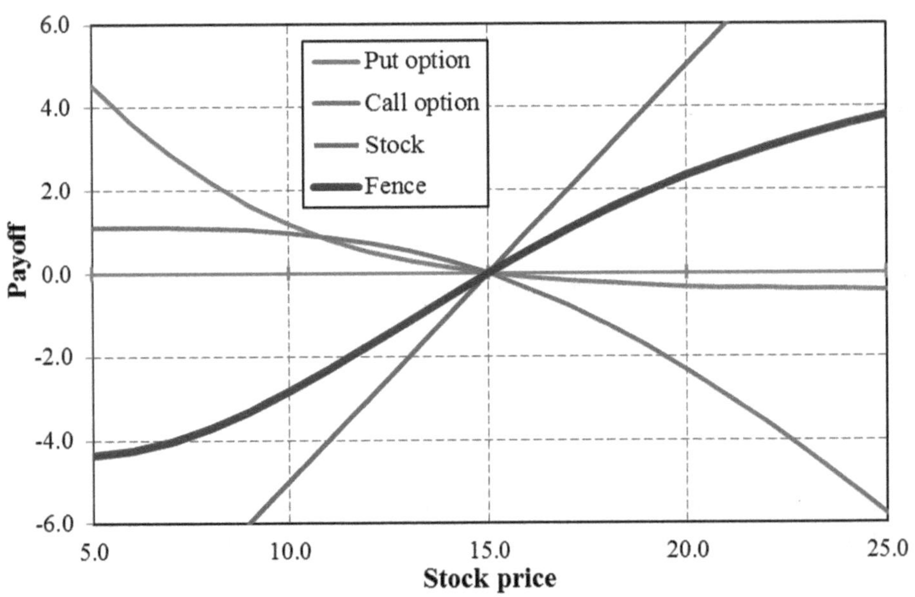

Fig. 8.27 A positive price-spread with call options

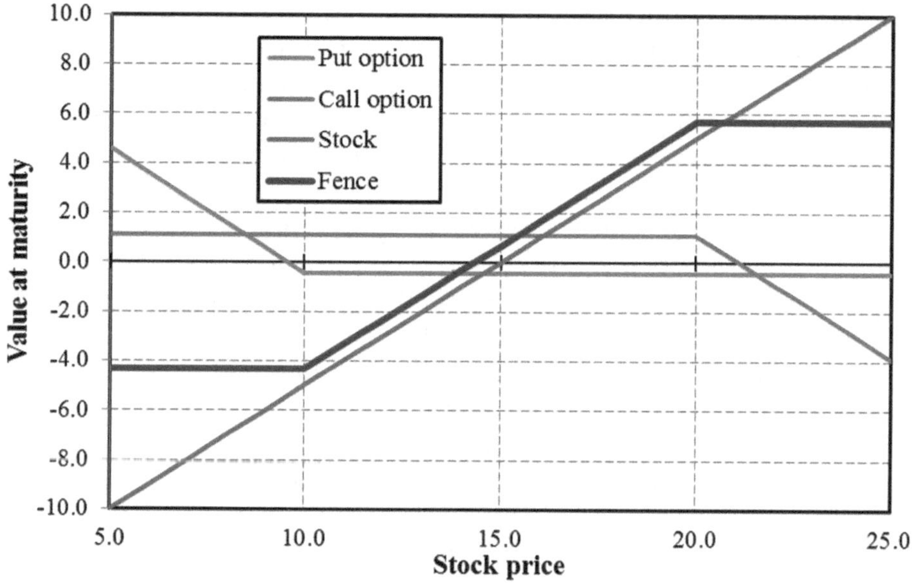

Fig. 8.28 The positive price-spread with call options at maturity

Profit
You profit in two ways;

1. the premium received for the sale of the call with near expiration and
2. from the upward movement of the stock price over the specified longer expiration.

Losses
Limited to the difference in strike prices +/− the initial cost/profit.

Margin Requirement
Yes, but with possible off-setting.

Comments
There is a risk that the issued option will be exercised. Initially the sale of the call reduces the cost of the spread, but each month you sell another call generating monthly income until the price of the stock gets close to the strike of the purchased call. Once the stock price has reached near the strike price of the long term purchased call you continue to profit from the stock/index's price increase without the need of selling more calls. Your potential profit is

unlimited, but your risk is limited and will vary during the time the spread is in effect. Since the spread is continually adjusted over time an accurate probability of profit cannot be determined prior to placing the trade. Margin requirements vary depending on the distance between strike prices. The rapid time decay of the closer expiration call sold and the slower time decay of the farther expiration call are your friend with this trade.

In Fig. 8.29 we illustrate a positive time-spread with two call options. We use a bought call option at strike 14 and maturity in a year and a sold call option at strike 16 with maturity in 6 months. The underlying price is 15, the volatility 40 % and the risk-free rate 2.0 %.

In Fig. 8.30 we illustrate the strategy after 6 months when the first option gains maturity.

This strategy can also be made by put options. In Fig. 8.31 we illustrate a positive time-spread with two put options. We use a sold put option at strike 14 and maturity in a year and a bought put option at strike 16 with maturity in 6 months. The underlying price is 15, the volatility 60 % and the risk-free rate 2.0 %.

In Fig. 8.32 we illustrate the strategy after 6 months when the first option gains maturity.

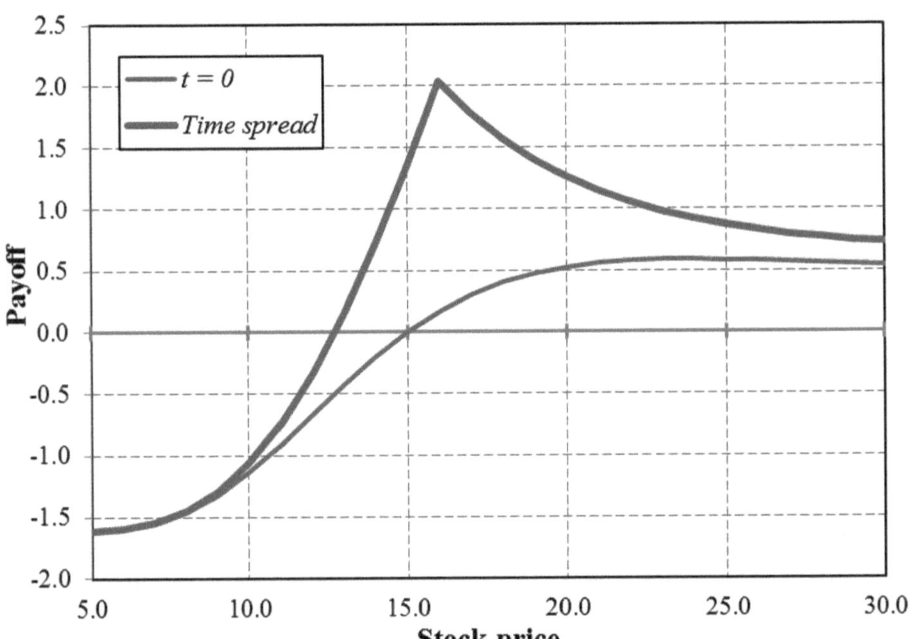

Fig. 8.29 A positive time-spread with call options

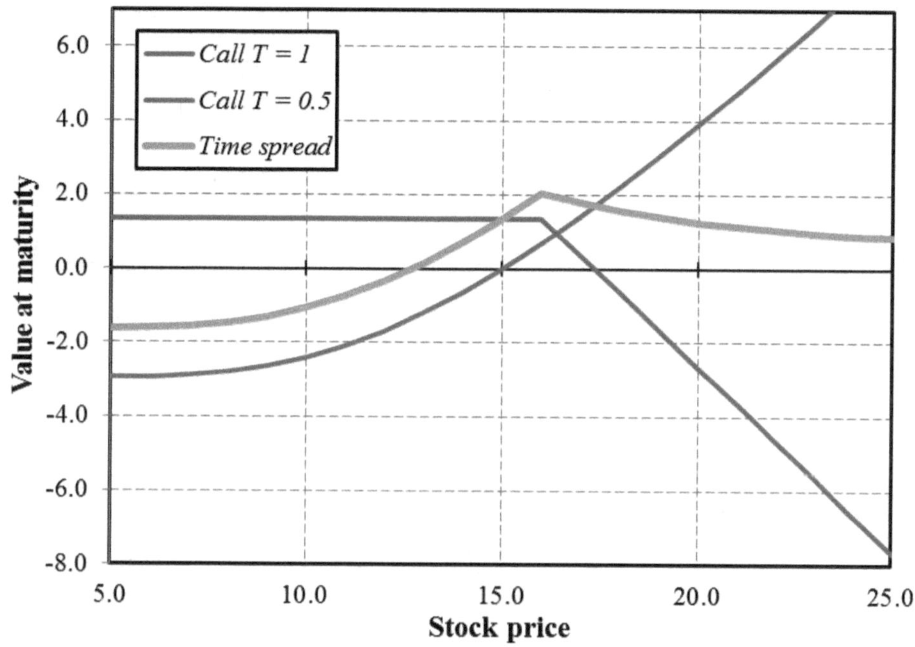

Fig. 8.30 The positive time-spread with call options on the first option maturity

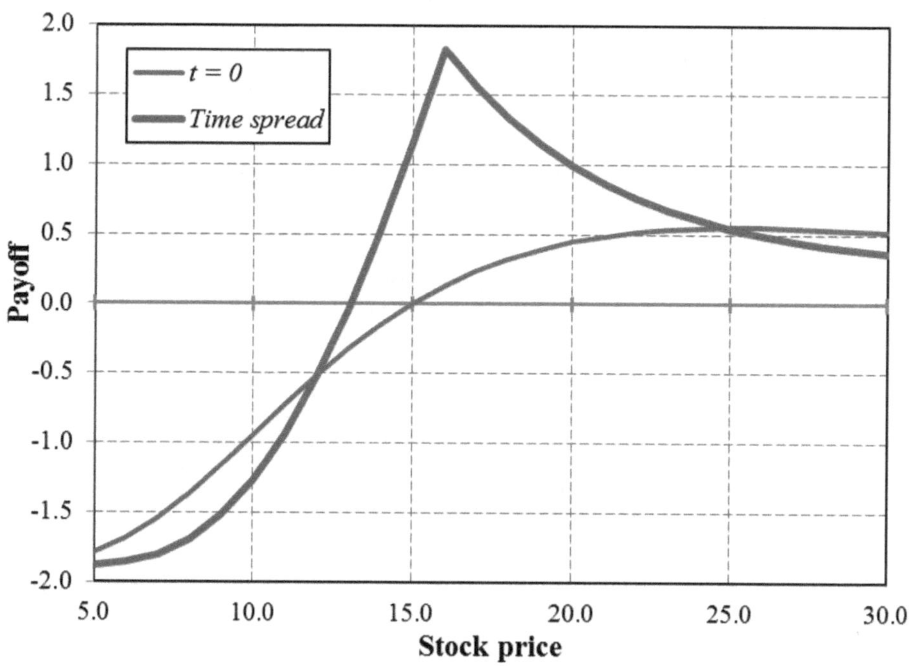

Fig. 8.31 A positive time-spread with put options

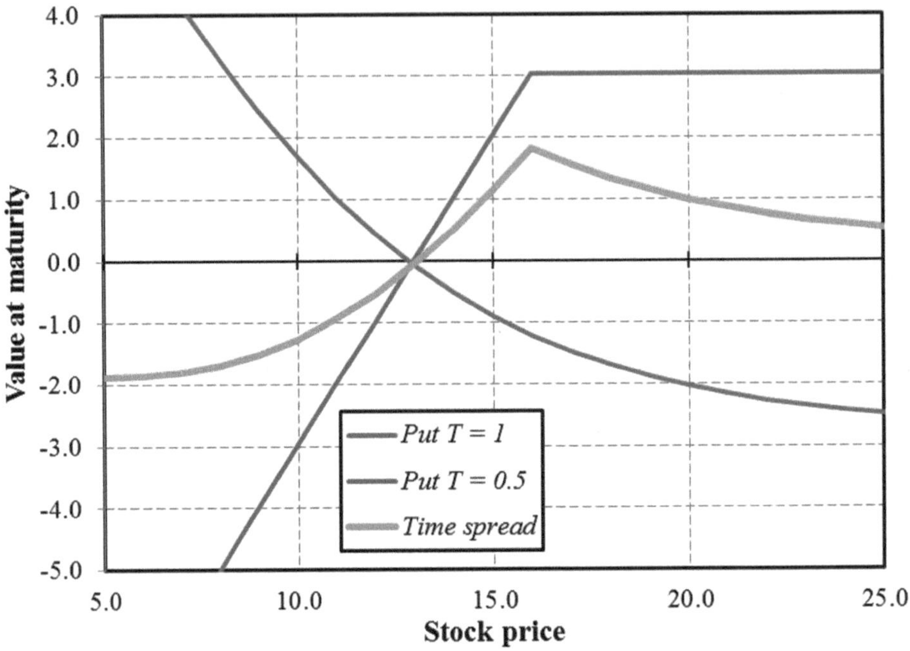

Fig. 8.32 The positive time-spread with put options on the first option maturity

8.4.7 Ratio Spread with Underlying's [1 2 0]

Market Belief
The investor believes in a limited increase in the underlying price.

Construction
Buy a call options with a low strike price and sell twice as many call options with a higher strike with the same maturity. You also buy the forward/future or own the underlying security.

Profit
Limited. The maximum profit is reached when the underlying price is above the upper strike price: The strike price of the issued call option minus the underlying price plus the difference between the strike prices plus possible profits or minus possible costs.

Losses
Losses are limited to the initial cost, that is, the underlying price plus costs or minus profits.

Break-Even
The break-even is reached at the underlying price plus the cost or minus profits.

Margin Requirement
Very limited since you own the underlying.

Trade Reasons

1. To get a better yield if the underlying reach the target price.
2. To get a profit on a neutral market.

Comments
This is a strategy if, at low risk, you want to increase your yield on owned stocks, in a neutral or weak increasing market. Over time, the yield is better than only an ownership in the stocks.

Follow Up
On an increase:

- Roll one of the issued call options to a higher strike price.
- Roll the call options to a higher strike price with later maturity.
- Buy more stocks or forwards.

On a decrease:

- Sell the owned call option.
- Sell the underlying.

In Fig. 8.33 we illustrate a ratio spread where we also own the underlying stock. The ration spread is made by buying a call option with a strike 10 and a sell two call options at 20. At the same time we own the underlying stock or forward/future. The initial stock price is 15, the risk-free interest rate 2 % and the volatility 60 %. Figure 8.34 illustrate the same situation at maturity.

8.4.8 Positive Back Spread [0 −1 1]

Back spreads is the common name when you hold more options than you is short. If you buy one more leg you get a ladder or a stair (see below).

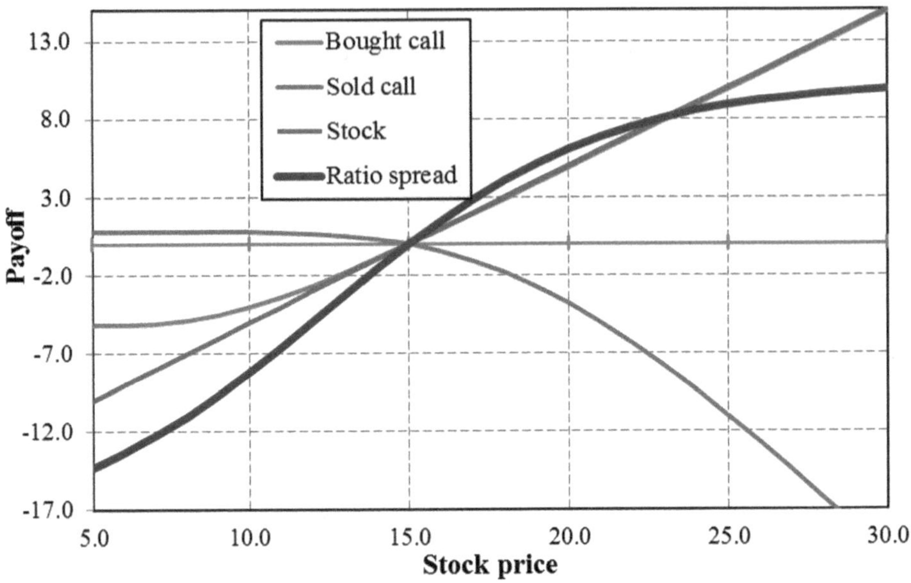

Fig. 8.33 A ratio-spread with call options

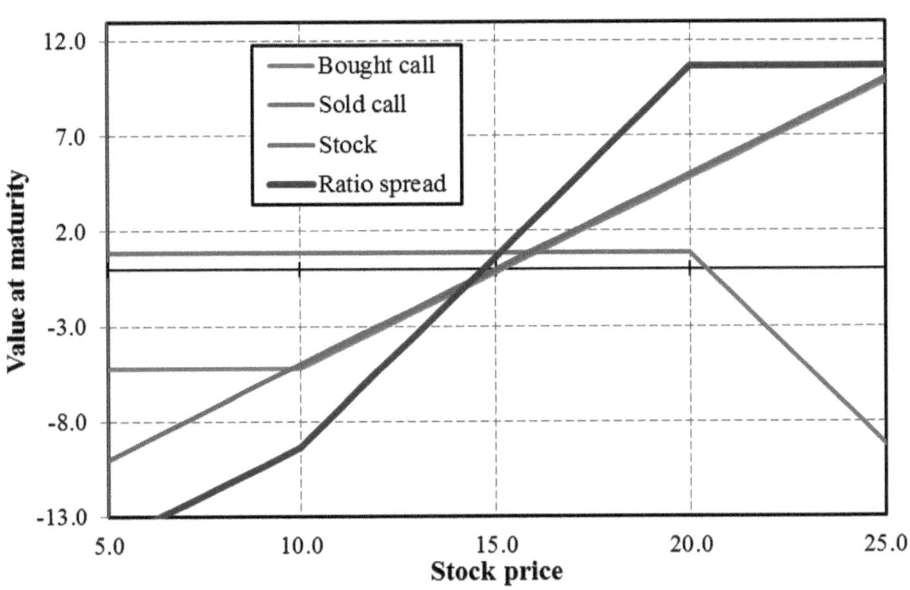

Fig. 8.34 The ratio-spread with call options at maturity

Market Belief
The investor believes in an increase of the underlying price, but wants a very good protection against a falling market.

Construction
Issue call options and buy twice as many with a higher strike price and same maturity

Profit
Unlimited.

Losses
Losses are limited to the difference between the strike prices plus possible cost or minus possible profit.

Break-Even
The break-even is the higher strike price plus the difference between the strike prices plus possible cost or minus possible profits.

Margin Requirement
Very limited.

Trade Reasons
Gives lower losses on a decrease compared with bought call options and a three-leg position.

Follow Up
On an increase:

- Issue to a higher strike price and buy back the issued call option.
- Sell the underlying forward.

On a decrease:

- Issue call options on a higher strike price and put options on a lower price.
- Issue call options on higher strike prices with later maturity.

In Fig. 8.35 we illustrate a positive back-spread with call options with strikes 10 and 20. The initial stock price is 15, the risk-free interest rate 2 % and the volatility 40 %. Figure 8.36 illustrate the same positive back-spread at maturity.

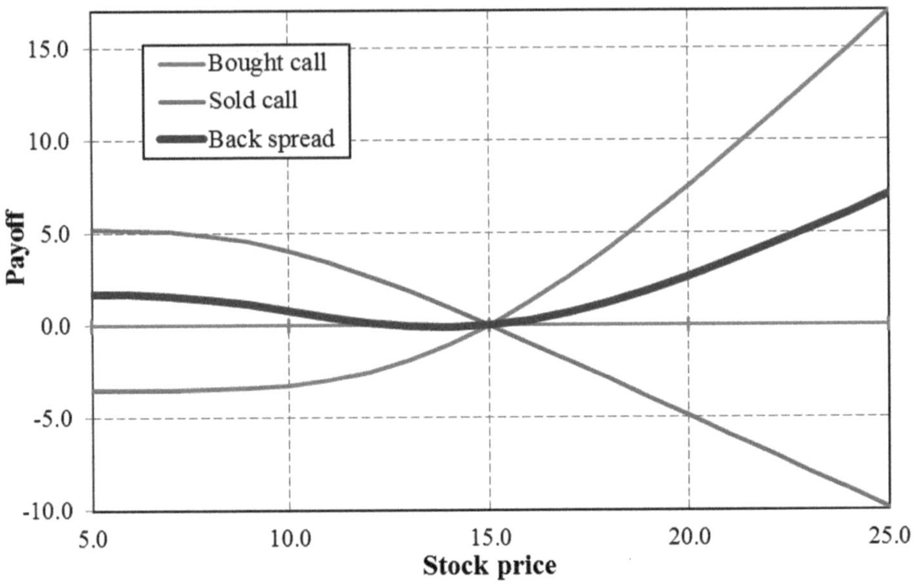

Fig. 8.35 A positive back-spread with call options

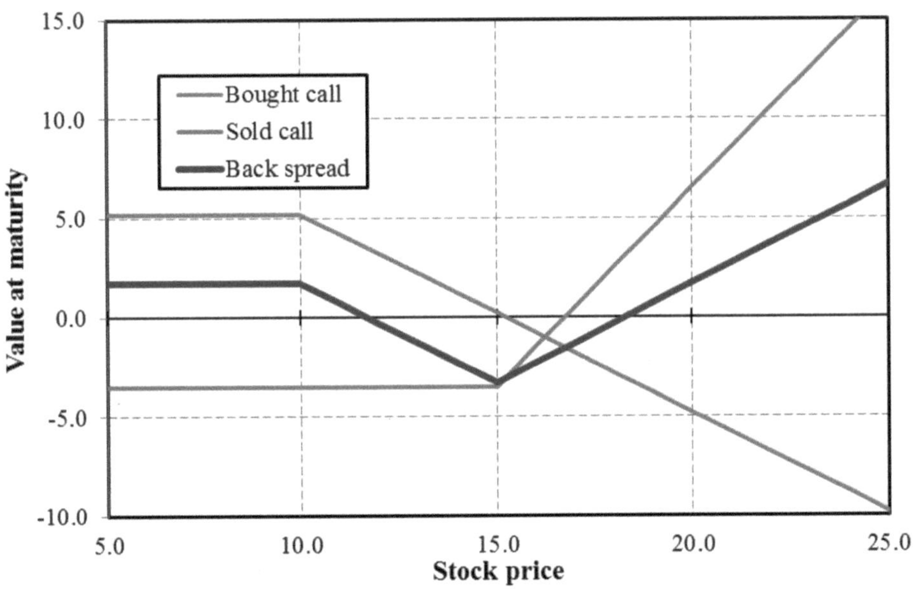

Fig. 8.36 The positive back-spread with call options at maturity

8.4.9 Long Synthetic Forward [1]

Market Belief
The investor believes in a strong increase of the underlying, but wants a more flexible position than buying the forward.

Construction
Issue a put option and buy a call option with same strike price.

Profit
Unlimited.

Losses
Unlimited: Strike price plus possible cost or minus possible income.

Break-Even
Strike price plus initial cost or minus initial income.

Margin Requirement
Always needed.

Trade Reasons

1. To create the same position as buying the underlying, but with a much lower cost.
2. This requires smaller movements than a long call option.
3. Gives higher potential yield than a long call option.

Follow Up
On an increase:

- Issue a call option to a higher strike price and use the income to buy back the issued put option.
- Lock-in the profit selling the forward.
- Roll the call option to a higher strike price and buy back the issued put option.

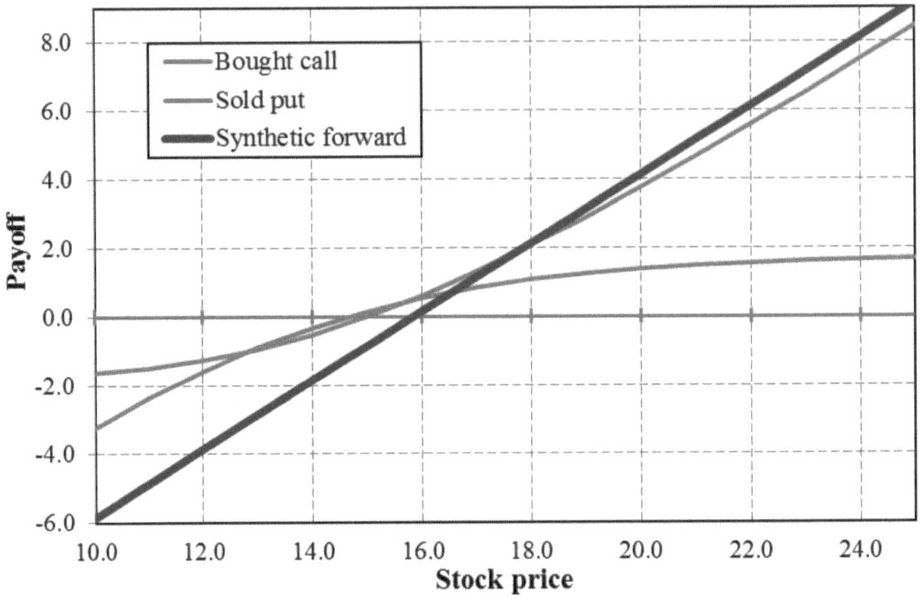

Fig. 8.37 A synthetic long forward/future or stock made by a long call and a short put option

On a decrease:

- Roll the issued put option to a lower strike price and maybe sell the owned call option.
- Roll the issued put option to a lower strike price with later maturity.
- Sell the owned call option and create a price-spread on a lower level.

In Fig. 8.37 we create long forward/future or stock by buying a call option and selling a put option ATM. The initial stock price is 15, the risk-free interest rate 2 % and the volatility 40 %. Figure 8.38 illustrate the same position at maturity.

8.4.10 Long Sloping Synthetic Forward [1 0 1]

Market Belief
The investor believes in a strong increase in the underlying price, but wants a more flexible position than buying the forward.

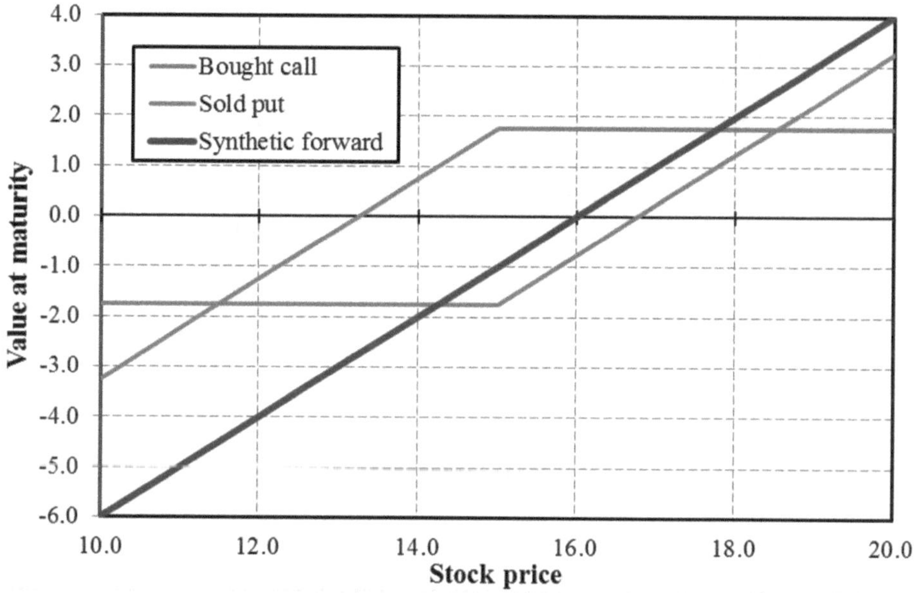

Fig. 8.38 The synthetic long forward/future or stock made by a long call and a short put option at maturity

Construction
Issue put options to a lower strike price and buy a call option with a higher strike price.

Profit
Unlimited.

Losses
Unlimited, put options strike price plus initial costs or minus initial income.

Break-Even
The strike price of the call option plus initial cost or the strike price of the put option minus initial income.

Margin Requirement
Always needed.

Trade Reasons

1. Lower risk than for a synthetic forward.
2. Higher potential than a positive stair.

Follow Up

On an increase:

- Issue a call option to a higher strike price and use the income to buy back the issued put option.
- Lock-in the earned profit by selling the forward.
- Roll the call option to a higher strike price and buy back the issued put option.

On a decrease:

- Roll the issued put option to a lower strike price and eventually sell the owned call option.
- Roll the issued put option to a lower strike price with later maturity.
- Sell the owned call options and create a price-spread on a lower level.

In Fig. 8.39 we create long sloping forward/future or stock by buying a call option at 16 and selling a put option at 13. The initial stock price is 15, the

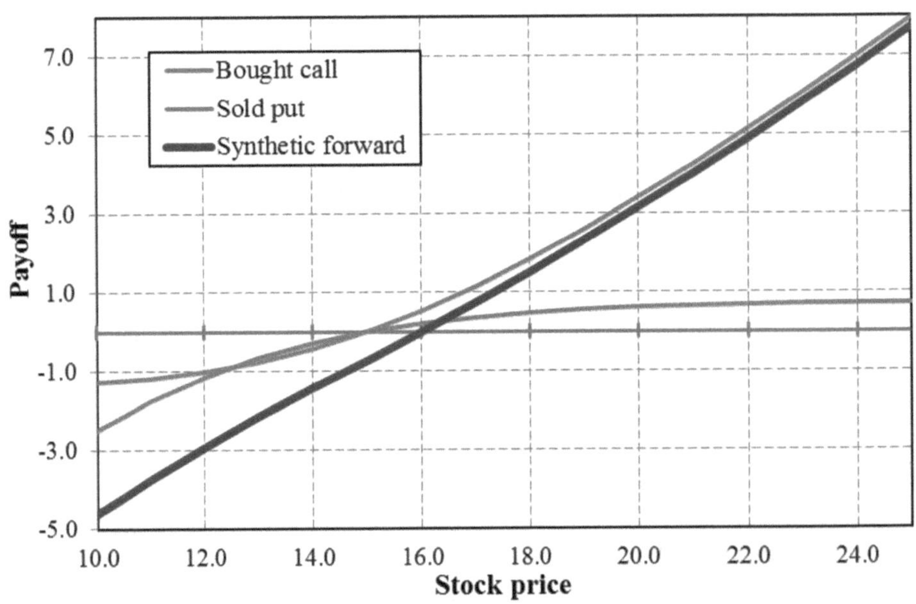

Fig. 8.39 A synthetic long sloped forward/future or stock made by a long call and a short put option

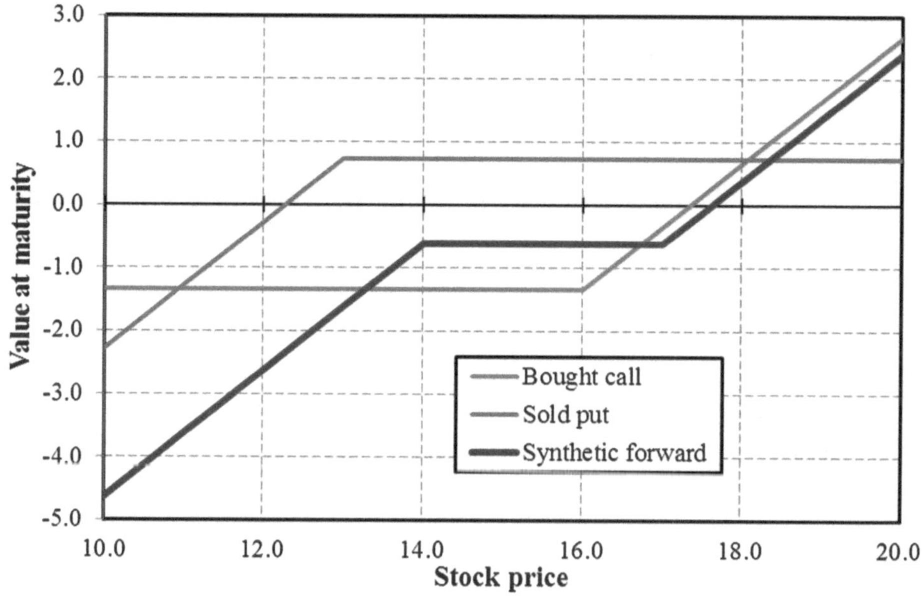

Fig. 8.40 The synthetic long sloped forward/future or stock made by a long call and a short put option at maturity

risk-free interest rate 2 % and the volatility 40 %. Figure 8.40 illustrate the same position at maturity.

8.4.11 Positive Stair [0 1 0 1 0]

Market Belief
The investor believes in an increase in the underlying price but want to be protected against a decrease.

Construction
Create two positive price-spreads, one with call options and another with put options.

Profit
Limited to the difference in the strike price of the call options minus initial cost or plus the initial profit.

Losses
Limited to the difference in the strike price of the put options plus the initial cost or minus the initial profit.

Break-Even
The lower call option strike price plus the initial cost or minus the initial profit.

Margin Requirement
Limited.

Trade Reasons

1. To get a lower risk than a synthetic forward or a sloped synthetic forward.
2. Lower cost than a positive price-spread.

Follow Up
On an increase:

- Roll the price-spread to a higher strike price.
- Sell the owned put option and move the issued call option to a higher strike price.

On a decrease:

- Issue more call options and create a ratio spread or a ladder.
- Sell the owned call options and roll the issued put option to a lower strike price with later maturity.

In Fig. 8.41 we have created a positive stair by buying a call option at 14 and selling another call option at 16. We also buy a put option at 10 and selling another put option at 12. The initial stock price is 15, the risk-free interest rate 2 % and the volatility 40 %. Figure 8.42 illustrate the same position at maturity.

8.4.12 A Ratio Spread with Call Options [0 1 −1]

Market Belief
This spread is used under the same conditions as a long call or the bull call debit spread. You should be very bullish on the stock/index and the expected

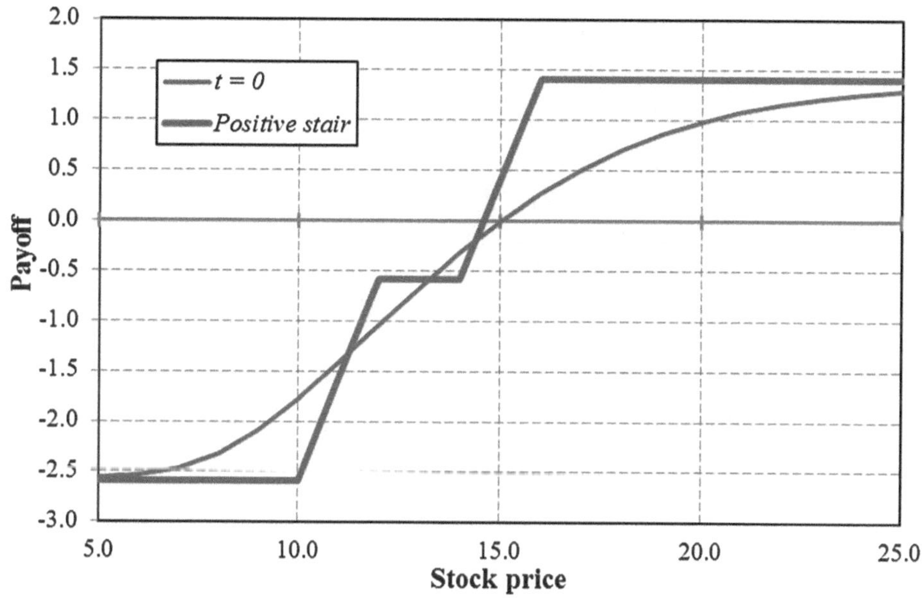

Fig. 8.41 A strategy called a positive stair made by two call options and two put option

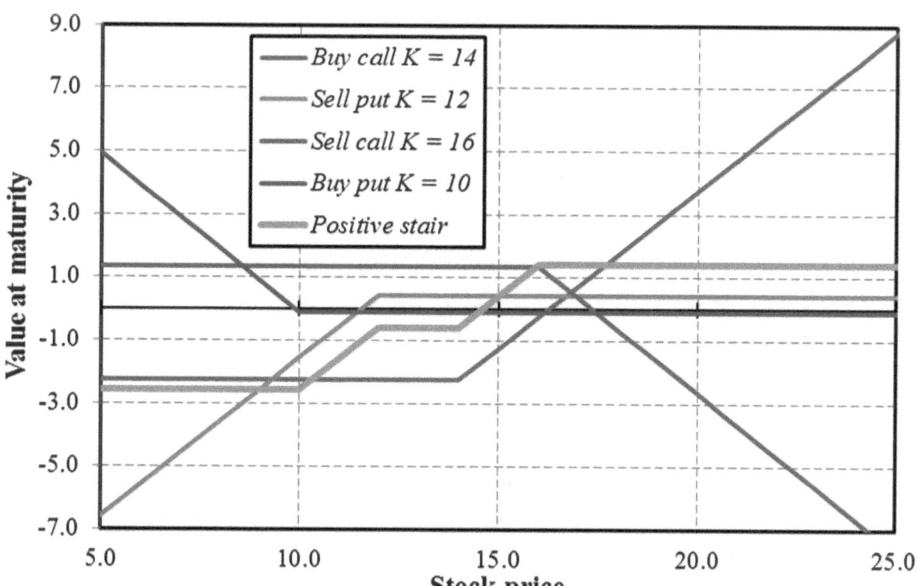

Fig. 8.42 This is how we can make a positive stair using two call options and two put options

range of the stock during the particular time period should extend significantly beyond the break-even points of the position.

Construction

You buy two of the higher strike call options that are near the current price and sone) reduces the cost of the two calls purchased often to the point or a free trade or credit to put on the spread and has unlimited profit potential, but the risk can be higher since you will be responsible for the difference in strike prices plus any premium paid or less any credit received.

Profit

Limited. The difference between the strike prices plus/minus the net profits/costs for the options. You get the maximum profit at the higher strike price.

Losses

Unlimited. The strategy gives a loss if there is a large increase in the underlying price.

Break-Even

As a result your break-even has two different points and you will lose money if the options expire at any point between the two break-even prices.

Margin Requirement

Always needed.

Trade Reasons

1. The strategy has lower cost than a bought price-spread.
2. The strategy gives only losses if there is a large increase in the underlying price.
3. There is a high probability to get a profit.

Comments

The strategy requires a careful follow-up. You can also use a 3:2 ratio (buy three, sell two) which will reduce the cost further or increase the credit received, but the break-even points will be extended even further requiring an even greater move in the stock price for a profit. The probability of profit should be greater than 40 % and rarely will exceed 60 %.

Follow Up
On an increase:

- Buy a call option with higher strike price.
- Move one of the issued call the options to a higher strike price and create a ladder.
- Buy the underlying forward if and when the prices pass the upper strike price.

On a decrease:

- Sell the owned call option.
- Sell the owned call option and buy back one of the issued call options.

In Fig. 8.43 we have created a ratio spread by buying a call option at 14 and selling two call options at 18. The initial stock price is 15, the risk-free interest rate 2 % and the volatility 40 %. Figure 8.44 illustrate the same position at maturity.

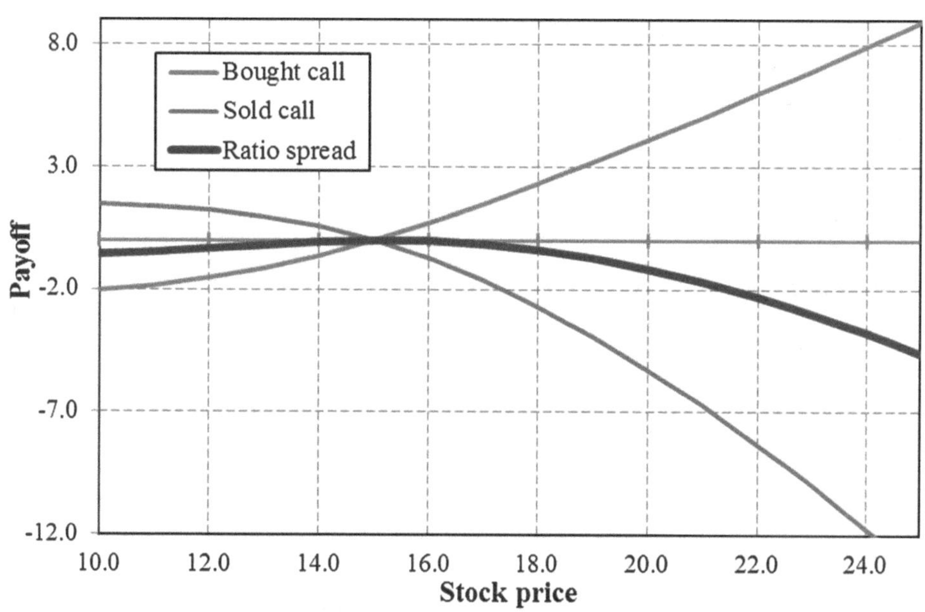

Fig. 8.43 A ratio spread made by two call options. We buy one at 14 and sell two at 18

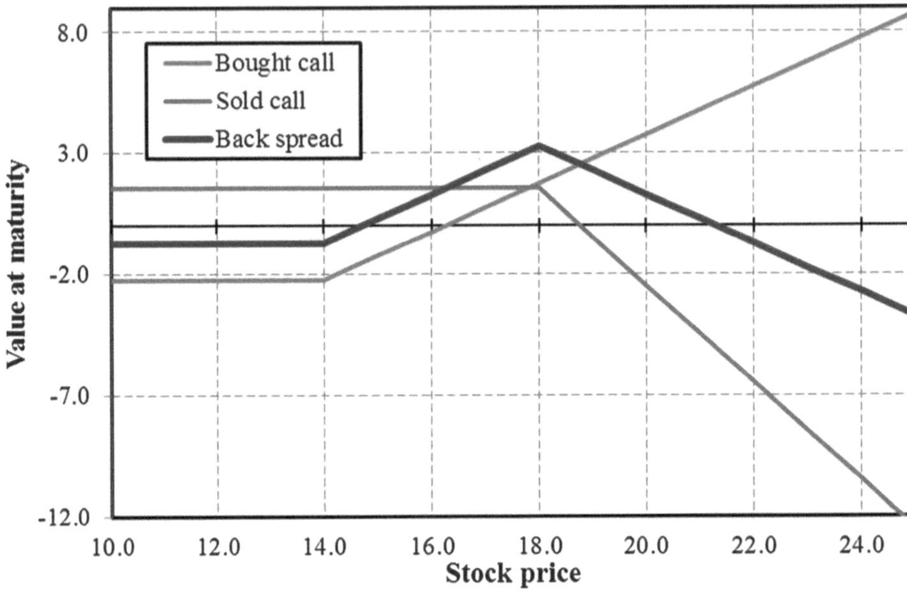

Fig. 8.44 The ratio spread at maturity, made by two call options. We buy one at 14 and sell two at 18

8.4.13 Positive Three-Leg Positions [1 0 1 0] and [0 1 0 1]

Market Belief

The investor believes in a strong increase in the underlying price, but at the same time needs a good protection on a decrease.

Construction

Sell put options with a low strike price, buy call options at a mid-strike and sell call options with a high strike price.

Profit

Limited to the difference in strike prices of the call options minus initial cost or plus possible profit.

Losses

Unlimited. The strike price of the put options plus initial cost.

Break-Even

The lower strike price on the put option plus possible cost.

Margin Requirement
Always needed.

Trade Reasons

1. The strategy gives no risk on the positive side as for the ratio spread.
2. The strategy gives lower price than a price-spread.
3. The strategy gives higher probability for a profit than a long call option.

Follow Up
On a profit:

- Move the put option to a higher strike price and use the income to move the issued call option to a higher strike price.
- Roll the price-spread to a higher strike price.

On losses:

- Roll the issued put option to a lower strike price and (maybe) sell the owned call option.
- Move the issued put option to a lower strike price with later maturity.
- Issue more call options and create a ratio spread or ladder.

In Fig. 8.45 we have created a three-leg strategy by selling a put option at 10, buying a call at 15 and selling another call options at 20. The initial stock price is 15, the risk-free interest rate 2 % and the volatility 40 %. Figure 8.46 illustrate the same position at maturity.

Another way to create a positive three-leg strategy is by using two put options and one call.

Market Belief
The investor believes in a strong increase in the underlying price, but at the same time needs a good protection on a decrease.

Construction
Issue put options and buy put options with a lower strike and use the income to buy call options.

Profit
Unlimited.

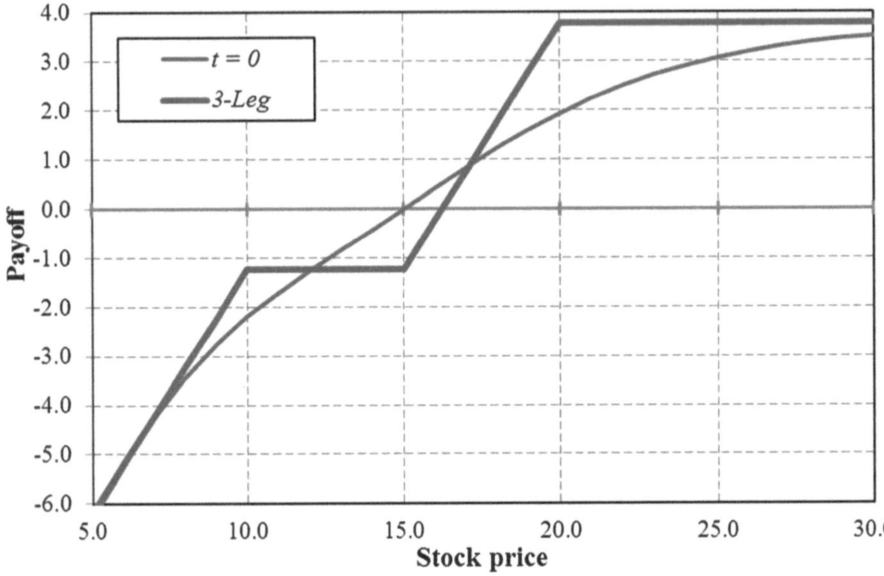

Fig. 8.45 A three-leg strategy made by one put and two call options

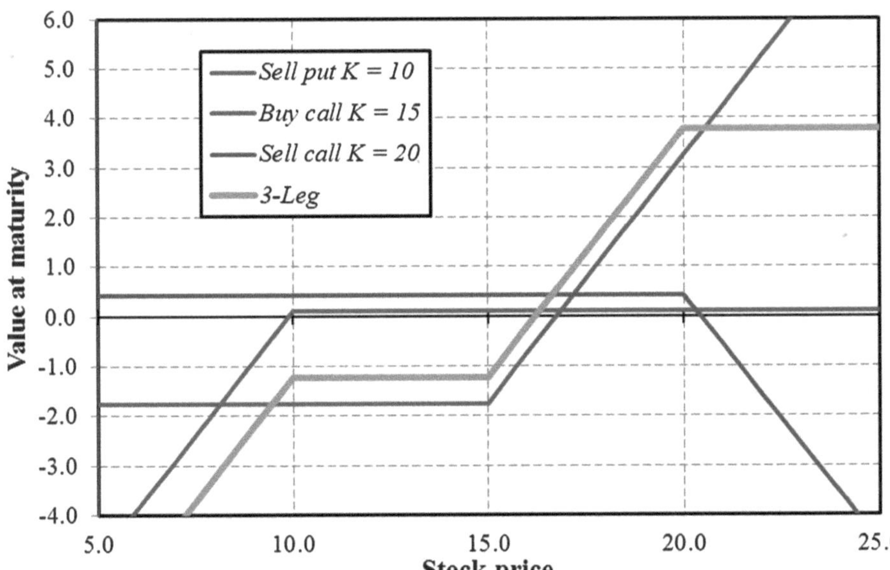

Fig. 8.46 The three-leg strategy at maturity made by one put and two call options

Losses
Limited. Maximum loss is the difference between the strike prices of the put options plus a possible net cost or minus a possible net profit.

Break-Even

Upwards: The strike price of the call options plus possible cost.
Downwards: The upper strike price on the put option minus possible income.

Margin Requirement
Limited.

Trade Reasons

1. The position gives a lower cost a bought call option and therefore a lower break-even.
2. The position gives no losses on limited increase as the back spread above.

Follow Up
With profit:

- Issue a call option on a higher price and use the income to close the positive price-spread.
- Roll the owned call option to a higher strike price and use the income to close the positive price-spread.

On losses:

- Sell the owned put option and roll the issued to a lower strike price.
- Sell the owned call option and create a price-spread on a lower level.

In Fig. 8.47 we have created a three-leg strategy by buying a put option at 10, and selling another at 15 and buying a call options at 20. The initial stock price is 15, the risk-free interest rate 2 % and the volatility 40 %. Figure 8.48 illustrate the same position at maturity.

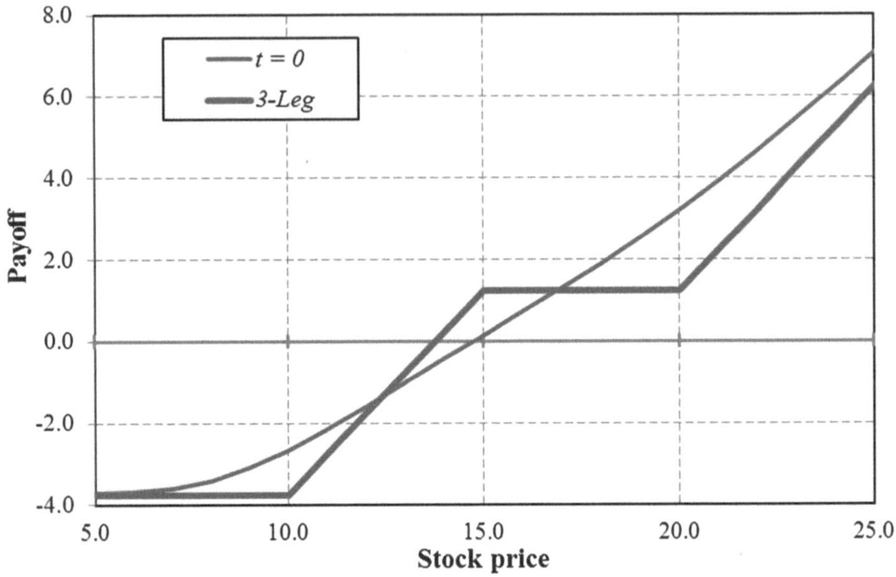

Fig. 8.47 A three-leg strategy made by one call and two put options

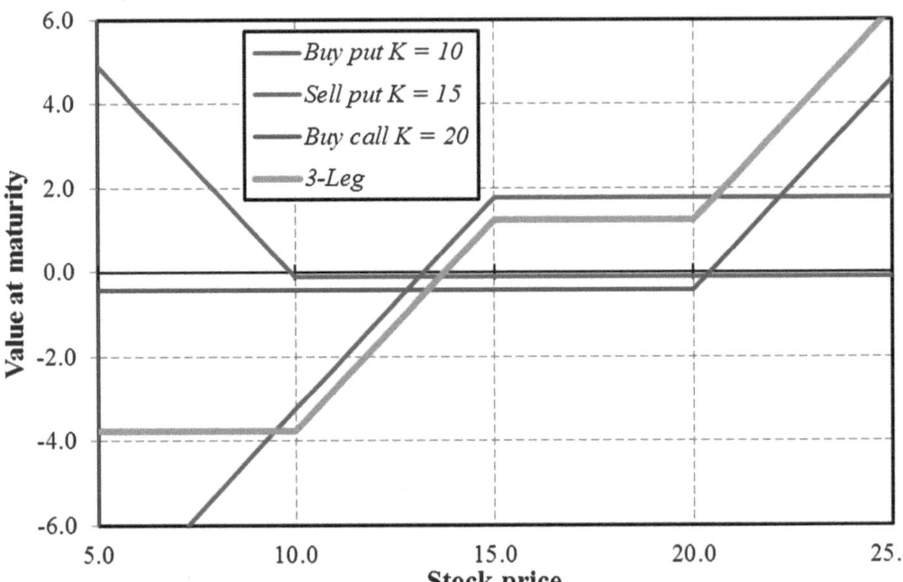

Fig. 8.48 The three-leg strategy at maturity made by one call and two put options

8.4.14 Positive Three-Leg Position with Ownership [1 0 1 0 *]

Market Belief

The investor believes in a limited increase of the underlying price but wants a good protection against a decrease.

Construction

Buy the underlying, issue a call option, buy put option and issue a put option on a lower strike.

Profit

Limited. The strike price of the call option minus the stock price minus initial cost or + initial income.

Losses

Unlimited.

Break-Even

Upwards: The underlying price plus initial cost.
Downwards: The underlying price minus initial income.

Margin Requirement

Very limited since you own the underlying.

Comments

If you start with a covered call protected with a long put option and thereafter want to get a better yield with issuing a put option on a low level, then you are in the following situation.

Trade Reasons

- To lower the risk of buying the underlying.
- With a low underlying price and when you are uncertain if the bottom is reached.

Follow Up
At profit:

- Sell the put option and move the call option to a higher price.
- Move the call option to a higher strike price with later maturity.

At losses:

- Sell the issued options.
- Move the issued put option to a lower strike price with later maturity.
- If you believes in a turn in price; sell the put option and buy more stocks on this low level.

In Fig. 8.49 we have created a three-leg strategy where we also own the underlying stock or forward/future. We sell a put option at 10, buying another at 15 and sell a call at 20. The initial stock price is 15, the risk-free interest rate 2 % and the volatility 40 %. Figure 8.50 illustrate the same position at maturity.

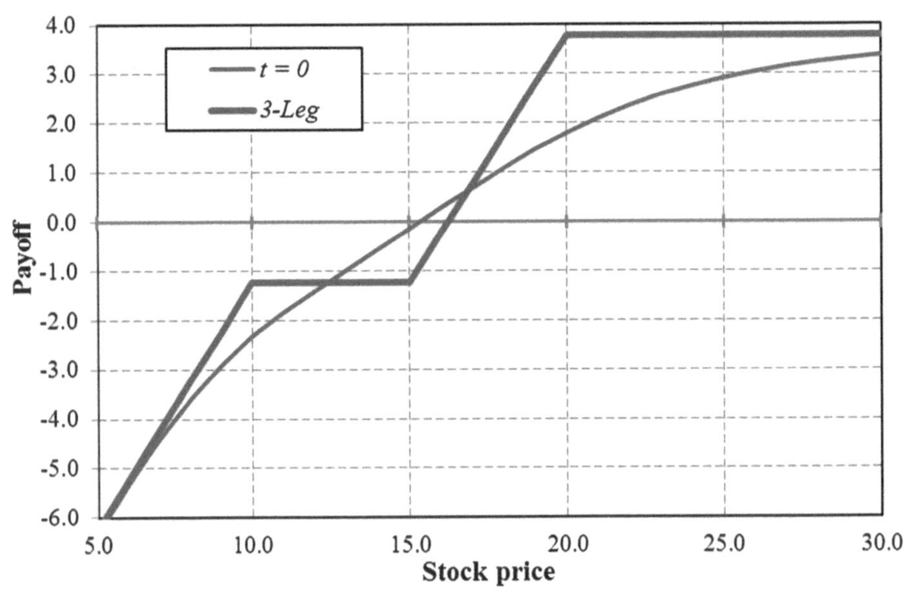

Fig. 8.49 A three-leg strategy made by one call, two puts and underlying

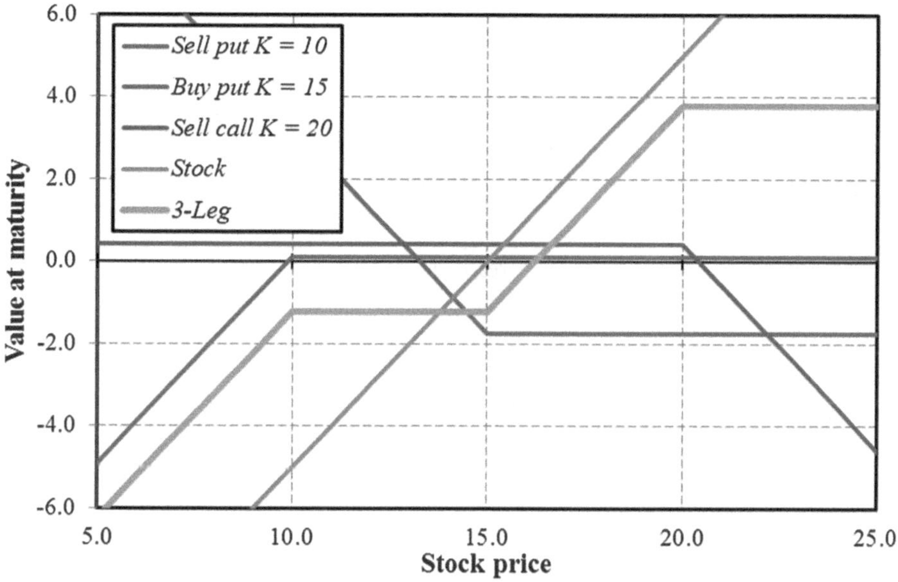

Fig. 8.50 The three-leg strategy from Fig.8.49 broken down to illustrate the time to maturity. As we can see, it's made one sold call, two puts and the underlying

8.5 Neutral Markets

In a neutral market we believe on small changes in underlying prices, i.e., low volatilities. We will now present some useful strategies under such market conditions

8.5.1 A Short Straddle [1 −1]

Market Belief
This spread is used when you believe that the stock/index will stay essentially unchanged with minimal price movement up or down in the near future. I.e., the investor believes in a market with low volatility.

Construction

1. Sell a put option and a call option with same strike price and maturity,
2. Buy the underlying and sell twice as many call options.

Profit

1. Limited to the premium, if the options are bought ATM.
2. The strike price minus the underlying price plus the premium.

Break-Even

1. b: the strike price minus the premium or
2. c: the strike price plus the premium.

Losses
Unlimited.

Margin Requirement
Always needed and high.

Comments
You profit if the price movement over the specified time period is less than the premium received from the sale of the call and put. The rapid time decay in the last month prior to expiration is your friend in this trade. The 5-day volatility should be higher than the 100-day volatility. Expiration should generally be less than 30 days of when the trade is placed. Probability of profit is generally less than 50 %. However, high margin requirements generally require having a larger trading account.

Trade Reasons
To get a payoff in a neutral market.

Follow Up
On increase (without ownership):

- Buy a call option with a lower strike price as protection.
- Buy a forward if the underlying increases above the upper level.

On decrease (without ownership):

- Buy a put option with lower strike price as protection.
- Sell a forward if the underlying decreases below the upper level.

Neutral (without ownership):

- Buy a call option with higher strike price and a put option with lower strike price. In that case you can lock the earned profit.
- If you can buy the call- and the put option with same strike price with later maturity you can lock the earned profit

At a profit (with ownership):

- Buy the same amount of call options to a higher strike price.
- Buy the same amount of call options to the same strike price with later maturity.

At a loss (with ownership):

- On an increase: buy more stocks or forwards.
- On an increase: buy call options with higher strike price.
- On a decrease: move the issued call options to a lower strike price.
- On a decrease: buy put options.

In Fig. 8.51 we have created a short straddle by selling a put and a call ATM, i.e., at strikes 15. The initial stock price is 15, the risk-free interest rate 2 % and the volatility 40 %. Figure 8.52 illustrate the same position at maturity. In Fig. 8.53 we make the same kind of strategy when we own the stock (or forward/future). Then we just sell two call options.

8.5.2 Short Strangle [1 0 −1]

If we take the straddle in Sect. 8.6.1 and spread the strikes apart, we create a strangle.

Market Belief
This spread is used when you believe that the price of the stock/index will stay within a specific range in the near future, that is, the investor believes in market with relative low volatility.

Construction
Sell a put option with a low strike price and a call option with a higher strike.

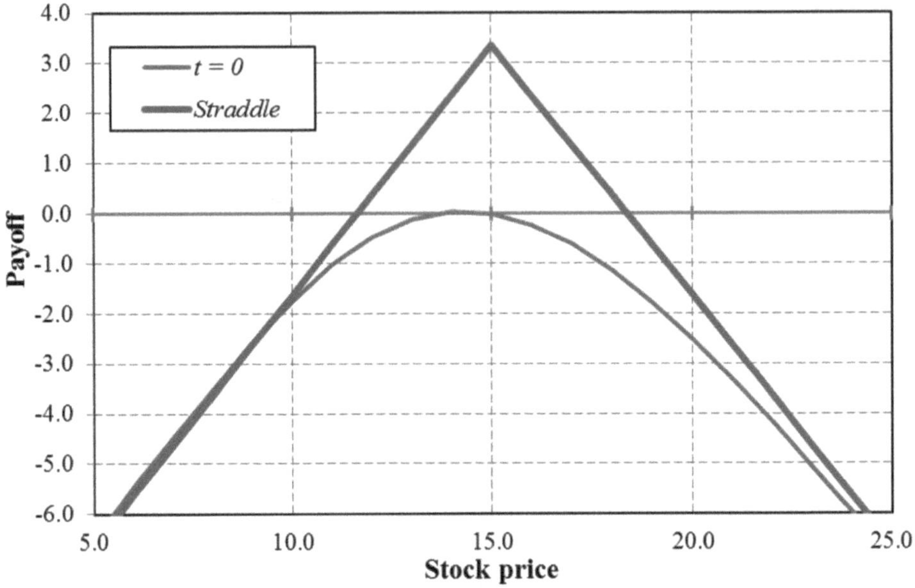

Fig. 8.51 A short straddle made by selling a call and a put at ATM

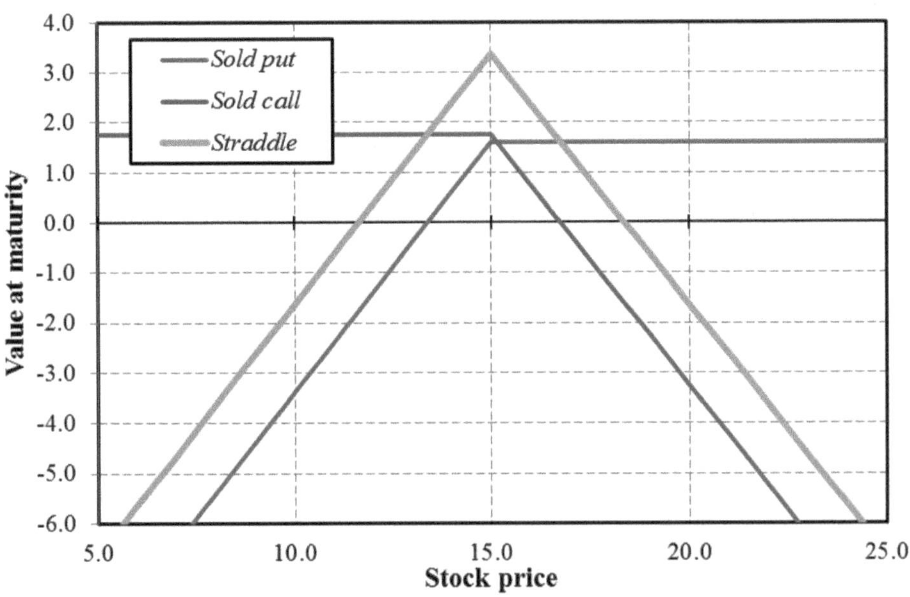

Fig. 8.52 The short straddle made by a call and a put at maturity

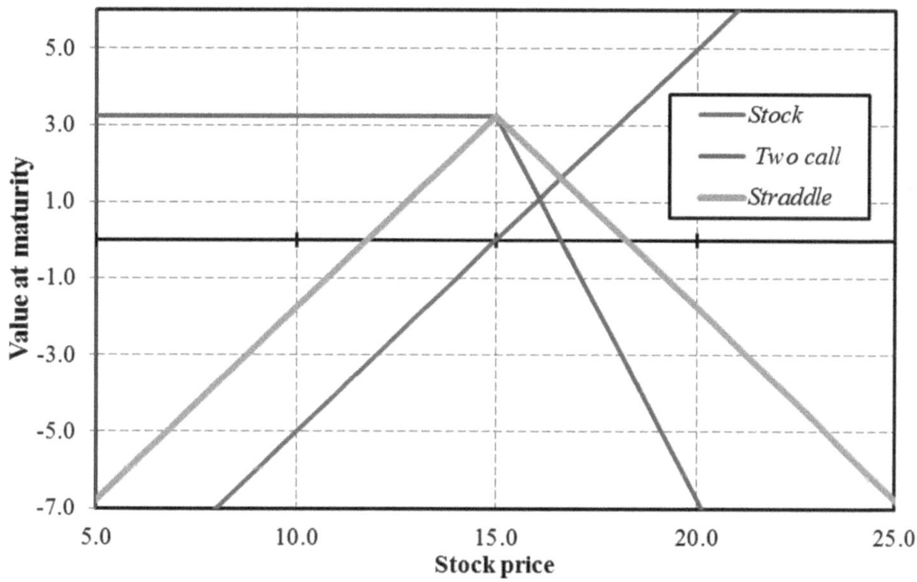

Fig. 8.53 The short straddle made by selling two calls and holding the underlying at maturity

Profit

You profit if the price movement over the specified time period stays within the range between the two strike prices or does not extend beyond either strike price more than the premium received from the sale of the call and put.

Break-Even

The point where the lower strike price minus the premium is reached and the point where the higher strike price plus the premium is reached.

Losses

Unlimited.

Margin Requirement

Always needed.

Comments

The rapid time decay in the last month prior to expiration is your friend in this trade. The 5-day volatility generally should be higher than the 100-day volatility. Expiration should generally be less than 30 days of when the trade is placed. Probability of profit is generally greater than 50 %. This is a very high probability trade to profit if entered correctly. However, high margin requirements generally require having a larger trading account.

Trade Reasons
To get a profit in a neutral or almost neutral market.

Follow Up
On an increase:

- Buy call options with lower strike price as protection.
- Buy the forward if the underlying reaches the level for the upward break-even.

On a decrease:

- Buy put options with lower strike price as protection.
- Sell the forward if the underlying falls below the lower level of break-even.

On neutral:

- If you are able to buy a call option with higher strike price and a put option with lower strike price so that the net profit is greater than the difference in strike price, then you have locked-in a profit.
- Buy call- and put option with same strike price with later maturity to lock-in the profit.

In Fig. 8.54 we have created a short strangle by selling a put and a call at strikes 12 and 17 respectively. The initial stock price is 15, the risk-free interest rate 2 % and the volatility 40 %. Figure 8.55 illustrate the same position at maturity.

8.5.3 A Long Butterfly [0 1 −1 0]

Market Belief
You believe that the stock price will fluctuate in a narrow range.

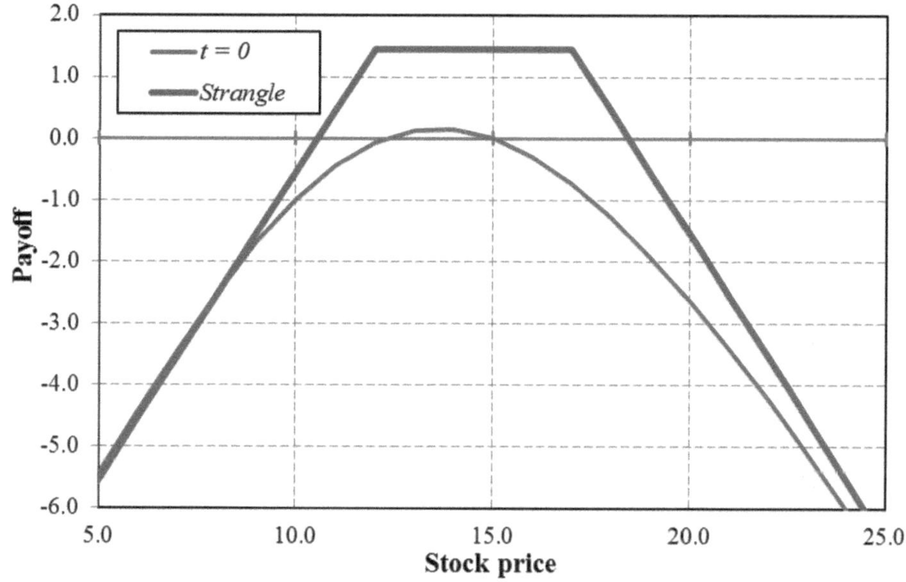

Fig. 8.54 A short strangle made by selling a call and a put

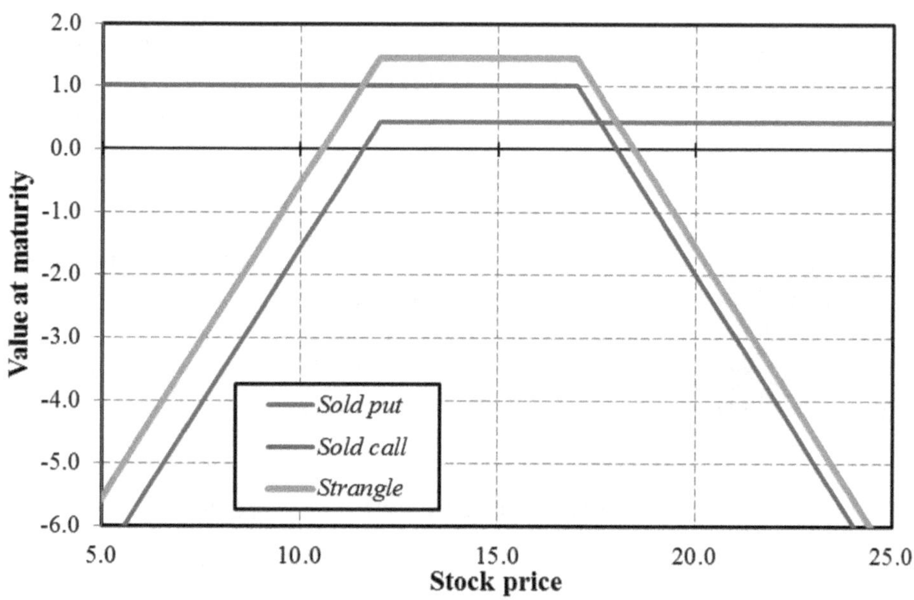

Fig. 8.55 The short strangle made by a call and a put at maturity

Construction

Call option with low strike bought and two call options with medium strike sold and call option with high strike bought. The same position can be created with puts.

Profit

Limited, reaching maximum at a high strike.

Break-Even

If call version used, downside break-even = low strike – net cost of spread, upside break-even is at high strike + net cost of spread.

Losses

Maximum loss realized if stock ends below low strike or above high strike and limited to net credit paid. For each point above low strike or below high strike, loss decreases by additional point.

Margin Requirement

Low.

Comments

This position can be difficult to buy and sell during a short time period. This position is a combined asset. As time passes, value of position increases/erodes toward expiration value. If volatility increases, increase/erosion slows; if volatility decreases, increase/erosion speeds up.

Trade Reasons

To get an income on a neutral or weak market with a minimized risk for losses.

In Fig. 8.56 we have created a long butterfly with call options. We made this by buying call options at 12 and 18 and selling two call options ATM, 15. The initial stock price is 15, the risk-free interest rate 2 % and the volatility 40 %. Figure 8.57 illustrate the same position at maturity.

8.5.4 A Neutral Time Spread or Calendar Spread

Market Belief

The investor believes in an initial week market with a strong increase in the future.

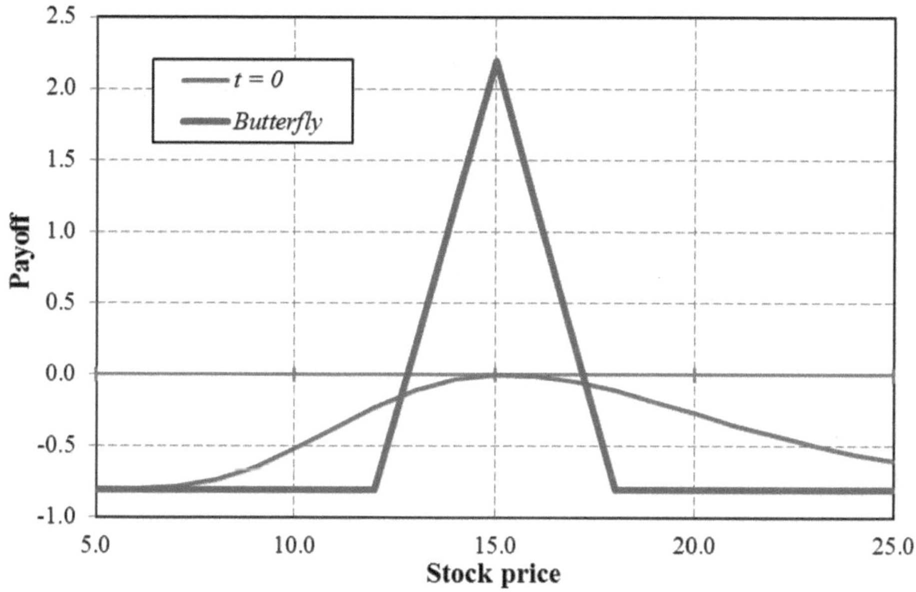

Fig. 8.56 A long butterfly by call options

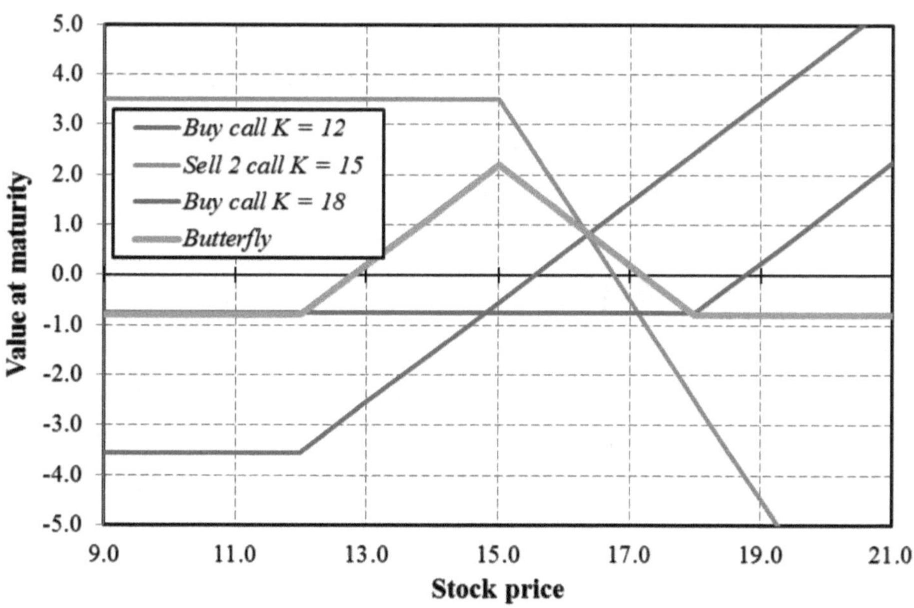

Fig. 8.57 A long butterfly by call options at maturity

Construction
Issue a call option with short time to maturity and buy another call option with the same strike price with later maturity. If the investor believes in the opposite then he/she can make the reverse strategy with put options.

Profit
Big, if the long option still is alive when the short one has expired. If the position are closed when the issued option have the maturity, the maximum profit is reached ATM.

Losses
Limited to the difference in the strike price plus the premium.

Margin Requirement
Yes, but limited.

Comments
The risk is that the issued option will be exercised. Commonly, ATM options are used. This strategy has a larger possible loss compared with a diagonal time-spread if the underlying price decreases, but less during an increase. The strategy is initializes when the shorter option has a month or less to maturity.

In Fig. 8.58 we have created a neutral time-spread, also called a calendar-spread using two call options. We have bought one with maturity one year and sold another with maturity g month. Both options have the strike given by 16. The initial stock price is 15, the risk-free interest rate 2 % and the volatility 40 %. Figure 8.59 illustrate the same strategy at maturity.

8.5.5 Covered Call or Synthetic Sold Put Option [1 0 *]

Market Belief
You are sure that the price of the stock you hold will not fall. Sell lower strike options if you are only somewhat convinced; sell higher strike options if you are confident stock will rise. If you think stock will stagnate, sell ATM options for maximum profit.

Construction
Issue call options. The number is calculated by delta.

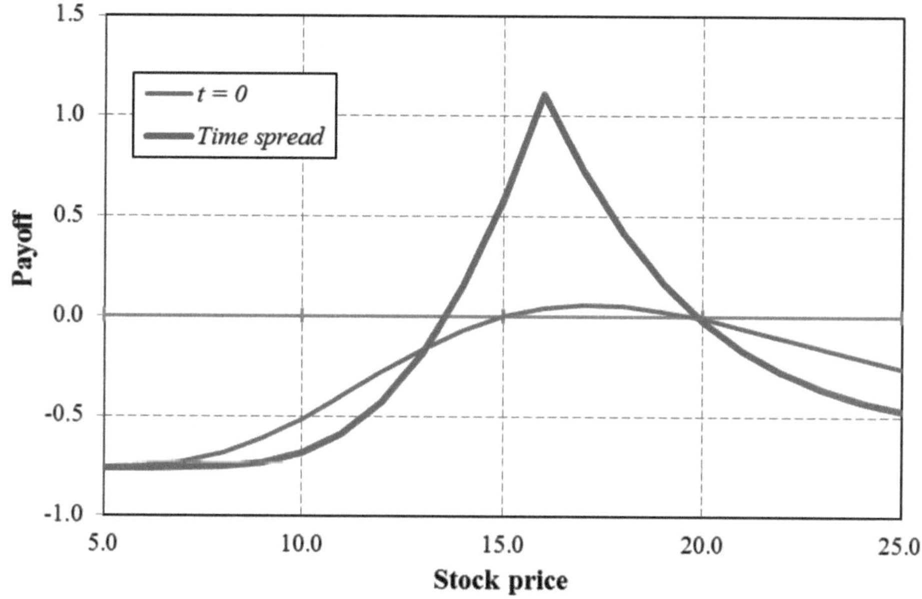

Fig. 8.58 A calendar spread by call options

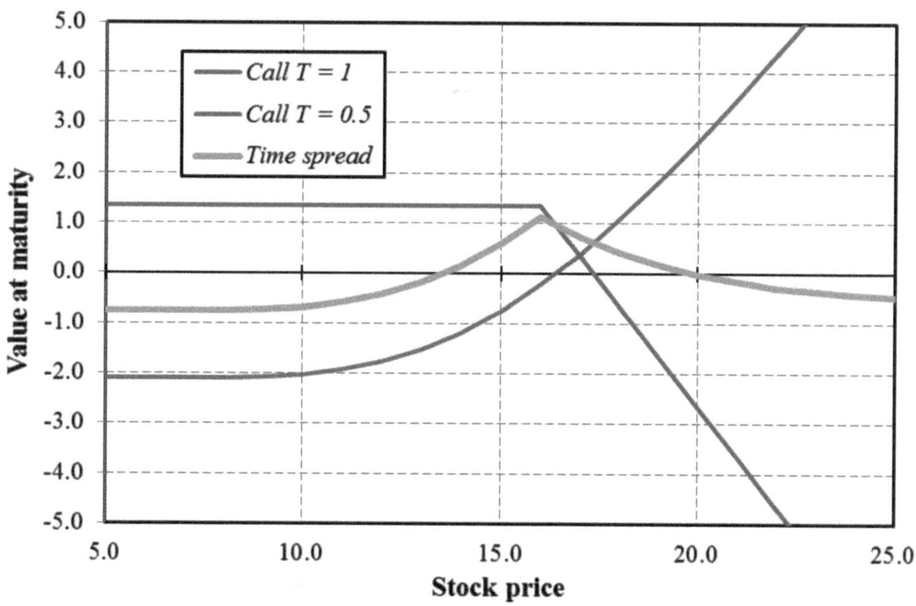

Fig. 8.59 The calendar spread by call options at maturity

Profit
Limited to the strike minus the market price plus the premium received. The maximum profit is the strike price minus the underlying price plus the premium.

Break-Even
Initial value minus the premium.

Losses
Similar to that incurred with ordinary stock ownership, only partially off-set by the option premium received. Main loss could be the opportunity loss if the market rises strongly.

Margin Requirement
Always needed.

Trade Reasons

1. To be compensated for an increase of the underlying.
2. To get an extra yield on a neutral market.

Follow Up
On profit:

- Move the call option to a higher strike price.
- Move the call option to a higher strike price with later maturity.
- Buy more stocks or forwards.

On losses:

- Move the call option to a lower strike price.
- Move the call option to a lower strike price with later maturity.
- Buy a put option.

Comments
This position is a growing asset. As time passes, value of position increases as the option loses its time value. Maximum rate of increasing profits occurs if option is at-the-money.

Warning: As an investor this can be dangerous since the issued option might be exercised, and you might not be allowed to sell the underlying stocks.

8.6 Volatile Markets

8.6.1 Long Straddle [−1 1]

Market Belief
You firmly believe that the stock moves far enough in either direction in the short-term. Buy higher/lower strike options if the position can encounter different probabilities of bullish or bearish movements of the stock; buy ATM options if those probabilities are almost equal.

Construction
Call option and put option are bought with the same strike, usually at-the-money, or buy the underlying and twice as many put options.

Profit
Increases as the stock rises or falls. At expiration, break-even points will be option exercise price +/− prices paid for options. For each point above upside break-even or below downside break-even, profit increases by an additional point.

Losses
Limited to the amount paid for options. Maximum loss realized if stock ends at option exercise. For each point above or below a, loss decreases by additional point.

Break-Even
The strike price minus the premium and the strike price plus the premium.

Margin Requirement
None

Comments
This position is a wasting asset. As time passes, value of position erodes toward expiration value. If volatility increases, erosion slows; if volatility decreases, erosion speeds up.

Trade Reasons

To get an income on a volatile market

Follow Up

On an increase (without ownership):

- Issue a call options with higher strike price and sell the owned put option.
- Sell the forward and the owned put option

On a decrease (without ownership):

- Issue a put option with higher strike price and sell the owned call option.
- Buy the forward and sell the call option

On a neutral market (without ownership):

- Issue a call options with a higher strike and a put options with a lower strike.
- Issue call- and put options with lower time to maturity.

On a profit (with ownership):

- On a decrease: sell half of the put options and buy more stocks on the new and low price.
- On an increase: issue a call option on a higher strike, sell the owned put options and buy half as many new on a higher strike price

On losses (with ownership) example:

- Issue 20 contracts of put options on a lower strike price and 10 contracts of call options to a higher strike than the underlying price.
- Issue 20 contracts of put options with same strike price with later maturity.

In Fig. 8.60 we have created a long strangle by buying a put and a call at strikes 15. The initial stock price is 15, the risk-free interest rate 2 % and the volatility 40 %. Figure 8.61 illustrate the same position at maturity. In Fig. 8.62 we have made the same strategy by holding the underlying and buy twice as many put options.

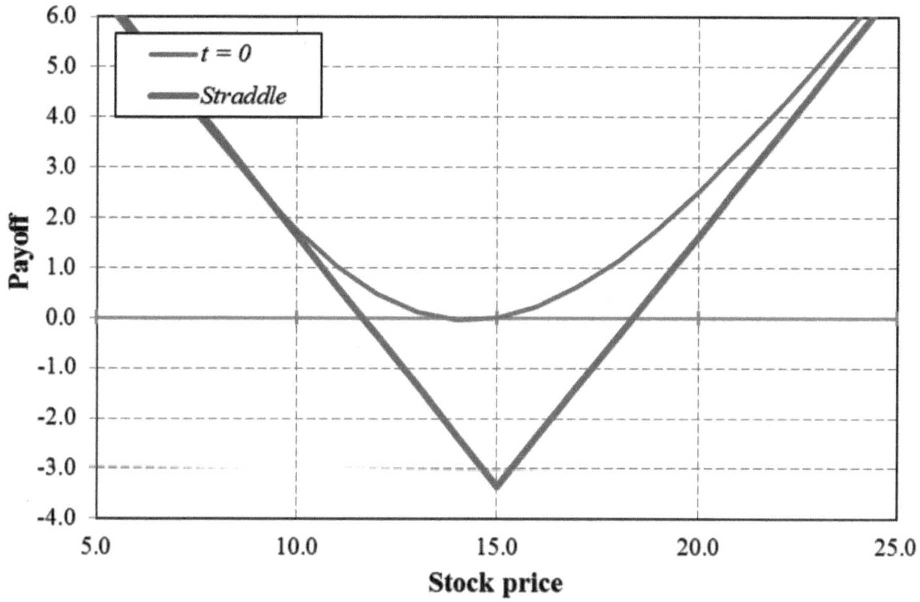

Fig. 8.60 A long straddle made by buying a call and a put at ATM

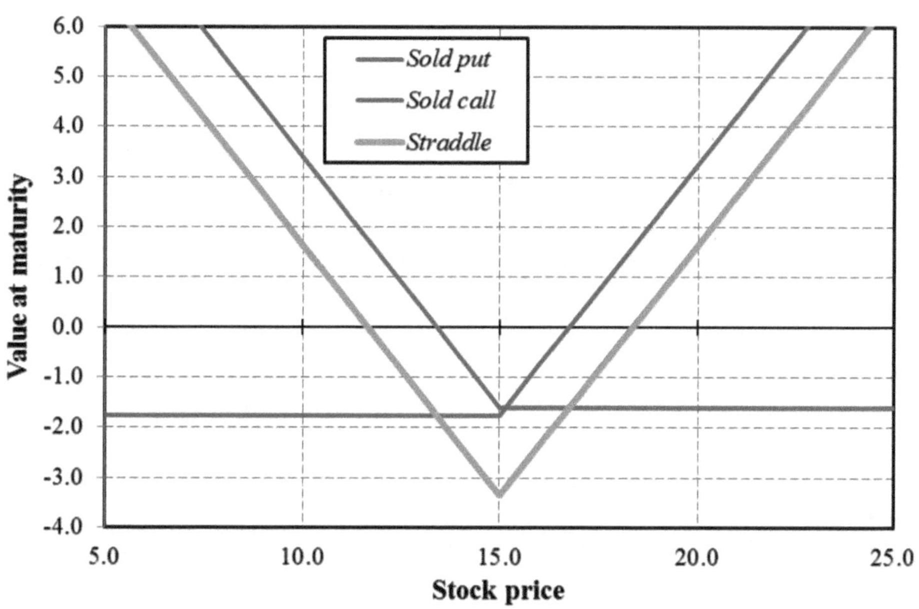

Fig. 8.61 The long straddle at maturity

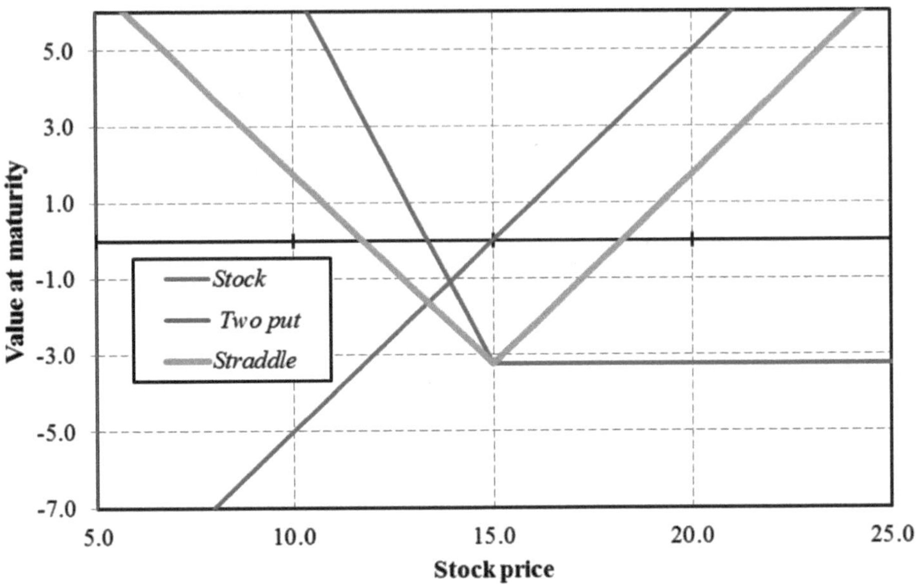

Fig. 8.62 The long straddle constructed by the underlying and two bought put options at maturity

8.6.2 Long Strangle [−1 1 0]

Market Belief

You strongly believe the stock will move far enough from the predefined range. This strategy is similar to the buy straddle but the premium paid here is less. Buy higher/lower strike options if the position can encounter different probabilities of bullish or bearish movements of the stock; buy ATM options if those probabilities are almost equal.

Construction

Put option is bought with a low strike and a call option is bought with a high strike.

Profit

Unlimited and increases as stock rises above the high or falls below the low strike. At expiration, break-even points will be the option exercise price for the low strike minus the prices paid for the options and option exercise price at the high strike plus the prices paid for options. For each point above upside break-even or below downside break-even, the profit increases by an additional point.

Break-Even
The point where the lower strike minus the premium is reached or where the higher strike plus the premium is reached.

Losses
Limited to amount paid for options. Maximum loss realized if stock ends between a and b. For each point above b or below a, loss decreases by additional point.

Margin Requirement
None.

Comments
This position is a wasting asset. As time passes, value of position erodes toward expiration value. If volatility increases, erosion slows; if volatility decreases, erosion speeds up.

Trade Reasons

- To get a profit on a big change in price independent of the direction.
- To get a lower cost than buying a straddle.

Follow Up
On an increase:

- Issue a call option with higher a strike price and sell the owned put option.
- Sell the forward and the owned put option.

On a decrease:

- Issue a put option with a higher strike price and sell the owned call option.
- Buy the forward and the owned call option

On a neutral market:

- Issue a call option with higher strike price and a put option with a lower strike.
- Issue a call- and a put option with lower time to maturity.

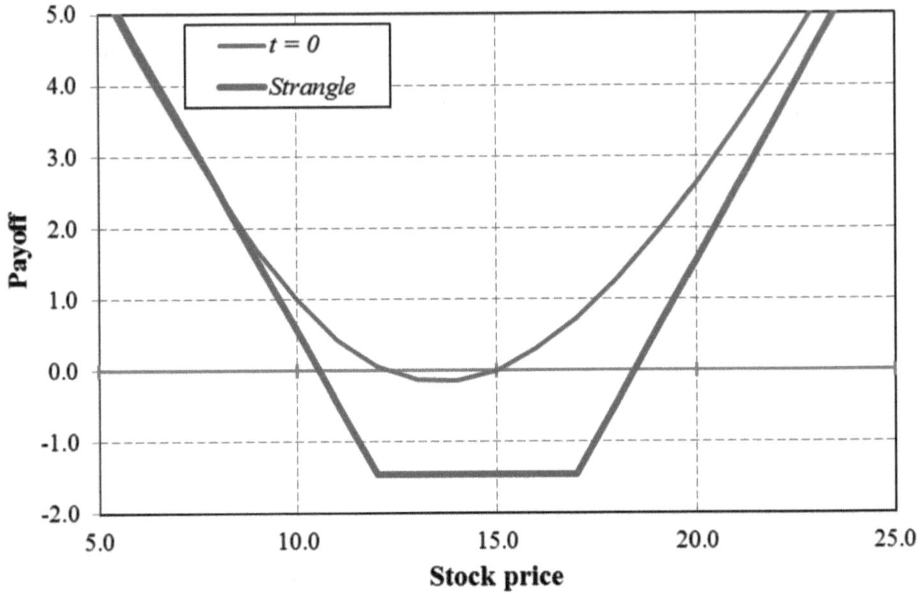

Fig. 8.63 A long strangle made by buying a call option at 17 and a put option at 12 where the ATM price is 15

In Fig. 8.63 we have created a long strangle by buying a put and a call at strikes 12 and 17 respectively. The initial stock price is 15, the risk-free interest rate 2 % and the volatility 40 %. Figure 8.64 illustrate the same position at maturity.

8.6.3 Short Butterfly [0 −1 1 0]

Market Belief
You believe that the stock price will move substantially.

Construction
Call option with low strike sold and two call options with medium strike bought and call option with high strike sold. The same position can be created with puts.

Profit
Limited to the initial credit received.

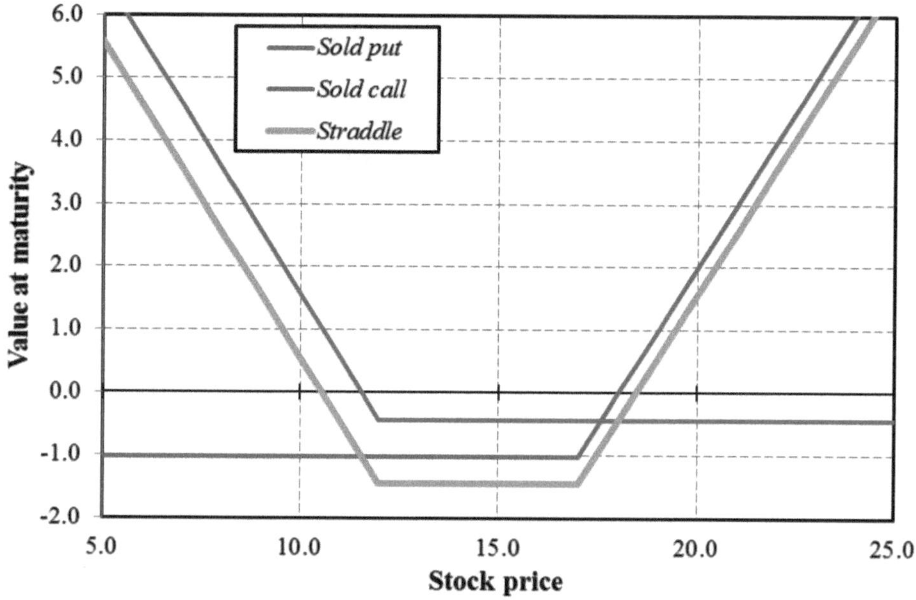

Fig. 8.64 A long strangle at maturity, made by buying a call option at 17 and a put option at 12 where the ATM price is 15

Losses
Limited to the difference between the lower and middle strikes minus the initial spread credit.

Margin Requirement
Low.

Comments
It can be difficult to realize and sell the position on a short time period. This position is a combined asset. As time passes, value of position increases/erodes toward expiration value. If volatility increases, increase/erosion slows; if volatility decreases, increase/erosion speeds up.

Trade Reasons
To get an income on volatile market to a low cost.

In Fig. 8.65 we have created a long butterfly with call options. We made this by selling call options at 12 and 18 and buying two call options ATM, 15. The initial stock price is 15, the risk-free interest rate 2 % and the volatility 40 %. Figure 8.66 illustrate the same position at maturity.

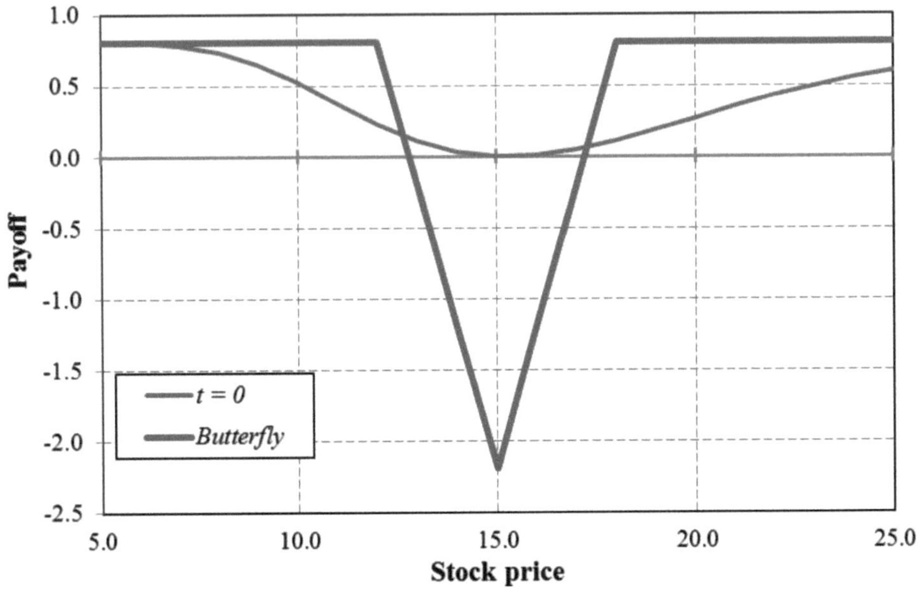

Fig. 8.65 A short butterfly by call options

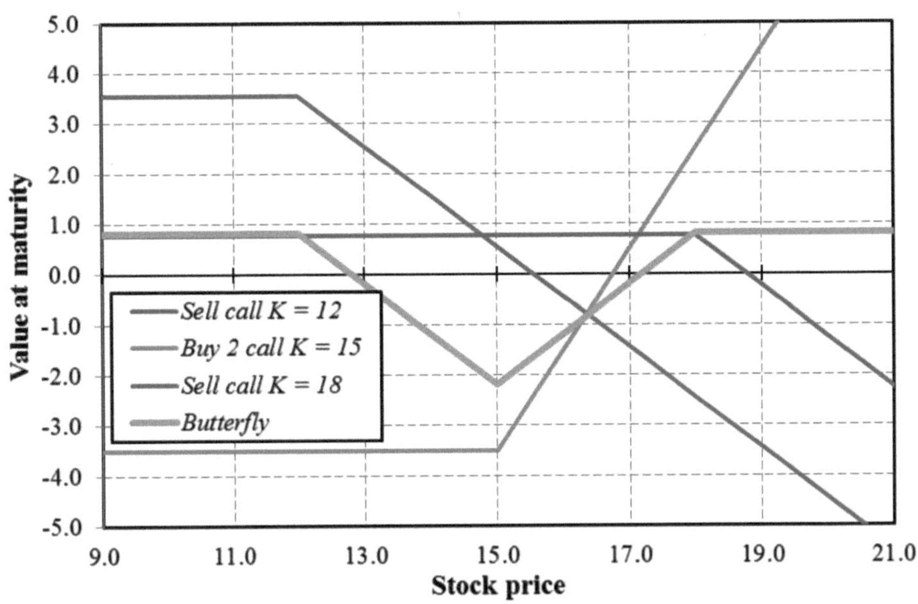

Fig. 8.66 A long butterfly by call options at maturity

8.7 Using Market Indexes in Pricing

As we know, in the Black–Scholes world we have a number of variables (or parameters) that we need to know to be able to calculate prices of stock options. These are:

- the initial stock price,
- the contracted strike price,
- time to maturity,
- the risk-free interest rate and
- the stock (or the implied) volatility.

The most important of these is volatility.

- The strike price is fixed if we don't have some kind of exotic option where the strike might change during the option's lifetime.
- The initial stock price is known, but the future value depends on the changes in the stock price during the lifetime of the option.
- The time to maturity is known if we don't have an exotic option where the maturity time may change.

This leave us with two parameters:

- the risk-free interest rate, that might change during the lifetime and
- volatility.

In the Black–Scholes formulas, the interest rate is used in the calculation of d_1 and d_2 and used to discount the strike price to a present value. Since lifetimes of options most of the time is short, less than a year, the option price will not change very much on changes of the interest rate. This can be seen by simulations or by calculating the rho, ρ i.e., the derivative of the option price with respect to the interest rate r.

This leave us with the most important variable of them all, volatility. The volatility is the variable that will has the greatest impact on the option price, except the underlying price. Also, the volatility is the variable that is the most difficult to estimate.

Exchanges and clearing houses use implied volatility to calculate the margin requirement, not the estimated historical stock volatility that is sometimes used in the Black–Scholes model. The reason is that volatility varies with the strike price. Options with strike prices far from the actual stock value—that is,

OTM or ITM do not have the same liquidity as options with strike prices near ATM.

We have also seen that the volatility is a measure of risk and is used to calculate the sharp ratio or the market price of risk. Therefore we would like to know the market belief on the volatility in general will be in the future. One way to do so is to calculate implied volatilities or study a volatility index, such as the CBOE, Volatility Index® (VIX® Index).

8.7.1 The CBOE VIX Index

The VIX Index was introduced in 1993 by Professor Robert E. Whaley in his paper "Derivatives on Market Volatility: Hedging Tools Long Overdue," in 1993 with two purposes in mind:

- to provide a benchmark of expected short-term market volatility
- to provide an index upon which futures and options contracts on volatility could be written

The VIX Index is computed every 15 seconds throughout the trading day to measure volatility. To compare with historical levels, values were computed back to the beginning of January 1986. This was particularly important since documenting the level of the market during the worst stock market crash since the Great Depression, namely the October 1987 crash. This data can provide useful benchmark information in during market turbulence in the future.

The CBOE also launched trading of VIX futures contracts at the CBOE Futures Exchange (CFE) in 2004 and VIX options at CBOE in 2006.

To understand the VIX, it is important to realize that it is the forward-looking, measure of volatility that the investors expect to see. It is not a backward-looking measure. Conceptually, VIX is like a bond's yield-to-maturity, i.e., a discount rate that equates a bond's price to the present market value when re-investing the coupons at the same rate (the ytm). Yield-to-maturity can also be interpreted as the expected future return of the bond over its remaining lifetime. In a similar manner, VIX is implied by the current prices of S&P 500 index options and represents expected future market volatility over the next 30 calendar days.

The original VIX index was based on the prices of S&P 100 (ticker OEX), not S&P 500 (ticker SPX), index option prices. In their early years, OEX options were the most actively traded index options in the USA Also, the original VIX was based on the prices of only eight ATM index calls and puts. Over the years, index option trading in the USA changed in a fundamental way. The SPX option market later became the most active index. Other factors included that the S&P 500 index is better known, futures contracts on S&P 500 are actively traded, and S&P 500 option contracts are European style (while the options on OEX are American style), which makes them easier to value. Also the trading motives of market participants changed. The index option market became dominated by portfolio insurers, who routinely bought and still buy, OTM and ATM index puts for insurance purposes. During the first ten months of 2008, for example, the average daily volume of SPX puts was over 70 % more than the SPX calls.

In September 2003, the CBOE changed the VIX calculation to account for these fundamental changes. First, they began to use SPX rather than OEX option prices. Second, they began to also include OTM options in the index computation since OTM put prices, in particular, contain important information regarding the demands for portfolio insurance. Including additional option series also helps make the VIX less sensitive to any single option price and hence less sensitive to manipulations.

In Fig. 8.67 we show how the VIX Index (i.e. volatility) changes with respect to the S&P 500 index. The Standard & Poor's 500 Index (S&P 500) is

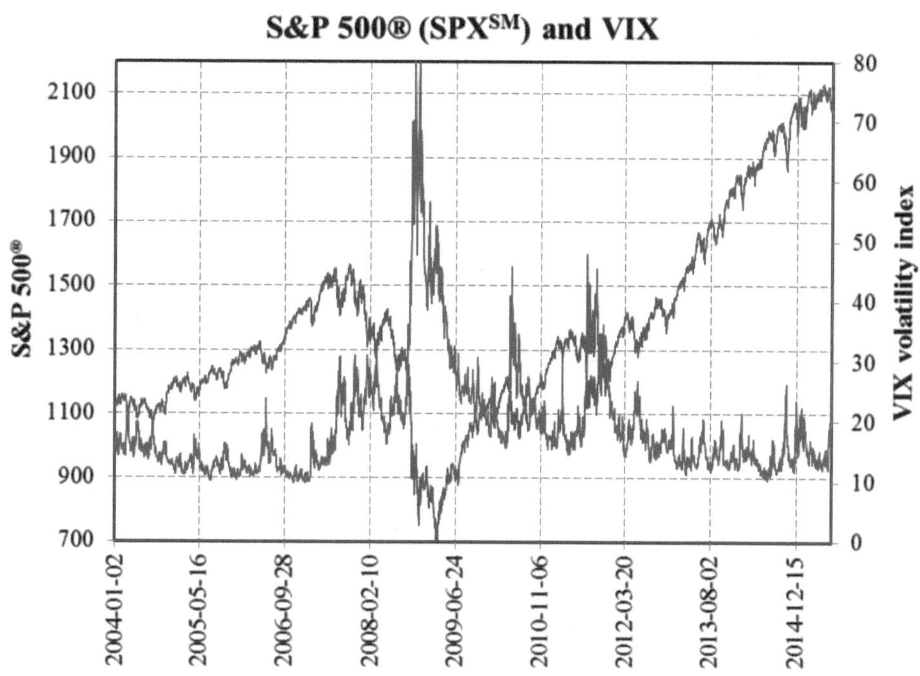

Fig. 8.67 The VIX index and its relationship to the S&P 500 (Source: CBOE)

designed to be a leading indicator of US equities. The index is based upon 500 large companies having common stock listed on the NYSE or NASDAQ. The components and their weightings are determined by S&P Dow Jones Indices. The index is one of the most followed equity indices, and many consider it one of the best representations of the US stock market.

The components of the S&P 500 are selected by a committee. This is similar to the Dow Jones Industrial Average. The committee selects the companies in the S&P 500 so they are representative of the industries in the United States economy. In order to be added to the index, a company must satisfy some liquidity-based size requirements. The securities must be publicly listed on either the NYSE or NASDAQ.

Sometimes known as the "fear gauge," the VIX Index generally stays below 20 in a steady to normal market. When VIX is above 20, it suggests the market is in distress and above 40 it's more like a financial crisis. This is shown in Fig. 8.67 where we can observe the financial crises in 2008–2009, in mid-2010 and during the autumn of 2011.

When the volatility moves up or down, hedgers buy index put options on the S&P 500 index as they are concerned about a potential drop in the market. The more investors demand, the higher the price. VIX is therefore an indicator that reflects the price of insurance on a portfolio.

The VIX index itself has a volatility index called VVIX. This index is an indicator of the expected volatility of a 30-day forward price on VIX. CBOE also calculates a term structure of VVIX for different VIX maturities. The values on VVIX's term structure is calculated from a portfolio of VIX options using a similar algorithm used to calculate the VIX itself. Approximate fair values of VIX futures prices and their standard deviations are derived from the VVIX term structure.

Each VIX can be viewed as a fear gauge for its underlying asset. The CBOE VIX Suite Heat Map highlights the daily variations of these. It uses a color spectrum from green to red to indicate the VIX indices that had the smallest to greatest percentage change from close to close. The latest CBOE VIX® Heat Map can be found at: http://www.cboe.com/micro/vix-and-volatility.aspx

In Fig. 8.68 we show the VVIX Index between 2007 and 2015 and in Fig. 8.69 we show the Volatility Heat Map for 2016-01-22 using the link above.

Beside the VIX and VVIX indexes (and many other volatility indexes), there are also a CBOE Skew Index called the Tail Index. Investors in the US stock market found this index especial interesting after the crash in October 1987 when they changed their view of S&P 500 returns. Investors realized then that

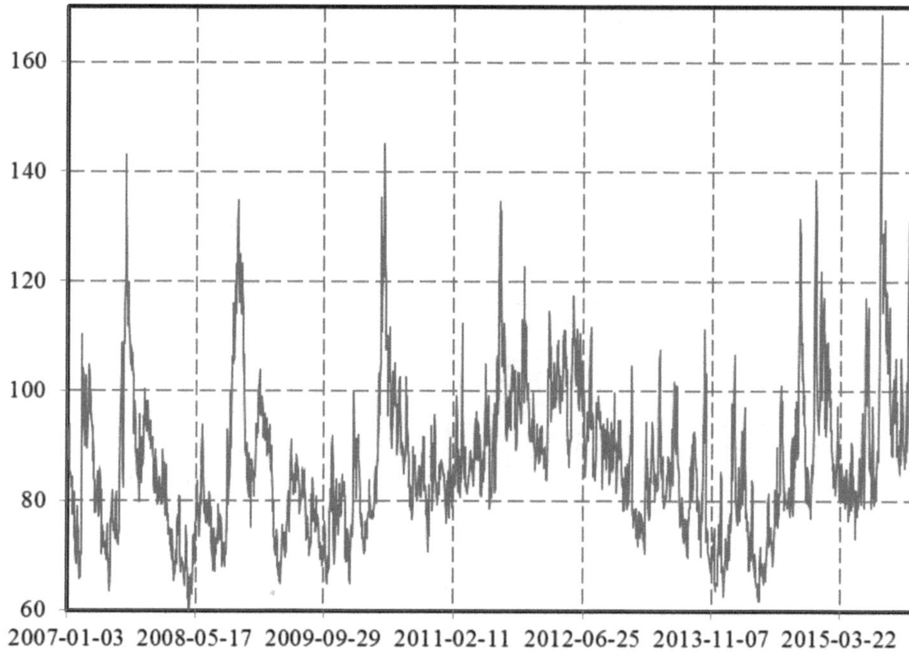

Fig. 8.68 The VVIX index from beginning of 2007 and the end of 2015 (Source: CBOE)

the returns of S&P 500, two or more standard deviations below the mean – is significantly greater than under an ordinary lognormal distribution.

The CBOE SKEW Index is derived from the priced in the left tail of the S&P 500, which explains the term Tail Index The left tail represent losses while the right tail profits. Similar to VIX, the price of S&P 500 tail risk is calculated from the prices of OTM options. SKEW typically ranges from 100 to 150. A SKEW value of 100 means that the perceived distribution of S&P 500 log-returns is normal, and the probability of outlier returns is therefore negligible. As SKEW rises above 100, the left tail of the distribution acquires more weight, and the probabilities of outlier returns become more significant. One can estimate these probabilities from the value of SKEW. Since an increase in perceived tail risk increases the relative demand for low strike puts, increases in SKEW also correspond to an overall steepening of the curve of implied volatilities, familiar to option traders as the volatility skew.

As illustrated in Fig. 8.70, in October 1987 the smile lost its symmetry and become biased towards the put side. The reason why the S&P 500 implied volatilities no longer smiles is the fact that investors began to prize low strike puts more than high strike calls. The standard deviation of returns is then insufficient to characterize risk and the probability of returns two or three

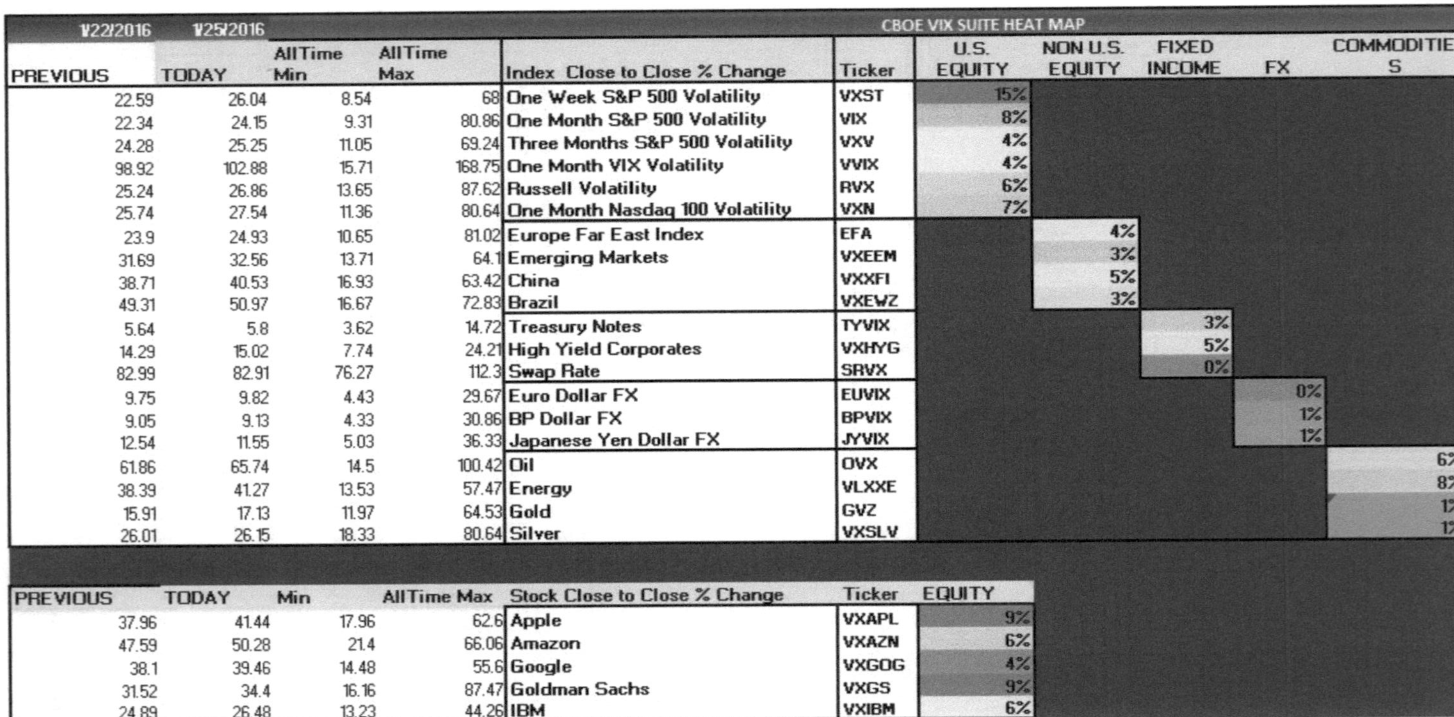

CBOE VIX SUITE HEAT MAP

1/22/2016 1/25/2016

PREVIOUS	TODAY	AllTime Min	AllTime Max	Index Close to Close % Change	Ticker	U.S. EQUITY	NON U.S. EQUITY	FIXED INCOME	FX	COMMODITIES
22.59	26.04	8.54	68	One Week S&P 500 Volatility	VXST	15%				
22.34	24.15	9.31	80.86	One Month S&P 500 Volatility	VIX	8%				
24.28	25.25	11.05	69.24	Three Months S&P 500 Volatility	VXV	4%				
98.92	102.88	15.71	168.75	One Month VIX Volatility	VVIX	4%				
25.24	26.86	13.65	87.62	Russell Volatility	RVX	6%				
25.74	27.54	11.36	80.64	One Month Nasdaq 100 Volatility	VXN	7%				
23.9	24.93	10.65	81.02	Europe Far East Index	EFA		4%			
31.69	32.56	13.71	64.1	Emerging Markets	VXEEM		3%			
38.71	40.53	16.93	63.42	China	VXXFI		5%			
49.31	50.97	16.67	72.83	Brazil	VXEWZ		3%			
5.64	5.8	3.62	14.72	Treasury Notes	TYVIX			3%		
14.29	15.02	7.74	24.21	High Yield Corporates	VXHYG			5%		
82.99	82.91	76.27	112.3	Swap Rate	SRVX			0%		
9.75	9.82	4.43	29.67	Euro Dollar FX	EUVIX				0%	
9.05	9.13	4.33	30.86	BP Dollar FX	BPVIX				1%	
12.54	11.55	5.03	36.33	Japanese Yen Dollar FX	JYVIX				1%	
61.86	65.74	14.5	100.42	Oil	OVX					6%
38.39	41.27	13.53	57.47	Energy	VLXXE					8%
15.91	17.13	11.97	64.53	Gold	GVZ					1%
26.01	26.15	18.33	80.64	Silver	VXSLV					1%

PREVIOUS	TODAY	Min	AllTime Max	Stock Close to Close % Change	Ticker	EQUITY
37.96	41.44	17.96	62.6	Apple	VXAPL	9%
47.59	50.28	21.4	66.06	Amazon	VXAZN	6%
38.1	39.46	14.48	55.6	Google	VXGOG	4%
31.52	34.4	16.16	87.47	Goldman Sachs	VXGS	9%
24.89	26.48	13.23	44.26	IBM	VXIBM	6%

Fig. 8.69 The volatility heat map on January 22, 2016 (Source: CBOE)

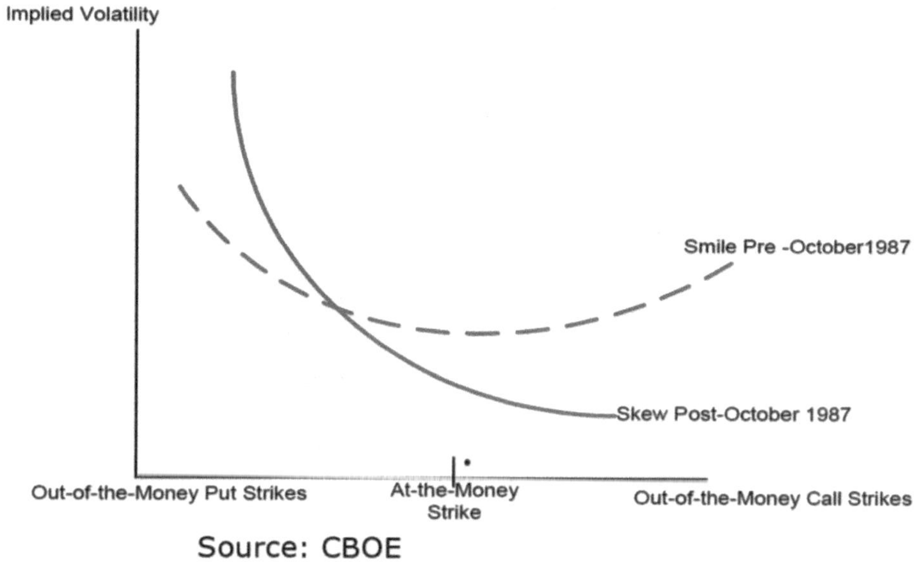

Fig. 8.70 The volatility smile changed shape to a skew in October 1987

standard deviations below the mean is not negligible, as it is under a normal distribution.

Figure 8.71 confirms that the S&P 500 distribution is far from normal. It carries "tail risk" where the frequency of outlier returns is greater than for a normal distribution and the distribution has a negative skew. This means that VIX, as a proxy for the standard deviation of the S&P 500 distribution, may not fully capture the perceived risk of a cash or derivative investment in the S&P 500 or in correlated assets. Similar to VIX, the SKEW is calculated from the price of a tradable portfolio of OTM S&P 500 (SPX^{SM}) options.

The SKEW is derived from the price of S&P 500 skewness. If we denoted by S, the coefficient of statistical skewness:

$$S = E\left[\left(\frac{R - \mu}{\sigma}\right)^3\right]$$

where we calculate the present value if the 30-day log-return of the S&P 500 minus the mean divided by the volatility. Since S tends to be negative, it is inconvenient to use it as an index. S is therefore transformed to SKEW by SKEW $= 100 - 10S$. With this definition, SKEW increases as S becomes more negative and the tail risk increases.

The SKEW between 1990 and 2010 is shown in Fig. 8.72.

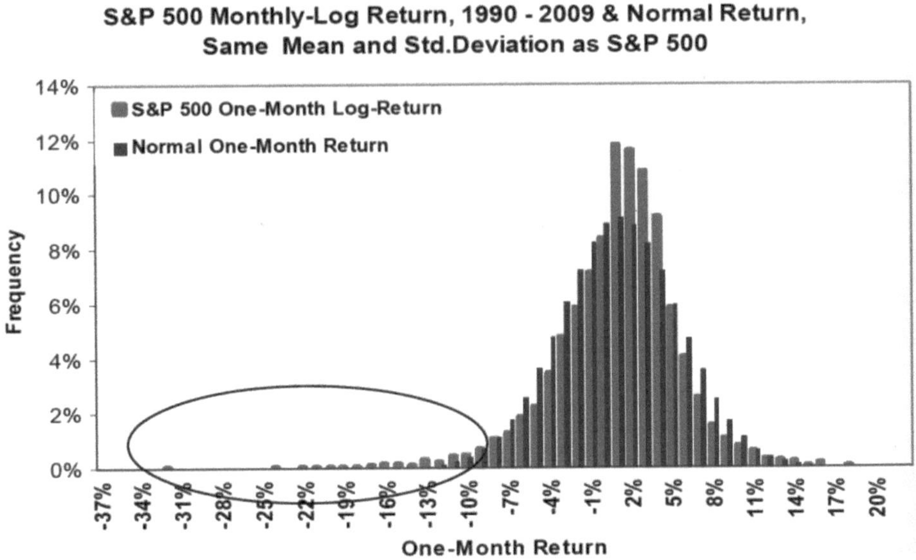

Fig. 8.71 The log return of S&P 500 shows a fat tail that can't be modelled with a normal distribution

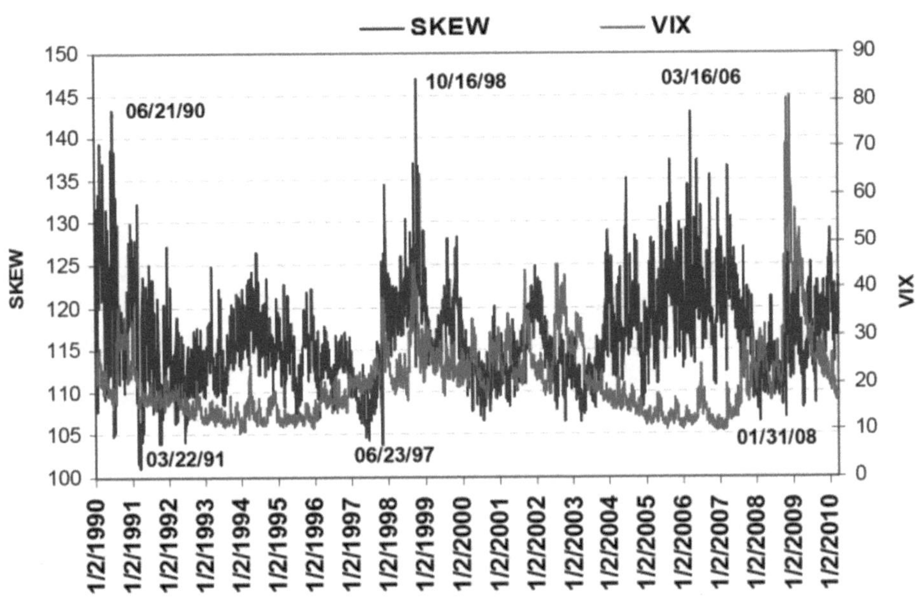

Fig. 8.72 The CBOE SKEW index for a period of 20 years

By using data such as the indexes above, traders can study the market view of volatility, the most important parameter to value derivative contracts on the stock market. The VIX calculation, step by step can be found in the White Paper: The CBOE Volatility Index – VIX at: https://www.cboe.com/micro/vix/part2.aspx.

8.8 Price Direction Matrix

	Call +/–	Put +/–
The value of the underlying decreases	+	–
The value of the underlying increases	–	+
Higher strike prices compared with a lower	–	+
Low strike prices compared with a higher	+	–
Long time to maturity	+	+
Short time to maturity	–	–
High volatility	+	+
Low volatility	–	–
High risk-free interest rate	+	–
Low risk-free interest rate	–	+
Dividends	–	+

As we can see above, the value of a long call option decrease if the interest rate increases. The reason is the following; Say that if we buy the option because we will buy the underlying in the future. Then we only have to pay an initial cost for the option and we can put the rest of our money at the bank. If the interest increases, then we get a better yield on the money on the bank. Therefore, the total position increase in value and so do the option.

8.9 Strategy Matrix

	Positive market	Neutral market	Negative market
Increasing volatility	Buy call option Positive price-spread Back spread Three leg position Protective put	Long straddle Long strangle Short neutral time-spread	Long put option Negative price-spread Back spread Three leg position
Neutral volatility	Buy underl./forward Buy synthetic forward Buy sloped synthetic Forward	DON'T TRADE	Short forward Short synthetic forward Short sloped synthetic forward

(continued)

	Positive market	Neutral market	Negative market
Decreasing volatility	Issue put options Positive price-spread Covered call Three leg position Ratio spread	Short straddle Short strangle Long neutral time-spread	Issue call option Negative price-spread Three leg position Ratio spread

Appendix: Some Source Codes

In VBA, (Visual Basic for Application) the code for a European call option with the binomial model can be written as:

```
Function eCall(S, K, T, r, sigma, n)
  u     = Exp(sigma*((T/n)^0.5))
  d     = 1/u
  r     = Exp(r*(T/n))
  rp_u  = (r - d)/(r*(u - d))
  rp_d  = 1/r - rp_u
  eCall = 0

  For i = 0 To n
  eCall = eCall + Application.Combin(n, i)*rp_u^i*rp_d^(n - i)* _
                 Application.Max(S*u^i*d^(n - i) - K, 0)
  Next i
End Function
```

A more general binomial model, which also handles American-type options, is given below. This code is written in VBA and handles some of the binomial models such as Cox–Ross–Rubenstein (with and without Black–Scholes smoothing) and the Leisen–Reimer model. The model also calculates the Greeks, delta, gamma and theta. To calculate rho and vega, two trees must be made with two different interest rates and volatilities, respectively

J.R.M. Röman, *Analytical Finance: Volume I*,
DOI 10.1007/978-3-319-34027-2

```
' Input mBin = Binomial model
'       mExe = Put or Call (PUT_ or CALL_)
'       mTyp = American or European

Function Binom(mBin As Long, mTyp As Long, mExe As Long, S As Double,
X As Double, T As Double, r As Double, v As Double, n As Long,
mDelta As Double, mGamma As Double, mTheta As Double)
  Dim d1 As Double, d2 As Double, a As Double, b As Double
  Dim aa As Double, vv As Double, U As Double, D As Double
  Dim udd As Double, u_d As Double, p As Double, Dt As Double
  Dim disc As Double, pe As Double, my As Double, m As Double
  Dim q As Double, ermqdt As Double, pdash As Double, ans As Double
  Dim idx As Long, i As Long, j As Long
  Dim lSt1(0 To 2) As Double
  Dim lC1(0 To 2) As Double
  ReDim lSt(0 To n + 2) As Double
  ReDim lC(0 To n + 2) As Double

  d1 = (Log(S / X) + (r + v * v / 2) * T) / (v * Sqr(T))
  d2 = d1 - v * Sqr(T)
  Dt = T / n
  disc = Exp(-r * Dt)

  If (mBin = CRR) Then
    U = Exp(v * Sqr(Dt))
    D = 1 / U
    udd = U / D
    u_d = U - D
    p = (1 / disc - D) / u_d
  ElseIf (mBin = PEG) Then
    pe = (Log(X / S)) / n
    U = Exp(pe + v * Sqr(Dt))
    D = Exp(pe - v * Sqr(Dt))
    udd = U / D
    u_d = U - D
    p = (1 / disc - D) / u_d
  ElseIf (mBin = JR) Then
    ' Remark pu = pd = 0.5!
    ' In this method the calc. of delta and gamma (rho+) is more
    ' difficult since S0 <> S0*u*d. Use ORC method in this case!
    my = r - 0.5 * v * v
    U = Exp(my * Dt + v * Sqr(Dt))
    D = Exp(my * Dt - v * Sqr(Dt))
    p = 0.5
    udd = U / D
    u_d = U - D
```

```
ElseIf (mBin = TIAN) Then
  m = Exp(r * Dt)
  vv = Exp(v * v * Dt)
  U = 0.5 * m * vv * (1 + vv + Sqr(vv * vv + 2 * vv - 3))
  D = 0.5 * m * vv * (1 + vv - Sqr(vv * vv + 2 * vv - 3))
  udd = U / D
  u_d = U - D
  p = (1 / disc - D) / u_d
ElseIf (mBin = LR) Then
  q = 0      ' No yield
  ermqdt = Exp((r - q) * Dt)
  d2 = BSDTwo(S, X, r, q, T, v)
  p = PPNormInv(d2, n)
  pdash = PPNormInv(d2 + v * Sqr(T), n)
  U = ermqdt * pdash / p
  D = ermqdt * (1 - pdash) / (1 - p)
  udd = U / D
  u_d = U - D
End If

' initialize stock prices at maturity  log(a^b) = b*log(a)
lSt(0) = S * D ^ n

For i = 1 To n
  lSt(i) = lSt(i - 1) * udd
Next i

' initialize option prices at maturity n
For i = 0 To n
  lC(i) = Max(0, bC(mExe, lSt(i), X))
Next i

idx = find_opt_index(lSt, X, n)

' step back through the tree
For j = n - 1 To 2 Step -1
  For i = 0 To j
    lC(i) = disc * (p * lC(i + 1) + (1 - p) * lC(i))
    lSt(i) = lSt(i) / D
    If (mTyp = AMERICAN) Then
      lC(i) = Max(lC(i), bC(mExe, lSt(i), X))
    End If
    If (BS_Smoothing) Then
      If (j = n - 1) Then
        If (i >= idx And i <= idx + 4) Then
```

```
          lC(i) = GBlackScholes(mExe, lSt(i), X, T / n, r, r, v)
        End If
      End If
    End If
  Next i
Next j

' save option value, used for calculating hedge parameters
mTheta = lC(1) / Dt ' For calculating Theta.

For i = 0 To 1
  lC1(i) = disc * (p * lC(i + 1) + (1 - p) * lC(i))
  lSt1(i) = lSt(i) / D

  If (mTyp = AMERICAN) Then
    lC1(i) = Max(lC1(i), bC(mExe, lSt1(i), X))
  End If
Next i

ans = disc * (p * lC1(1) + (1 - p) * lC1(0))
If (mTyp = AMERICAN) Then ans = Max(ans, bC(mExe, S, X))

mGamma = ((lC(2) - lC(1)) / (lSt(2) - lSt(1)) - _
    (lC(1) - lC(0)) / (lSt(1) - lSt(0))) / (0.5 * (lSt(2) - lSt(0)))
mDelta = (lC1(1) - lC1(0)) / (lSt1(1) - lSt1(0))
mTheta = (mTheta - ans / Dt) / 2
If (mBin = TIAN) Then mTheta = mTheta / 2
Binom = ans
End Function
```

To find the index closest to the strike for the Black–Scholes smoothing, we use the following function:

```
Function find_opt_index(lSt() As Double, mX As Double, n As Long)
  Dim idx As Long, i As Long

  idx = 0

  For i = 0 To n - 1
    If (lSt(i) >= mX) Then Exit For
    idx = idx + 1
  Next i
  find_opt_index = idx - 2
End Function
```

The boundary for call and put, is given by:

```
Function bC(aIsCall As Long, S As Double, X As Double) As Double
  If aIsCall Then
    bC = S - X
  Else
    bC = X - S
  End If
End Function
```

We also use two functions from the Black–Scholes model:

```
Function BSDTwo(S As Double, X As Double, r As Double, q As Double, _
                T As Double, v As Double) As Double
  BSDTwo = (Log(S / X) + (r - q - 0.5 * v * v) * T) / (v * Sqr(T))
End Function
  double BSDTwo(double S, double X, double r, double q, double T,

Function BlackScholes(mCall As Long, S As Double, X As Double, _
                      T As Double, r As Double, b As Double, _
                      v As Double) As Double
  Dim d1 As Double, d2 As Double

  d1 = (Log(S / X) + (b + v * v / 2) * T) / (v * Sqr(T))
  d2 = d1 - v * Sqr(T)

  If (mCall) Then
    BlackScholes = S*Exp((b - r) * T)*CND(d1) - X*Exp(-r*T)*CND(d2)
  Else
    BlackScholes = X*Exp(-r*T)*CND(-d2) - S*Exp((b - r)*T)*CND(-d1)
  End If
End Function
```

In the Leisen–Reimer model we use the Peizer–Pratt inversion formula:

```
Function PPNormInv(z As Double, n As Long) As Double
  Dim c1 As Double

  n = 2 * Int(n / 2) + 1 ' == odd(n);
  c1 = Exp(-((z / (n + 1 / 3 + 0.1 / (n + 1))) * _
           (z / (n + 1 / 3 + 0.1 / (n + 1)))) * (n + 1 / 6))
  PPNormInv = 0.5 + Sgn(z) * Sqr((0.25 * (1 - c1)))
End Function
```

The function CND to calculate the cumulative normal eistribution is given in VBA below. This function below gives the normal distribution function with a maximum error of 10^{-8}:

```
do Function CND(X As Double) As Double
  Dim sign As Long
  Dim x2, q0, q1, q2 As Double

  If (X < 0) Then
    X = -X
    sign = -1
  ElseIf (X > 0) Then
    sign = 1
  Else ' (x = 0.0)
    CND = 0.5
    Exit Function
  End If

  If (X > 20) Then
    If (sign < 0) Then
      CND = 0
    Else
      CND = 1
    End If
    Exit Function
  End If

  X = X*0.707106781186547
  x2 = X*X
  If (X < 0.46875) Then
    q1 = 3209.37758913847 + x2*(377.485237685302 + x2* _
        (113.86415415105 + x2*(3.16112374387057 + x2* _
        0.185777706184603)))

    q2 = 2844.23683343917 + x2*(1282.61652607737 + x2* _
        (244.024637934444 + x2*(23.6012909523441 + x2)))
    CND = 0.5*(1 + sign*X*q1/q2)
  ElseIf (X < 4) Then
    q1 = X*(8.88314979438838 + X*(0.56418849698867 + X* _
        2.15311535474404E-08))
    q1 = X*(881.952221241769 + X*(298.6351381974 + X* _
        (66.1191906371416 + q1)))
    q1 = 1230.339354798 + X*(2051.07837782607 + X*(1712.04761263407 _
      + q1))
```

```
    q2 = X*(117.693950891312 + X * (15.7449261107098 + X))
    q2 = X*(3290.79923573346 + X*(1621.38957456669 + X* _
          (537.18110186201 + q2)))
    q2 = 1230.33935480375 + X*(3439.36767414372 + X* _
        (4362.61909014325 + q2))

    CND = 0.5*(1 + sign*(1 - Exp(-x2)*q1/q2))
  Else
    q0 = 1/x2
    q1 = 6.58749161529838E-04 + q0*(1.60837851487423E-02 + q0* _
        (0.125781726111229 + q0*(0.360344899949804 + q0* _
        (0.305326634961232 + q0*1.63153871373021E-02))))

    q2 = 2.33520497626869E-03 + q0*(6.05183413124413E-02 + q0* _
        (0.527905102951428 + q0*(1.87295284992346 + q0*´
        (2.56852019228982 + q0))))
    CND = 0.5*(1 + sign*(1 - Exp(-x2)/X*(0.564189583547756 - _
    q0*q1/q2)))
  End If
End Function

Function nd(X As Double) As Double
  nd = 1/Sqr(2*3.141592654)*Exp(-X*X/2)
End Function
```

Below we give an example of how to write a C++ program to solve option prices with the Crank–Nicholson method. The model also calculates the Greeks, delta, gamma and theta. To calculate rho and vega, two grids must be made with two different interest rates and volatilities, respectively, The function bC is the same as for the binomial model.

```
double CrNi(double S, double K, double T, double sig, double r,
            int N, int Nj,  double div, int mIsCall,
            int mIsAmerican, double D, double &mDelta,
            double &mGamma, double &mTheta)
{
  double dt  = T/N;
  double dx  = sig*sqrt(D*dt);
  double nu  = r - div - 0.5*sig*sig;
  double edx = exp(dx);
  double pu  = -0.25*dt*((sig/dx)*(sig/dx) + nu/dx);
  double pm  = 1.0 + 0.5*dt*(sig/dx)*(sig/dx) + 0.5*r*dt;
```

```
  double pd  = -0.25*dt*((sig/dx)*(sig/dx) - nu/dx);
  double *St, *C[2], lambda_U, lambda_L;

  St   = new double [2*Nj + 3];
  C[0] = new double [2*Nj + 3];
  C[1] = new double [2*Nj + 3];

  // Initialize the asset prices at maturity.
  St[0] = S*exp(-Nj*dx);
  for (int j = 1; j <= 2*Nj; j++) {
    St[j] = edx*St[j-1];
}

  // Initialize the option values at maturity.
  for (int j = 0; j <= 2*Nj; j++)
    C[0][j] = max(0.0, bC(mIsCall, St[j], K));

  // Compute derivative boundary condition
  if (mIsCall) {
    lambda_U = St[2*Nj] - St[2*Nj-1];
    lambda_L = 0.0;
}
  else {
    lambda_L = -(St[1] - St[0]);
    lambda_U = 0.0;
}

   // Step backwards through the lattice
   for (int i = N - 1; i >= 0; i--) {
     solve_CN(C, pu, pm, pd, lambda_L, lambda_U, Nj);
     if (i == 0) mTheta = (C[0][Nj] - C[1][Nj])/dt;

     // Apply early exercise condition
     for (int j = 0; j <= 2*Nj; j++) {
       if (mIsAmerican)
         C[0][j] = max(C[1][j], bC(mIsCall, St[j], K));
       else
         C[0][j] = C[1][j];
}
}
```

```
  mDelta = (C[1][Nj+1]  - C[1][Nj-1])/(St[Nj+1] - St[Nj-1]);
  mGamma = ((C[0][Nj+2] - C[0][Nj]) /(St[Nj+2] - St[Nj]) -
           (C[0][Nj]   - C[0][Nj-2])/(St[Nj]   - St[Nj-2]))/
           (0.5*(St[Nj+2] - St[Nj-2]));

  return C[0][Nj];
}

double solve_CN(double **C, double pu, double pm, double pd,
                double lambda_L, double lambda_U, int Nj)
 {
   double *pmp; // = array();
   double *pp;  // = array();

   pp  = new double [2*Nj + 3];
   pmp = new double [2*Nj + 3];

   // Substitute boundary condition at j = -Nj into j = -Nj + 1
   pmp[1] = pm + pd;
   pp [1] = -pu*C[0][2] - (pm - 2.0)*C[0][1] -
            pd*C[0][0] + pd*lambda_L;

   // Eliminate the upper diagonal
   for (int j = 2; j < 2*Nj; j++) {
     pmp[j] = pm - pu*pd/pmp[j-1];
     pp [j] = - pu*C[0][j+1] - (pm - 2.0)*C[0][j]
             - pd*C[0][j-1] - pp[j-1]*pd/pmp[j-1];
 }

   // Use boundary condition at j = Nj and equation at j = Nj - 1
   C[1][2*Nj] = (pp[2*Nj-1] + pmp[2*Nj-1]*lambda_U)/
              (pu + pmp[2*Nj-1]);
   C[1][2*Nj-1] = C[1][2*Nj]  - lambda_U;

   // Back substitution
   for (int j = 2*Nj - 2; j >= 1; j--)
     C[1][j] = (pp[j] - pu*C[1][j+1])/pmp[j];

   C[1][0] = C[1][1] - lambda_L;
   delete [] pp;
   delete [] pmp;
}
```

A VBA code for the Black–Scholes model (i.e., with continuous dividends) and the Greeks are given below:

```
Function BlackScholes(mCall As Integer, s As Double, x As Double, _
                      T As Double, r As Double, b As Double, _
                      v As Double) As Double
  Dim d1 As Double, d2 As Double

  d1 = (Log(s/x) + (b + v*v/2)* T)/(v*Sqr(T))
  d2 = d1 - v * Sqr(T)

  if (mCall) Then
    if (T > 0.0)
      BlackScholes = s*Exp((b - r)*T)*CND(d1)
                   - x*Exp(-r * T)*CND(d2)
    Else
      BlackScholes = WorksheetFunction.Max(s - x, 0)
  Else
    if (T > 0.0)
      BlackScholes = x*Exp(-r*T)*CND(-d2)
                   - s*Exp((b - r)*T)*CND(-d1)
    Else
      BlackScholes = WorksheetFunction.Max(x - s, 0)
End Function

Function Delta(mCall As Integer, s As Double, x As Double, _
               T As Double, r As Double, b As Double, _
               v As Double) As Double
  Dim d1 As Double

  If (T > 0) Then
    d1 = (Log(s/x) + (b + v*v/2)*T)/(v*Sqr(T))
  Else
    d1 = (Log(s/x) + (b + v*v/2)*T)/(v*Sqr(T))

  If (mCall) Then
    Delta = Exp((b - r)*T)*CND(d1)
  Else
    Delta = Exp((b - r)*T)*(CND(d1) - 1)
  End If
End Function

Function Gamma(s As Double, x As Double, T As Double, _
               r As Double, b As Double, v As Double) As Double
  Dim d1 As Double

  d1 = (Log(s/x) + (b + v*v/2)*T)/(v*Sqr(T))
  Gamma = nd(d1)*Exp((b - r)*T)/(s*v*Sqr(T))
End Function
```

```
Function Vega(s As Double, x As Double, T As Double, r As Double,_
              b As Double, v As Double) As Double
  Dim d1 As Double

  d1 = (Log(s/x) + (b + v*v/2)*T)/(v*Sqr(T))
  Vega = s*Exp((b - r)*T)*nd(d1)*Sqr(T)
End Function

Function Theta(mCall As Integer, s As Double, x As Double, _
               T As Double, r As Double, b As Double, _
               v As Double) As Double
  Dim d1 As Double
  Dim d2 As Double

  d1 = (Log(s/x) + (b + v*v/2)*T)/(v*Sqr(T))
  d2 = d1 - v*Sqr(T)

  If (mCall) Then
    Theta = -s*Exp((b - r)*T)*nd(d1)*v/(2*Sqr(T)) - _
      (b - r)*s*Exp((b - r)*T)*CND(d1) - r*x*Exp(-r*T)*CND(d2)
  Else
    Theta = -s*Exp((r - b)*T)*nd(d1)*v/(2*Sqr(T)) + _
      (b - r)*s*Exp((b - r)*T)*CND(-d1) + r*x*Exp(-r*T)*CND(-d2)
  End If
End Function

Function Rho(mCall As Integer, s As Double, x As Double, _
             T As Double, r As Double, b As Double, _
             v As Double) As Double
  Dim d1 As Double
  Dim d2 As Double

  d1 = (Log(s/x) + (b + v*v/2)*T)/(v*Sqr(T))
  d2 = d1 - v*Sqr(T)

  If (mCall) Then
    If (b = 0) Then
      Rho = -T*BlackScholes(1, s, x, T, r, b, v)
    Else
      Rho = T*x*Exp(-r*T)*CND(d2)
    End If
  Else
    If (b = 0) Then
      Rho = -T*BlackScholes(0, s, x, T, r, b, v)
    Else
```

```
      Rho = -T*x*Exp(-r*T)*CND(-d2)
    End If
  End If
End Function
```

A VBA implementation of the RollGeske–Whaley model is given below

```
Public Const BIGNUM As Long = 100000000
Public Const EPSILON As Double = 0.00001

Function RollGeskeWhaley(s As Double, x As Double, v As Double, _
                        r As Double, T As Double, D As Double, _
                        TD As Double) As Double
  Dim SX As Double, ci As Double, HighS As Double, LowS As Double
  Dim a1 As Double, a2 As Double, b1 As Double, b2 As Double,
  Dim c As Double, i As Double

  SX = s - D * Exp(-r * TD)

  ' Not optimal to exercise.....
  If (D <= x*(1 - Exp(-r*(T - TD)))) Then
    RollGeskeWhaley = BlackScholes(mCall, SX, x, T, r, r, v)
    Exit Function
  End If
  ci = BlackScholes(mCall, SX, x, T - TD, r, r, v)
  HighS = s

  Do While ((ci - HighS - D + x) > 0 And (HighS < BIGNUM))
    HighS = 2*HighS
    ci = BlackScholes(mCall, HighS, x, T - TD, r, r, v)
  Loop

  If (HighS < BIGNUM) Then
    RollGeskeWhaley = BlackScholes(mCall, SX, x, T, r, r, v)
    Exit Function
  End If

  LowS = 0
  i = HighS * 0.5
  ci = BlackScholes(mCall, i, x, T - TD, r, r, v)

  ' Find the critical Stock Price with Bisection.
  Do While (Abs(ci - i - D + x) > EPSILON And _
          (HighS - LowS) > EPSILON)
```

```
    If ((ci - i - D + x) < 0) Then
      HighS = i
    Else
      LowS = i
    End If

    i = (HighS + LowS)/2
    ci = BlackScholes(mCall, i, x, T - TD, r, r, v)
  Loop

  a1 = (Log(SX/x) + (r + v*v/2)*T)/(v*Sqr(T))
  a2 = a1 - v*Sqr(T)
  b1 = (Log(SX/i) + (r + v*v/2)*TD)/(v*Sqr(TD))
  b2 = b1 - v*Sqr(TD)

  c = SX*(CND(b1) + M(a1, -b1, Sqr(TD / T))) - _
     x*Exp(-r*T)*M(a2, -b2, -Sqr(TD/T)) - _
     (x - D)*Exp(-r*TD)*CND(b2)
  RollGeskeWhaley = c
End Function
```

Where the function for the bivariate normal distribution is given below. The bivariate normal distribution is defined as:

$$M(x,y,\lambda) = \frac{1}{2\pi\sigma_x\sigma_y\sqrt{1-\lambda^2}}\exp\left\{\frac{(x-\mu_x)^2}{\sigma_x^2} - \frac{2\lambda(x-\mu_x)(y-\mu_y)}{\sigma_x\sigma_y} + \frac{(y-\mu_y)^2}{\sigma_y^2}\right\},$$

where μ_x and μ_y is the mean of x and y, σ_x and σ_y the standard deviations. The correlation function λ is defined by

$$\lambda = \mathrm{corr}(x,y) = \frac{\sigma_{xy}}{\sigma_x\sigma_y}.$$

Hereby, the probability measure $p(x, y)$ is given by

$$p(x,y) = \int_{-\infty}^{x}\int_{-\infty}^{y} f(\xi,\zeta,\lambda)d\xi d\zeta.$$

A VBA implementation is given below

```
' Calculates the Cumulative Probability in Bivariate Normal dist.
' with |e(x)| < 1.0e-7

 Function M(a As Double, b As Double, c As Double) As Double
   Dim sa As Double, sb As Double, r1 As Double, r2 As Double
   Dim D As Double

   sa = 1
   sb = 1

   If (a <= 0 And b <= 0 And c <= 0) Then
     M = MM(a, b, c)
   ElseIf (a*b*c <= 0) Then
     If (b > 0 And c > 0) Then
       M = CND(a) - MM(a, -b, -c)
     ElseIf (a > 0 And c > 0) Then
       M = CND(b) - MM(-a, b, -c)
     ElseIf (a > 0 And b > 0) Then
       M = CND(a) + CND(b) - 1 - MM(-a, -b, c)
     End If
   Else
     If (a < 0) Then sa = -1
     If (b < 0) Then sb = -1
     r1 = (c*a - b)*sa/Sqr(a*a - 2*a*b*c + b*b)
     r2 = (c*b - a)*sb/Sqr(a*a - 2*a*b*c + b*b)
     D = (1 - sa*sb)/4
     M = MM(a, 0, r1) + MM(b, 0, r2) - D
   End If
 End Function

Function MM(a As Double, b As Double, c As Double) As Double
  Dim aa As Double, bb As Double
  Dim i As Integer, j As Integer
  Dim am(4) As Double
  Dim bm(4) As Double

  am(0) = 0.325303
  am(1) = 0.4211071
  am(2) = 0.1334425
  am(3) = 0.006374323

  bm(0) = 0.1337764
  bm(1) = 0.6243247
  bm(2) = 1.3425378
  bm(3) = 2.2626645
```

```
  MM = 0
  aa = a/Sqr(2*(1 - c*c))
  bb = b/Sqr(2*(1 - c*c))

  For i = 0 To 4
    For j = 0 To 4
      MM = MM + am(i)*am(j)*f(bm(i), bm(j), aa, bb, c)
    Next j
  Next i
  MM = MM*Sqr(1 - c*c)/3.121592654
End Function

Function f(x As Double, y As Double, a As Double, b As Double, _
          c As Double) As Double
  f = Exp(a*(2*x - a) + b*(2*y - b) + 2*c*(x - a)*(y - b))
End Function
```

A VBA implementation of Barone–Adesi–Whaley is given below:

```
Function BaroneAdesiWhaleyCall(s As Double, x As Double, _
                               v As Double, r As Double, _
                               T As Double, b As Double) As Double
  Dim sk As Double, n As Double, K As Double, d1 As Double
  Dim q2 As Double, a2 As Double

  If (b >= r) Then
    BaroneAdesiWhaleyCall = BlackScholes(mCall, s, x, T, r, b, v)
    Exit Function
  End If

  sk = kc(x, T, r, b, v)
  n = 2*b/(v*v)
  K = 2*r/(v*v*(1 - Exp(-r*T)))
  d1 = (Log(sk/x) + (b + v*v/2)*T)/(v*Sqr(T))
  q2 = (-(n - 1) + Sqr((n - 1)*(n - 1) + 4*K))/2
  a2 = (sk/q2)*(1 - Exp((b - r)*T)*CND(d1))

  If (s < sk) Then
    BaroneAdesiWhaleyCall = BlackScholes(mCall, s, x, T, r, b, v)_
                            + a2*(s/sk) ^ q2
    Exit Function
  End If

  BaroneAdesiWhaleyCall = s - x
End Function
```

```
Function BaroneAdesiWhaleyPut(s As Double, x As Double, _
                              v As Double, r As Double, _
                              T As Double, b As Double) As Double
  Dim sk As Double, n As Double, K As Double, d1 As Double
  Dim q1 As Double, a1 As Double

  sk = kp(x, T, r, b, v)
  n = 2*b/(v*v)
  K = 2*r/(v*v*(1 - Exp(-r*T)))
  d1 = (Log(sk/x) + (b + v*v/2)*T)/(v*Sqr(T))
  q1 = (-(n - 1) - Sqr((n - 1)*(n - 1) + 4*K))/2
  a1 = -(sk/q1)*(1 - Exp((b - r)*T)*CND(-d1))

If (s > sk) Then
    BaroneAdesiWhaleyPut = BlackScholes(mPUT, s, x, T, r, b, v) _
                          + a1*(s/sk) ^ q1
    Exit Function
  End If

  BaroneAdesiWhaleyPut = x - s
End Function

Function kc(x As Double, T As Double, r As Double, b As Double, _
            v As Double) As Double
  ' Calculation of seed value, Si
  Dim M As Double, q2u As Double, Su As Double, h2 As Double
  Dim Si As Double, d1 As Double, q2 As Double, LHS As Double
  Dim RHS As Double, bi As Double, E As Double, K As Double

  M = 2*r/(v*v)
  q2u = 1 + 2*M
  Su = x/(1 - 1/q2u)
  h2 = -(b*T + 2*v*Sqr(T))*x/(Su - x)
  Si = x + (Su - x)*(1 - Exp(h2))

  If (T = 0) Then T = 0.000000001
  K = 2*r/(v*v*(1 - Exp(-r*T)))
  d1 = (Log(Si/x) + (b + v*v/2)*T)/(v*Sqr(T))
  q2 = 1 + 2*K
  LHS = Si - x
  RHS = BlackScholes(mCall, Si, x, T, r, b, v) + _
    (1 - Exp((b - r)*T)*CND(d1))*Si/q2
  bi = Exp((b - r)*T)*CND(d1)*(1 - 1/q2)
  bi = bi + (1 - Exp((b - r)*T)*CND(d1)/(v*Sqr(T)))/q2
  E = 0.000001
```

```
  ' Newton-Raphson algorithm for finding critical price Si
  Do While (Abs(LHS - RHS)/x > E)
    Si = (x + RHS - bi*Si)/(1 - bi)
    d1 = (Log(Si/x) + (b + v*v/2)*T)/(v*Sqr(T))
    LHS = Si - x
    RHS = BlackScholes(mCall, Si, x, T, r, b, v) + _
      (1 - Exp((b - r)*T)*CND(d1))*Si/q2
    bi = Exp((b - r)*T)*CND(d1)*(1 - 1/q2)
    bi = bi + (1 - Exp((b - r)*T)*CND(d1)/(v*Sqr(T)))/q2
  Loop
  kc = Si
End Function

Function kp(x As Double, T As Double, r As Double, b As Double, _
           v As Double) As Double
   ' Calculation of seed value, Si
   Dim M As Double, q1u As Double, Su As Double, h1 As Double
   Dim Si As Double, d1 As Double, q1 As Double, LHS As Double
   Dim RHS As Double, bi As Double
   Dim E As Double, K As Double, n As Double

   n = 2*b/(v*v)
   M = 2*r/(v*v)
   q1u = (-(n - 1) - Sqr((n - 1)*(n - 1) + 4*M))/2
   Su = x/(1 - 1/q1u)
   h1 = (b*T - 2*v*Sqr(T))*x/(x - Su)
   Si = Su + (x - Su)*Exp(h1)

   If (T = 0) Then T = 0.000000001
   K = 2*r/(v*v*(1 - Exp(-r*T)))
   d1 = (Log(Si/x) + (b + v*v/2)*T)/(v*Sqr(T))
   q1 = (-(n - 1) - Sqr((n - 1)*(n - 1) + 4*K))/2
   LHS = x - Si
   RHS = BlackScholes(mPUT, Si, x, T, r, b, v) - _
     (1 - Exp((b - r)*T)*CND(-d1))*Si/q1
   bi = -Exp((b - r)*T)*CND(-d1)*(1 - 1/q1)
   bi = bi - (1 + Exp((b - r)*T)*CND(-d1)/(v*Sqr(T)))/q1
   E = 0.000001

   ' Newton Raphson algorithm for finding critical price Si
   Do While (Abs(LHS - RHS)/x > E)
     Si = (x - RHS + bi*Si)/(1 + bi)
     d1 = (Log(Si/x) + (b + v*v/2)*T)/(v*Sqr(T))
     LHS = x - Si
```

```
        RHS = BlackScholes(mPUT, Si, x, T, r, b, v) - _
          (1 - Exp((b - r)*T)*CND(-d1))*Si/q1
        bi = -Exp((b - r)*T)*CND(-d1)*(1 - 1/q1)
        bi = bi - (1 + Exp((b - r)*T)*CND(-d1)/(v*Sqr(T)))/q1
    Loop
    kp = Si
End Function
```

A VBA implementation of the Bjerksund-Stensland model is given below:

```
Function BjerksundStenslandCall(s As Double, x As Double, _
                               v As Double, r As Double, _
                               T As Double, b As Double) _
                                   As Double
  Dim Beta As Double, BInfinity As Double, B0 As Double
  Dim ht As Double, i As Double
  Dim alpha As Double, ss As Double

  If (b >= r) Then // Never optimal to exersice before maturity
    BjerksundStenslandCall = BlackScholes(mCall, s, x, T, r, b, v)
  Else
    Beta = (0.5 - b/(v*v)) + Sqr((b/(v*v) - 0.5)*(b/(v*v) - 0.5)_
        + 2*r/(v*v))
    BInfinity = Beta/(Beta - 1)*x
    B0 = Max(x, r/(r - b)*x)
    ht = -(b*T + 2*v*Sqr(T))*B0/(BInfinity - B0)
    i = B0 + (BInfinity - B0)*(1 - Exp(ht))
    alpha = (i - x)*(i ^ (-Beta))

    If (s >= i) Then
      BjerksundStenslandCall = s - x
      Exit Function
    End If

    ss = alpha*(s^Beta) - alpha*phi(s, T, Beta, i, i, r, b, v)
    ss = ss + (phi(s, T, 1, i, i, r, b, v) _
      - phi(s, T, 1, x, i, r, b, v))
    ss = ss - (x*phi(s, T, 0, i, i, r, b, v) _
      - x*phi(s, T, 0, x, i, r, b, v))
    BjerksundStenslandCall = ss
  End If
End Function
```

```
Function BjerksundStenslandPut(s As Double, x As Double,
                              v As Double, r As Double, _
                              T As Double, b As Double) As Double
  BjerksundStenslandPut = _
    BjerksundStenslandCall(x, s, v, r - b, T, -b)
End Function

Function phi(s As Double, T As Double, gamma As Double, _
           h As Double,_i As Double, r As Double, b As Double,_
           v As Double) As Double
  Dim lambda As Double, D As Double, kappa As Double, f As Double

  lambda = (-r + gamma*b + 0.5*gamma*(gamma - 1)*(v*v))*T
  D = -(Log(s/h) + (b + (gamma - 0.5)*(v*v))*T)/(v*Sqr(T))
  kappa = 2*b/((v*v)) + (2*gamma - 1)
  f = CND(D) - ((i/s)^kappa)*CND(D - 2*Log(i/s)/(v*Sqr(T)))
  phi = Exp(lambda)*(s^gamma)*f
End Function
```

References

Albanese, C., & Campolieti, G. (2006). *Advanced derivatives pricing and risk management*. San Diego: Elsevier Academic Press.

Barone-Adesi, G., et.al. (1999). VaR without correlations for nonlinear portfolios. *Journal of Futures Markets, 19*, 583–602.

Barucci, E. (2003). *Financial markets theory*. New York: Springer.

Berkowitz, J., & O'Brien, J. (2002). How accurate are value-at-risk models at commercial banks? *Journal of Finance, 57*, 1093–1111.

Björk, T. (1998). *Arbitrage theory in continuous time*. Oxford: Oxford University Press.

Black, F. (1976). The pricing of commodity contracts. *Journal of financial economics, 3* (1), 167–179.

Black, F., & Scholes, M. (1973). The pricing of options and corporate liabilities. *The Journal of Political Economy, 8*, 637–654.

Campbell, S. D. (2005). A review of backtesting and backtesting procedures. Finance and economics, discussion series Divisions of Research and Statistics and Monetary Affairs. Federal Reserve Board, Washington, DC.

Clewlow, L., & Strickland, C. (2000). *Implementing derivative models*. London: Wiley.

Conroy, R. M. (2003). *Binomial option pricing*. Darden Graduate School of Business Administration, University of Virginia. UVA-F-0943

Cox, J., Ross, S. A., & Rubenstein, M. (1979). Option pricing: A simplified approach. *Journal of Financial Economics, 7*(1979), 145–166.

Curran, M. (1994). Valuing Asian and Portfolio options by conditioning on the geometric mean price. *Management Science, 40*(12), 1705–1711.

Dana, R.-A., & Jeanblanc, M. (2007). *Financial markets in continuous time*. Berlin: Springer.

J.R.M. Röman, *Analytical Finance: Volume I*,
DOI 10.1007/978-3-319-34027-2

Davis, M., & Etheridge, A. (2006). *Louis Bachelier's theory of speculation: The origins of modern finance*. Princeton: Princeton University Press.

Derman, E. (1996). *Model risk*. Goldman Sachs Quantitative Strategies Research Notes.

DeRosa, D. F. (2000). *Options on foreign exchange*. New York: Wiley.

Drezner, Z. (1978). Computation of the bivariate normal integral. *Mathematics of Computation, 32*, 277–279.

Glasserman, P. (2000). *Monte Carlo methods in financial engineering*. New York: Springer.

Glasserman, P. (2003). *Monte Carlo methods in financial engineering*. New York: Springer.

Glasserman, P. (2004). *Monte Carlo methods in financial engineering*. New York: Springer.

Haug, E. (1997). *The complete guide to option pricing formulas*. New York: McGraw-Hill.

Ho, T. S. Y., & Lee, S. B. (2005). *Securities valuations*. Oxford: Oxford University Press.

Hull, J. *Option futures and other derivatives. (9th Edition)* Prentice-Hall. ISBN-13: 978-0133456318

Hunt, P. J., & Kennedy, J. E. (2004). *Financial derivative in theory and practise*. Chichester: Wiley.

Ingersoll, J. E. (1987). *Theory of financial decision making*. Totowa: Rowman & Littlefield.

Jarrow, R., & Rudd, A. (1983). *Option pricing* (pp. 183–188). Homewood: Dow Jones-Irwin.

Jorion, P. (2001). *Value at risk: The new benchmark for managing financial risk*. New York: McGraw-Hill.

Kerry B. (2005). *A cource in derivative securities*. New York: Springer.

Kijima, M. (2003). *Stochastic processes with application to finance*. London: Chapman & Hall.

Kwok, Y.-K. (2008). *Mathematical models of financial derivatives*. New York: Springer.

Leisen, D. (1998). Pricing the American put option: A detailed convergence analysis for binomial models. *Journal of Economic Dynamics and Control, 22*, 1419–1444.

Leisen, D., & Reimer, M. (1996). Binomial models for option valuation – Examine and improving convergence. *Applied Mathematical Finance, 3*, 319–346.

London, J. (2005). *Modeling derivatives in C++*. London: Wiley.

Longstaff, F. A. (1989). Pricing options with extendible maturities. *Journal of Finance, 45*(3), 935–957.

Martin, J. S. (2001). *Applied math for derivatives*. London: Wiley.

Merton, R. C. (1973). Theory of rational option pricing. *The Bell Journal of Economics and Management Science, 4*(1), 141–183.

Merton, C. R. (1976). Option pricing when stock returns are discontinuous. *Journal of Financial Economics,* 3, 167–179.

Munk, C. (2003). *Fixed income analysis: Securities, pricing and risk management.* http://citeseerx.ist.psu.edu/viewdoc/download?doi=10.1.1.129.9912&rep=rep1&type=pdf

Neftci, S. N. (2000). *An introduction to the Mathematics of Financial Derivatives.* New York: Academic Press.

Neftci, S. N. (2004). *Principles of financial engineering.* San Diego: Academic Press.

Pratt, J. W. (1968). A normal approximation for binomial, F, Beta, and other common, related tail probabilities, II. *The Journal of the American Statistical Association, 63*, 1457–1483.

Rebonato, R. (2001). Managing model risk. In C. Alexander (Ed.), *Handbook of risk management.* FT-Prentice Hall.

Reiner, E., & Rubinstein, M. (1991). Breaking down the barriers. *Risk, 4*, 28–35.

Rendleman, R. J., & Bartter, B. J. (1980). The pricing of options on debt securities. *The Journal of Financial and Quantitative Analysis, 15*, 11–24.

Ross, S. M. (1983). *Stochastic processes.* New York: Wiley.

Rubinstein, M. (2000). On the relation between binomial and trinomial option pricing models. *The Journal of Derivatives, 8*(2), 47–50.

Schweizer, M. (2012). *On Bermudan options.* ETH Zurich.

Shreve, Steven. (2004). *Stochastic calculus for finance II.* New York: Springer.

Shreve, Steven. (2005). *Stochastic calculus for finance I.* New York: Springer.

Sidenius, J. (2000). LIBOR market models in practice. *Journal of Computational Finance, 3*(3), 75–99.

Stoklosa, J. (2007). *Studies of barrier options and their sensitivities.* The University of Melbourne.

Tian, Y. (1993). A modified lattice approach to option pricing. *Journal of Futures Markets, 13*(5), 564–577.

Trigeorgis, L. (1991). A log-transformed binomial numerical analysis for valuing complex multi-option investments. *Journal of Financial and Quantitative Analysis, 26*(3), 309–326.

Wilmott, P. (2000). *On quantitative finance.* Chichester: Wiley.

Index

Note: Page numbers followed by "n" denote footnotes.

J.R.M. Röman, *Analytical Finance: Volume I*,
DOI 10.1007/978-3-319-34027-2

R

S

FSC
www.fsc.org
MIX
Papier | Fördert gute Waldnutzung
FSC® C083411

Kolibri 360 GmbH
Ferdinand-Jühlke-Straße 7
99095 Erfurt, Deutschland
produktsicherheit@kolibri360.de